VOLUME ONE HUNDRED AND EIGHT

# ADVANCES IN PROTEIN CHEMISTRY AND STRUCTURAL BIOLOGY

## Stress and Inflammation in Disorders

VOLUME ONE HUNDRED AND EIGHT

# ADVANCES IN PROTEIN CHEMISTRY AND STRUCTURAL BIOLOGY

## Stress and Inflammation in Disorders

Edited by

**ROSSEN DONEV**
*Biomed Consult Ltd*
*Swansea, United Kingdom*

ELSEVIER

ACADEMIC PRESS
An imprint of Elsevier

Academic Press is an imprint of Elsevier
50 Hampshire Street, 5th Floor, Cambridge, MA 02139, United States
525 B Street, Suite 1800, San Diego, CA 92101-4495, United States
The Boulevard, Langford Lane, Kidlington, Oxford OX5 1GB, United Kingdom
125 London Wall, London, EC2Y 5AS, United Kingdom

First edition 2017



ISBN: 978-0-12-812388-1
ISSN: 1876-1623

For information on all Academic Press publications
visit our website at https://www.elsevier.com/books-and-journals



*Publisher:* Zoe Kruze
*Acquisition Editor:* Alex White
*Senior Editorial Project Manager:* Helene Kabes
*Production Project Manager:* Surya Narayanan Jayachandran
*Senior Cover Designer:* Miles Hitchen

Typeset by SPi Global, India

# CONTENTS

# CONTRIBUTORS

**Ilenia Agliarulo**
Università di Napoli Federico II, Napoli, Italy

**Elísabet Alcocer-Gómez**
Research Laboratory, University of Sevilla, Sevilla, Spain

**Gjumrakch Aliev**
"GALLY" International Biomedical Research Consulting LLC, San Antonio, TX; School of Health Sciences and Healthcare Administration, University of Atlanta, Johns Creek, GA, United States; Institute of Physiologically Active Compounds Russian Academy of Sciences, Chernogolovka, Russia

**Maria Rosaria Amoroso**
Università di Napoli Federico II, Napoli, Italy

**Rosario Avolio**
Università di Napoli Federico II, Napoli, Italy

**Anupam Bishayee**
College of Pharmacy, Larkin Health Sciences Institute, Miami, FL, United States

**Beatriz Castejón-Vega**
Research Laboratory, University of Sevilla, Sevilla, Spain

**Valentina Condelli**
Laboratorio di ricerca preclinica e traslazionale, IRCCS-CROB, Centro di Riferimento Oncologico della Basilicata, Rionero in Vulture, Italy

**Mario D. Cordero**
Research Laboratory, University of Sevilla, Sevilla, Spain

**Xiaoyun Dai**
Yong Loo Lin School of Medicine, National University of Singapore, Singapore, Singapore

**Eduardo I. Díaz-Barba**
Laboratorio de Mitocondria-Estrés Oxidativo y Patología, División de Neurociencias, Centro de Investigación Biomédica de Occidente, Instituto Mexicano del Seguro Social; Centro Universitario de Ciencias Exactas e Ingenierías, Universidad de Guadalajara, Guadalajara, Jalisco, Mexico

**Franca Esposito**
Università di Napoli Federico II, Napoli, Italy

**Joaquim Fernández-Solà**
Hospital Clínic, Institut de Recerca Biomèdica August Pi i Sunyer (IDIBAPS), Universitat de Barcelona, Barcelona, Spain

**Luis J. Flores-Alvarado**
Centro Universitario de Ciencias Exactas de la Salud, Universidad de Guadalajara, Guadalajara, Jalisco, Mexico

**J. Francisco Santoscoy-Gutiérrez**
Laboratorio de Mitocondria-Estrés Oxidativo y Patología, División de Neurociencias, Centro de Investigación Biomédica de Occidente, Instituto Mexicano del Seguro Social, Guadalajara, Jalisco, Mexico

**Aristea S. Galanopoulou**
Laboratory of Developmental Epilepsy, Albert Einstein College of Medicine; Montefiore/Einstein Epilepsy Center, Montefiore Medical Center, Bronx, NY, United States

**C. George Priya Doss**
School of Biosciences and Technology, VIT University, Vellore, Tamil Nadu, India

**Héctor González-Usigli**
Hospital de Especialidades, Centro Médico Nacional de Occidente, Guadalajara, Jalisco, Mexico

**Kreeti Kajal**
School of Biosciences and Technology, VIT University, Vellore, Tamil Nadu, India

**Mohammad Amjad Kamal**
Metabolomics and Enzymology Unit, Fundamental and Applied Biology Group, King Fahd Medical Research Center, King Abdulaziz University, Jeddah, Saudi Arabia; Enzymoics and Novel Global Community Educational Foundation, Hebersham, NSW, Australia

**Alan Prem Kumar**
Yong Loo Lin School of Medicine; Cancer Science Institute of Singapore, National University of Singapore; National University Cancer Institute, National University Health System, Singapore, Singapore; Curtin Medical School, Faculty of Health Sciences, Curtin University, Perth, WA, Australia; University of North Texas, Denton, TX, United States

**Matteo Landriscina**
Laboratorio di ricerca preclinica e traslazionale, IRCCS-CROB, Centro di Riferimento Oncologico della Basilicata, Rionero in Vulture; Università degli Studi di Foggia, Foggia, Italy

**Francesca Maddalena**
Laboratorio di ricerca preclinica e traslazionale, IRCCS-CROB, Centro di Riferimento Oncologico della Basilicata, Rionero in Vulture, Italy

**Junji Magae**
Magae Bioscience Institute, Tsukuba, Japan

**R. Magesh**
Faculty of Research and Bio Medical Sciences, Sri Ramachandra University, Chennai, Tamil Nadu, India

**Danilo Swann Matassa**
Università di Napoli Federico II, Napoli, Italy

**Jason Chua Min-Wen**
NUS Graduate School for Integrative Sciences and Engineering, National University of Singapore; Stem Cell & Regenerative Biology, Genome Institute of Singapore, Agency for Science Technology and Research, Singapore, Singapore

**Mario Mireles-Ramírez**
Hospital de Especialidades, Centro Médico Nacional de Occidente, Guadalajara, Jalisco, Mexico

**Srishti Mishra**
Yong Loo Lin School of Medicine, National University of Singapore, Singapore, Singapore

**Solomon L. Moshé**
Laboratory of Developmental Epilepsy, Albert Einstein College of Medicine; Montefiore/ Einstein Epilepsy Center, Montefiore Medical Center, Bronx, NY, United States

**Genaro Gabriel Ortiz**
Laboratorio de Mitocondria-Estrés Oxidativo y Patología, División de Neurociencias, Centro de Investigación Biomédica de Occidente, Instituto Mexicano del Seguro Social, Guadalajara, Jalisco, Mexico

**Fermín P. Pacheco Moisés**
Centro Universitario de Ciencias Exactas e Ingenierías, Universidad de Guadalajara, Guadalajara, Jalisco, Mexico

**Ilaria Peluso**
Centre for Food and Nutrition, Council for Agricultural Research and Economics, Rome, Italy

**Anna Planavila**
Institut de Biomedicina de la Universitat de Barcelona (IBUB), Universitat de Barcelona; CIBER Fisiopatología de la Obesidad y Nutrición (CIBEROBN), Barcelona, Spain

**Paloma Rivero-Moragrega**
Laboratorio de Mitocondria-Estrés Oxidativo y Patología, División de Neurociencias, Centro de Investigación Biomédica de Occidente, Instituto Mexicano del Seguro Social, Guadalajara, Jalisco, Mexico

**Sugandhi Saini**
School of Biosciences and Technology, VIT University, Vellore, Tamil Nadu, India

**Yousef Sawikr**
Faculty of Medicine, University of Benghazi, Benghazi, Libya

**Gautam Sethi**
Magae Bioscience Institute, Tsukuba, Japan; School of Biomedical Sciences, Curtin Health Innovation Research Institute, Curtin University, Perth, WA, Australia

**Oleksii Shandra**
Laboratory of Developmental Epilepsy, Albert Einstein College of Medicine, Bronx, NY, United States

**Ng Shyh-Chang**
Stem Cell & Regenerative Biology, Genome Institute of Singapore, Agency for Science Technology and Research, Singapore, Singapore

**R. Siva**
School of Biosciences and Technology, VIT University, Vellore, Tamil Nadu, India

**Víctor J. Sánchez-González**
Centro Universitario de los Altos, Universidad de Guadalajara, Guadalajara, Jalisco, Mexico

**Angélica L. Sánchez-López**
Laboratorio de Mitocondria-Estrés Oxidativo y Patología, División de Neurociencias, Centro de Investigación Biomédica de Occidente, Instituto Mexicano del Seguro Social, Guadalajara, Jalisco, Mexico

**Lorenzo Sánchez-Romero**
Laboratorio de Mitocondria-Estrés Oxidativo y Patología, División de Neurociencias, Centro de Investigación Biomédica de Occidente, Instituto Mexicano del Seguro Social, Guadalajara, Jalisco, Mexico

**P. Sneha**
School of Biosciences and Technology, VIT University, Vellore, Tamil Nadu, India

**Himani Tanwar**
School of Biosciences and Technology, VIT University, Vellore, Tamil Nadu, India

**D. Thirumal Kumar**
School of Biosciences and Technology, VIT University, Vellore, Tamil Nadu, India

**Francesc Villarroya**
Institut de Biomedicina de la Universitat de Barcelona (IBUB), Universitat de Barcelona; CIBER Fisiopatología de la Obesidad y Nutrición (CIBEROBN), Barcelona, Spain

**Charles Emmanuel Jebaraj Walter**
Faculty of Biomedical Sciences, Technology & Research, Sri Ramachandra University, Chennai, India

**Benjamin Chua Yan-Jiang**
NUS Graduate School for Integrative Sciences and Engineering, National University of Singapore; Stem Cell & Regenerative Biology, Genome Institute of Singapore, Agency for Science Technology and Research, Singapore, Singapore

**Nagendra Sastry Yarla**
Institute of Science, GITAM University, Visakhapatnam, Andhra Pradesh, India

# PREFACE

Inflammation is the body's response to stress—whether from diet, lifestyle, or environment. Inflammation can be caused by many different factors: low-grade chronic food allergies; dysbiosis that is an imbalance of bacteria and fungi in gastrointestinal tract causing immune system to overreact to bacteria in the gut; chronic psychological, emotional, or physical stress; environmental toxicity from air, water, food pollutants, and toxic metals; diet and lifestyle—imbalance between fat, sugar, and protein intake, chronic dehydration, lack of sleep, etc. can all increase inflammation in body. All these factors result in inflammatory processes in our body via contributing to the forming of free radicals, an oxygen-containing molecule containing one or more unpaired electrons, making it highly reactive with other molecules. This process is known as oxidative stress. Although this important field has been considerably advanced in last few decades, it is still of a great difficulty to create the entire picture of the complexly related processes occurring upon triggering of inflammation. As a result, it is a big challenge to identify the target(s) which would be most beneficial to target by therapeutics without causing significant adverse effects. These challenges justify further in-depth studies in this very promising and dynamic field.

The current volume of the *Advances in Protein Chemistry and Structural Biology* (APCSB) is focused on *Stress and Inflammation in Disorders*. The first chapter in this volume elucidates the role of oxidative stress in central nervous system and the implication of reactive oxygen species, which include both oxygen-free radicals and nonradical oxygen derivative, in neuropathology. Neurodegenerative diseases are a heterogeneous group of disorders characterized by the progressive and irreversible destruction of specific neuronal populations. Although that loss of neurons is complex and multifactorial, there is substantial evidence that oxidative stress is a critical factor in the etiology of the major neurodegenerative diseases such as Alzheimer's disease, Parkinson's disease, stroke or vascular ischemia, and multiple sclerosis. All this is discussed in the first chapter of this volume of the APCSB. The second chapter in the volume critically analyzes and discusses the mechanisms involved in neuroinflammation and the possible role of some nutraceuticals in the prevention and therapy of Alzheimer's disease by targeting inflammatory processes. The third chapter focuses on the role of inflammation in epileptic encephalopathies and more specifically in the development and identification of therapeutic targets for West syndrome (WS), an infantile epileptic encephalopathy that manifests with infantile spasms, hypsarrhythmia, and poor neurodevelopmental outcomes. With this targeted review authors discuss the evidence pro and against a number of key questions: Does activation of inflammatory pathways in the brain cause epilepsy in WS and does it contribute to the associated comorbidities and progression? Can activation of certain inflammatory pathways be a compensatory or protective event? Does activation of brain inflammatory signaling pathways contribute to the transition of WS to Lennox–Gastaut syndrome? Are there any lead candidates or unexplored targets for future therapy development for WS targeting inflammation?

The fourth and fifth chapters address the involvement of brain-derived neurotrophic factor (BDNF) and methylenetetrahydrofolatereduc tase (MTHFR) in prevalent mood disorders, such as major depression and bipolar disorder. The role of polymorphisms found associated with these mood disorders for the functionality of BDNF and MTHFR is addressed by applying a computational pipeline that includes the use of various in silico tools, e.g., molecular docking and molecular dynamics simulation.

The thematic volume of the APCSB continues with chapter six which is a comprehensive review on the role of the stress-induced inflammasome in human diseases. Inflammasome complex activation is an important function mediated by the immune system. The inflammasome has emerged as a stress sensor due to its activation upon different stress conditions. The implication of this sensor complex in a wide spectrum of diseases such as cardiovascular, neurodegenerative, psychiatric, and metabolic diseases and its consequences for development of new therapeutic strategies has been discussed.

Metabolic reprogramming is one of the most frequent stress-adaptive response of cancer cells to survive environmental changes and meet increasing nutrient requirements during their growth. Chapter seven summarizes the interconnection between metabolism, chemoresistance, inflammation, and epithelial-to-mesenchymal transition, and the central role of tumor necrosis factor-associated protein 1 in the regulation of these processes. This extensive review schedules new lights on molecular networks at the basis of ovarian cancer. The eight chapter is focused on the use of ascochlorin and its derivatives in cancer therapy. Authors analyze the chemopreventive and therapeutic properties of ascochlorin and its derivatives from inflammatory perspective. Identification of multiple molecular targets modulated by these novel anticancer agents is also thoroughly discussed.

A number of recent basic research studies have identified proteins produced by the heart, referred to as cardiokines, which may function in signaling either locally or peripherally and are deregulated during cardiac stress and inflammation. The search for new biomarkers for heart failure diagnosis and therapeutics has yielded a gradually increasing number of identified cardiokines, which estimate between 30 and 60 cardiokines in total. The last chapter in this thematic volume of the APCSB focuses in detail on the above topic and elucidates on the development of novel or better diagnostic tools and treatment modalities for cardiac diseases.

The aim of this volume is to promote further basic and translational research, and design of new therapeutic strategies in the field of inflammation as a basis for a number of disorders and diseases. Advancing our knowledge on this very exciting and essential topic would allow further design of successful treatments for a wide range of diseases and disorders such as psychiatric and neurological (e.g., major depression, bipolar disorder, autism spectrum disorders, Alzheimer's disease, Parkinson's disease, multiple sclerosis, schizophrenia, cerebral ischemia, epileptic encephalopathies) as well as different cancer types and cardiovascular diseases.

Dr. Rossen Donev
Biomed Consult
United Kingdom

CHAPTER ONE

# Oxidative Stress: Love and Hate History in Central Nervous System

**Genaro Gabriel Ortiz*,[1], Fermín P. Pacheco Moisés[†], Mario Mireles-Ramírez[‡], Luis J. Flores-Alvarado[§], Héctor González-Usigli[‡], Víctor J. Sánchez-González[¶], Angélica L. Sánchez-López*, Lorenzo Sánchez-Romero*, Eduardo I. Díaz-Barba*,[†], J. Francisco Santoscoy-Gutiérrez*, Paloma Rivero-Moragrega***

*Laboratorio de Mitocondria-Estrés Oxidativo y Patología, División de Neurociencias, Centro de Investigación Biomédica de Occidente, Instituto Mexicano del Seguro Social, Guadalajara, Jalisco, Mexico
[†]Centro Universitario de Ciencias Exactas e Ingenierías, Universidad de Guadalajara, Guadalajara, Jalisco, Mexico
[‡]Hospital de Especialidades, Centro Médico Nacional de Occidente, Guadalajara, Jalisco, Mexico
[§]Centro Universitario de Ciencias Exactas de la Salud, Universidad de Guadalajara, Guadalajara, Jalisco, Mexico
[¶]Centro Universitario de los Altos, Universidad de Guadalajara, Guadalajara, Jalisco, Mexico
[1]Corresponding author: e-mail address: genarogabriel@yahoo.com

## Contents



*Advances in Protein Chemistry and Structural Biology*, Volume 108
ISSN 1876-1623
http://dx.doi.org/10.1016/bs.apcsb.2017.01.003

## Abstract

Molecular oxygen is essential for aerobic organisms in order to synthesize large amounts of energy during the process of oxidative phosphorylation and it is harnessed in the form of adenosine triphosphate, the chemical energy of the cell. Oxygen is toxic for anaerobic organisms but it is also less obvious that oxygen is poisonous to aerobic organisms at higher concentrations of oxygen. For instance, oxygen toxicity is a condition resulting from the harmful effects of breathing molecular oxygen at increased partial pressures. Reactive oxygen species (ROS) are chemically reactive molecules containing oxygen that are formed as a natural byproduct of the normal metabolism of oxygen and have important roles in cell signaling and homeostasis. However, in pathological conditions ROS levels can increase dramatically. This may result in significant damage to cell structures. Living organisms have been adapted to the ROS in two ways: they can mitigate the unwanted effects through removal by the antioxidant systems and can advantageously use them as messengers in cell signaling and regulation of body functions. Some other physiological functions of ROS include the regulation of vascular tone, detection, and adaptation to hypoxia. In this review, we describe the mechanisms of oxidative damage and its relationship with the most highly studied neurodegenerative diseases.

## 1. INTRODUCTION

While it is true that oxygen is an essential element for life, it can also be harmful. How can this be? Free radicals are oxygen molecules in the body that are produced by ultraviolet radiation, smoke, pollution, acts of breathing and eating, and other factors (Jan et al., 2015). Free radicals are atoms or molecular species with unpaired electrons in the outermost bonding orbital, which make them extremely unstable. Thus, they can react with other molecules very quickly, and they may assign or free electron to absorb this substance it reacts. Donor molecules not only are damaged but just may also be converted into free radicals. Other features that are free radicals, they can contain one or more electrons couple having a variable electronic load; i.e., may be neutral, positive, or negative (Fig. 1).

Oxygen free radicals are a natural consequence of the evolutionary process. Reactive oxygen species (ROS) includes both oxygen free radicals and nonradical oxygen derivative. Therefore, all oxygen radicals are ROS, but not all ROS are free radicals of oxygen. Superoxide anion ($O_2^{\bullet -}$) and hydrogen peroxide ($H_2O_2$) are highly selective in their reactions with biological molecules, while hydroxyl ($OH^{\bullet}$) is not selective. Due to oxygen consumption, single-celled organisms were able to generate much more chemical energy in the form of adenosine triphosphate (ATP) than anaerobes, which allowed

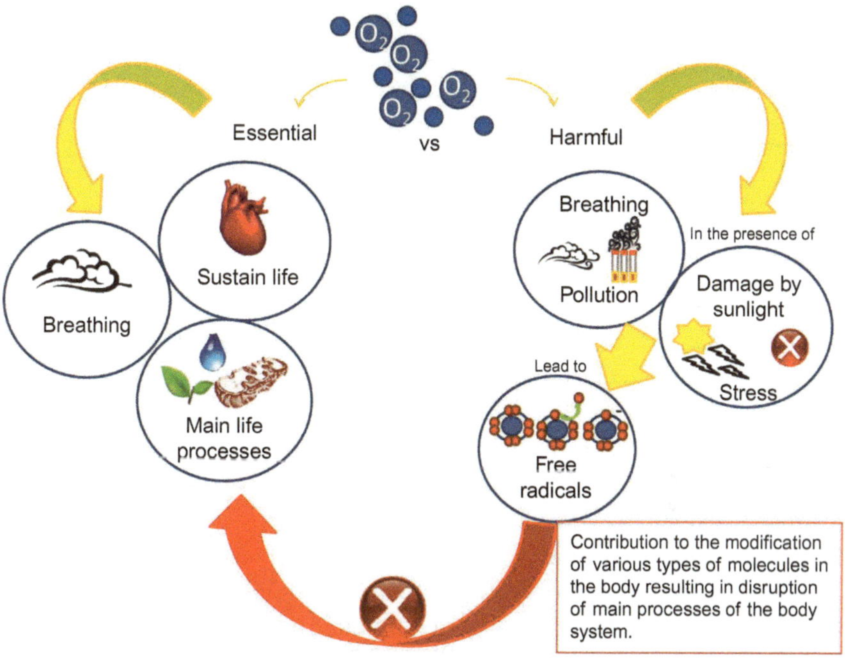


**Fig. 1** Oxygen is an essential molecule to sustain life as well as it is needed for main processes and breathing; however, it can also be harmful, that is, in the presence of some factors like smoke, pollution, stress, or sunlight, molecules of oxygen can be converted into free radicals, which are extremely unstable molecules. Contributing to alteration of various types of molecules, consequently resulting in the disruption of the normal functioning of many body systems affecting the functional capacity of the individual.

them to establish relations with neighboring cells and better defend the aggressive environment in which they lived. This allowed the creation of increasingly complex multicellular organisms until the mammal (Rahman, 2007).

Most of the oxygen we breathe goes to the mitochondria, the intracellular structure responsible for among other things generate about 95% of ATP consumed by the cell. During this process, some oxygen molecules are not metabolized properly; partially reduced $O_2^{\bullet\sim}$ is produced in an amount of 5 kg/year in man. $O_2^{\bullet\sim}$ is a ROS which, besides damaging the mitochondria and biomacromolecules, can be transformed into other reactive species such a $H_2O_2$ and, particularly $OH^{\bullet}$. Oxygen radicals can also be generated at cytosol and neuropil and other cellular structures, although generally to a lesser extent due to lower oxygen utilization outside the mitochondria (Barja & Herrero, 2000).

The chain reactions driven by ROS have the potential to damage each cell component including DNA. Over time, this damage can affect the resilience of the cells permanently, our skin ages prematurely and we become more vulnerable to getting sick (Lobo, Patil, Phatak, & Chandra, 2010). But free radicals are not always negative as they participate in a number of functions that are fundamental to ensure good dynamics and optimum level of our health, performing:

- Phagocytosis: a destruction by cells called phagocytes of the solid particles, organized or inert.
- Formation of eicosanoids: molecules that are the product of oxygenation of essential fatty acids, which play an important role in hemostasis, renal function, and control of gastric secretion.

Neurodegenerative diseases of the nervous system associated with the presence of free radicals are: Alzheimer, Parkinson, cerebral ischemia, multiple sclerosis (MS) among others. Although free radicals have been in the spotlight as the cause or effect of various diseases, new evidence might show that have beneficial effects on the pathophysiology. For a long time, we have been told that free radicals were responsible for cancer to cause changes in the DNA of cells, and also decrease the number of mitochondria, causing diseases and cell aging. However, there is enough work suggesting that free radicals are a measure that has our body to counter the effects of aging and protect us from diseases and especially infections (Schultz, Chao, & Mcginnis, 2009).

Nobody doubts the vital benefits of oxygen, however, when the tissues within produce "free radicals" can occur that instead of causing serious attacks benefit to the cellular structure. Even old age is nothing but slow oxidation in the body. Oxygen, so fundamental to the life of aerobic organisms, to the point that without it life on earth would be impossible, is not all good for our cells as we might suppose. From scientific information in recent years, it has accumulated a wealth of data on cellular oxygen toxicity and its consequences for the survival of organisms. In other words, the oxygen paradoxically allows life but also destroys (McCord, 2000).

Oxygen is used by aerobic organisms as an oxidizing agent, as necessary for the survival energy, growth, and multiplication of these comes from processes where metabolic fuels such as carbohydrates, lipids, and proteins are burned, becoming mainly carbon dioxide ($CO_2$) and water ($H_2O$). Thus, the reduced state of carbon, for example, present in a carbohydrate, is transformed into $CO_2$, releasing a considerable amount of energy, while the oxygen $O_2$ in its oxidized form, is reduced to O2-4. This process of metabolic fuel oxidation is not performed in a single step within our cells, for a very simple and logical reason: if this transformation so happens, most of the energy that is just the element needed by the cell, it would be lost as heat producing only an increase in system temperature. The solution to this problem, thermodynamic probably occurred in the early stages of the evolution of aerobic organisms, was to conduct the oxidation of metabolic fuels in a quantum way, that is, step by step, achieving maximum recovery energy in chemical form and wasting at least as heat (Davies, 1995).

## 1.1 Biological Importance of Oxygen Radicals (ROS)

Oxygen as an element is highly toxic to many forms of life, and aerobic living organisms are reliable to use it as a terminal electron acceptor achieving greater efficiency in energy production from metabolic fuels. However, oxygen can harm even aerobic cells.

Living organisms have been adapted to the ROS in two ways: they can mitigate the unwanted effects through removal by the antioxidant systems and can advantageously use them as messengers in cell signaling and regulation of body functions. Among the ROS advantageous physiological functions are the regulation of vascular tone, detection, and adaptation to hypoxia, and even the same oxidative stress response (Juránek, Nikitovic, Kouretas, Hayes, & Tsatsakis, 2013).

The amounts of ROS are determined to balance between by the production and removal, and a change in that balance in behalf of accumulation (oxidative stress) in turn generates adaptive responses in antioxidant systems maintaining what is called "redox homeostasis" (Fig. 2). Under basal conditions, there is a small concentration of ROS; however, when this level is exceeded signaling pathways sensitive to redox state are activated. This change in the balance due to the endogenous production of ROS or regulated generating environmental conditions of oxidative stress, but in both cases, if the increase in the concentration of ROS is transient and/or low magnitude, antioxidants soon systems are reliable to restore the initial state. On the other hand, under conditions certain increases the production of ROS stronger and more persistently and antioxidant responses may not be sufficient to restore the balance to the original level, generating a new equilibrium level, where ROS concentrations are higher and the gene expression pattern is modified due to sustained stimulation of signaling pathways sensitive to redox state, including the nuclear factor kappa beta (NF-κB) and MAP kinase (MAPK). Excessive amounts of free radicals

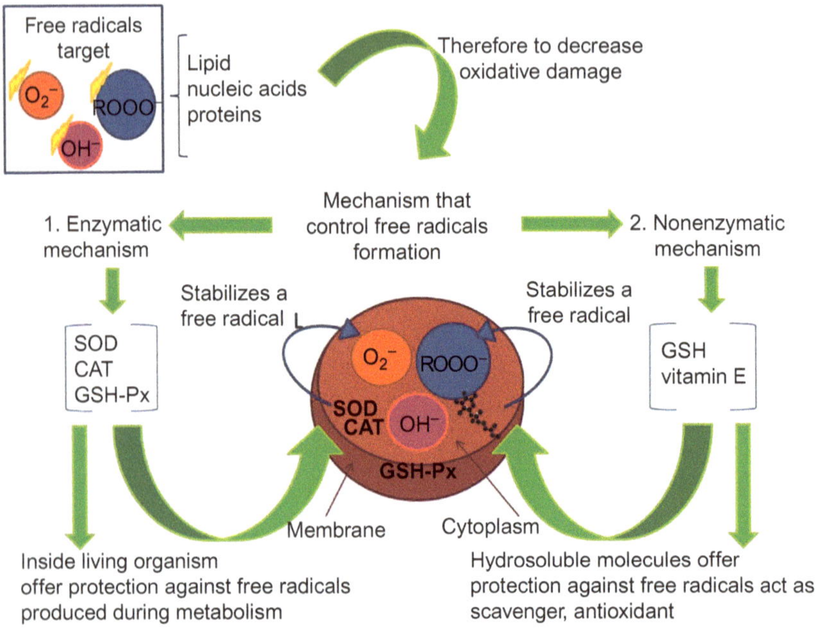


**Fig. 2** (1) Enzymatic mechanism controls the formation and fate of the radicals or toxic substances generated by the same cell; this mechanism is constituted mainly by superoxide dismutase (SOD), catalase (CAT), and glutathione peroxidase (GSH-Px). (2) Nonenzymatic mechanism. The most important components are glutathione (GSH) and vitamin E. The most important characteristics are that act as a "scavenger" of free radicals and antioxidant.

are involved in the pathogenesis of many diseases, including Alzheimer, Parkinson, ischemia of brain, MS, etc. That it is also known physical activity itself is an additional source of free radicals (Halliwell, 2006).

## 1.2 Training During Oxidative Metabolism: The "Oxygen Paradox"

For many years, it was believed that increased mitochondrial oxygen flow to meet the demands of ATP during exercise was sufficient to explain the excessive production of high ROS in mitochondria. However, it is now known that muscle hypoxia that develops during the work can also result in the production of ROS, although in small quantities. Below you will deepen a little more to this apparent paradox. Guzy et al. reported that mitochondria can act as the oxygen sensor and in response to hypoxia produces superoxide anion controlled manner in the complex III of the electron transport chain (mitochondrial intermembrane space), which is released into the cytosol where it stabilizes hypoxia-inducible factor (HIF-1α) may mediate so indirectly through the latter to the hypoxia adaptive responses (Barja & Herrero, 2000).

The hypoxia-inducible factors (HIF-1 and HIF-2) are the mediators of many responses to hypoxia, including the transcriptional activation of genes of erythropoietin (EPO), vascular endothelial growth factor (VEGF), glycolytic enzymes, transferrin, and myoglobin. Among the HIF-1α is constitutively transcribed and translated. However, under normoxia and hyperoxia after it is degraded by a group hydroxylation of proline hydroxylase. The activity of this group of enzymes is maximum $FiO_2$ 45%, which facilitates recognition by the ubiquitin–proteasome system (Fig. 3). Hydroxylation and degradation of HIF-1α under hypoxic conditions are inhibited by generating ROS, and, although amounts of additional ROS are generated during also hyperoxia is thought that they are not sufficient for inhibiting proline hydroxylase they are at a level close to maximum STI activity ($FiO_2$ 45%). This explains why only hypoxia (and not hyperoxia) results in the stabilization of HIF-1α (Majmundar, Wong, & Simon, 2010).

Given the above, it can rethink the theory of mitochondrial ROS production during exercise. No longer just try to mass action where the increase in activity in the electron transport chain would lead to a rise in the production of ROS, but also can be a "controlled" free radical production in the mitochondria act as oxygen sensor which would be reliable to detect the decrease in muscle $pO_2$ (mitochondrial). That occurs during exercise, in response to which produce and release more superoxide into the cytosol, generating adaptive responses to hypoxia induced the exercise by the stabilization of HIF-1α. In this regard, human acute exercise load increases the concentrations of both the HIF-1α protein, such as EPO and VEGF mRNA in the muscle cell. That it is worth emphasizing the "adaptive and controlled" superoxide ion production. During exercise-induced hypoxia would be advantageous within a certain range of concentrations, beyond which the beneficial adaptive responses mediated by HIF-1α is not would observe, and predominate, however, the undesired effects on different cellular components (oxidative stress) (Majmundar et al., 2010).

In the same way that the oxidation of carbohydrates, lipids, and proteins are gradually made, in recent years it has been discovered that molecular oxygen is also being reduced in stages, forming in each well-defined products whose properties physical and chemical researchers have worried the area. These intermediaries reduction not only allow many cellular functions

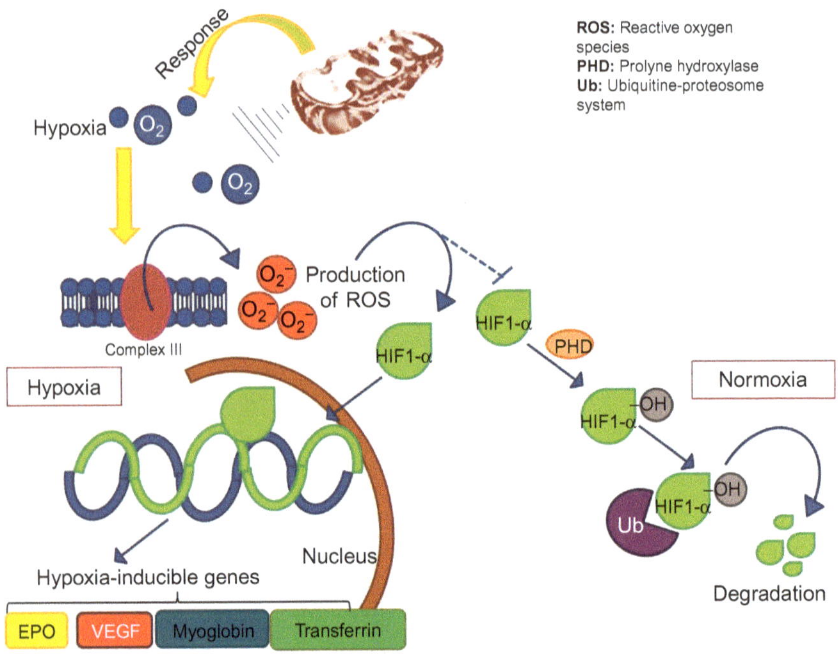


**Fig. 3** Free radicals target lipid, nucleic acids, and proteins leading to an impaired function of those molecules and decrease flow of the properties, causing disruption and physiological damage. Therefore to decrease damage in cells of living organisms, there are mechanisms that control free radical formation: enzymatic and nonenzymatic mechanism. Mitochondria can respond to an absence of oxygen environment (hypoxia) leading to a production of reactive oxygen species in complex III of the electron transport chain, resulting in the stabilization of a factor that mediates hypoxia adaptive response (HIF1-α), the hypoxia-inducible factors (HIF-1 and HIF-2) are the mediators of many responses to hypoxia, including the transcriptional activation of genes of erythropoietin (EPO), vascular endothelial growth factor VEGF, glycolytic enzymes, transferrin, and myoglobin. Among the HIF-1α is constitutively transcribed and translated. Under normoxia conditions (HIF1-α) factor can be degraded by a group hydroxylation of proline hydroxylase, which facilitates recognition by the ubiquitin–proteasome system, leading to degradation protein.

are carried out, but also of a potential danger to the cell. The molecular oxygen to become water must receive four electrons, but cannot simultaneously because energy barriers that must be overcome each electron to be incorporated into the molecule. Thus, different transient reduction intermediates, some of which exhibit characteristics of free radicals, are formed (Kardeh, Ashkani-Esfahani, & Alizadeh, 2014).

The electrons in the atoms occupy regions of space known as orbital's. Each orbital contains a maximum of two electrons. A free radical is defined simply as any chemical species capable of independent existence and containing one or more unpaired electrons. Free radicals can react with other molecules in different ways. And a free radical can donate their unpaired electron to another molecule. You can also grab an electron from another molecule to reach a situation of stability. In all these reactions, the free radical becomes the current molecule reacts in turn a free radical, and therefore, a fairly common feature in free radical reactions is that this chain of processes: A radical results in the formation of another radical. Only two free radicals found the process stops.

The formation of oxygen free radicals within cells of an organism is not a biological accident as these active forms are required and used by the cell to perform many of their functions. For example, involved in the synthesis of molecules such as collagen, which protein component provides strength to the tissues; in the formation of cholesterol; in the synthesis of prostaglandins, hormones that act on very small quantities and regulate such varied functions such as body temperature, blood clotting, or work delivery. Free radicals are also important in the formation of nucleic acids and enzyme systems that allow metabolize drugs or toxins such as alcohol, for example. Of all the mechanisms that are involved somehow free radicals, perhaps the best known is that which relates to the bactericidal process polymorphonuclear leukocytes, i.e., cells belonging to the group of white blood cells that are responsible for destroying bacteria or other offenders cell. The white blood cells form at its inner free radicals that diffuse extracellular damaging the biochemical function of bacteria while the phagocyte, incorporating them into the cell and ultimately destroying enzymatically highly active (Pham-Huy, He, & Pham-Huy, 2008).

From this perspective, one can think that the efficiency of cell function will depend on the perfect balance to be maintained between the rate of formation of necessary free radicals to their functions and the quantity of waste that lack of control produce toxic effects. This perfect balance depends largely on the existence of a complex intracellular mechanism against oxygen toxicity. This mechanism controls the formation and fate of the radicals generated by the same cell or toxic substances. The protection mechanism is constituted by two types of components: enzymatic and nonenzymatic. Within the former, superoxide dismutase (SOD), catalase (CAT), and glutathione peroxidase (GSH-Px) are very important. These enzymes, which are present in the cytoplasm (SOD) in the membranes (GSH-Px) or in specialized cell organelles (CAT), acting trapping superoxide and hydroxyl radicals (SOD and CAT) or destroying organic peroxides produced by the destruction fatty acids during the process (GSH-Px) (Kang, Narabayashi, Sata, & Takeshige, 1983).

The most important nonenzymatic components are glutathione (GSH) and vitamin E. Glutathione a tripeptide found in all cells and has important characteristics as "scavenger" of free radicals. It can act as a cellular antioxidant, and as it has been shown that different drugs that decrease the intracellular concentration of glutathione favor lipoperoxydation (LPO) damage mainly at the level of the cell membrane. Such is the case of the effect of an overdose of acetaminophen, an analgesic for widespread use, but which in high doses induces LPO and liver damage and a large decrease in the concentration of glutathione. A similar example is the excessive consumption of alcohol, liver damage that originated through a similar mechanism (Halliwell & Gutteridge, 1995). Vitamin E is one of the nonenzymatic components of the cellular defense system. By the fact of not being synthesized by animal organisms, they must purchase the power through vegetable oils and fats. This vitamin has a long history of research since the 1930s, when it was discovered; it was linked with fertility, virility, and even longevity. In short, it is a powerful antioxidant that is a free radical scavenger and hence its importance in cellular metabolism. Vitamin E has been used for therapeutic purposes; its effect may be closely related to their protective characteristics of the cell structure.

Mitochondrial respiration comprising a coordinated four electron reduction of oxygen to water, the donation of electrons by the coenzyme NADH or succinate to mitochondrial complexes I and II, respectively, of the chain of mitochondrial electron transport. The mitochondrial transport system is not perfect, and superoxide ions are produced whose enzymatic disproportionation leads to the formation of hydroperoxide. Peroximal oxidation of fatty acids generates hydroperoxides as byproducts. Microsomal enzymes such as cytochrome P450 responsible for metabolizing xenobiotics, usually reduces oxygen to superoxide. The activity of phagocytic cells during its attack on pathogens produces a mixture of oxidants and free radicals including superoxide, hydroperoxide, peroxynitrite, and hypochlorite. Regarding lipids is known that a hydroperoxyl radical withdraws a hydrogen atom of the double bond of a neighboring unsaturated lipid to form a hydroperoxide and an alkyl radical which in turn reacts with oxygen to regenerate a hydroperoxyl lipid radical capable of initiating of new oxidative process. An important consequence of the oxidation of membranous lipids is impaired and therefore the flow properties thereof that can cause disruption of membrane bound proteins (Absi, Ayala, Machado, & Parrado, 2000).

Oxidative damage to nucleic acids may be of very complex features. The electrochemical properties of 8-oxo-guanine (oxo8gua) and 8-oxodihidroxi-deoxyguanosine (oxo8dG) allowed the coupling of systems extremely sensitive electrochemical detection HPLC, which allowed the study of their formation, accumulation, and excretion in living creatures. Identifying specific enzymatic repair oxidative damage has provided proof of the significance of oxidative damage to DNA as well as tools for manipulating load damage in vivo by genetic knock out. Protein oxidation has not been well characterized, although they have been able to document specific damages including oxidation of sulfhydryl groups, reduction of disulfides, oxidative adducts of amino acid residues, reaction with aldehydes, peptide fragmentation, etc.

However, today we know that free radicals are signaling molecules that meet physiological functions. It is therefore necessary to know what those molecules are, how they form, and what effects occur, to act against them preventing cell damage while maintaining its cellular signaling functions. For instance, it has recently been shown that the elevation of free radicals in certain areas of the hypothalamus to regulate appetite is related to the same control and prevention of obesity. Consequently, one must be careful when it comes to control free radicals in the body, because we can alter physiological mechanisms of important regulation, causing more damage than we wanted to prevent.

Contrary to what one might think, not all free radicals are harmful, since there are some that are produced by the immune system (one that protects us from external aggression) to kill bacteria and fungi. They serve purpose are neutralized by the body, by activating enzymes called CAT and superoxide, which "disarm" free radicals generated to prevent unbalanced state (Castellanos, Sobrino, & Castillo, 2006). Moreover, these elements enter the body naturally also through processes like breathing; to better understand the above, it is considered that the body use oxygen to obtain energy from food and in turn provide it to all the organs perform their biochemical functions. The procedure involves the blood, where important protein called hemoglobin (containing iron), through which the vital fluid can absorb 50 times more oxygen than water. In this complex process, essential to life is called oxidation, and for him free to kill bacteria and provide some protection to the body radicals are generated popular. However, factors such as pollution, certain household chemicals, certain drugs, snuff, X-rays, and pesticides can increase production and thereby generate significant health problems, such as cancer (Poljsak & Milisav, 2013).

# 2. OXIDATIVE STRESS AND NEUROPATHOLOGY

Neurodegenerative diseases are a heterogeneous group of disorders gradually characterized by the progressive and irreversible destruction of specific neuronal populations. That loss of anatomically related physiologically or neuronal systems is complex and multifactorial. Although the etiology of the major neurodegenerative diseases (including Alzheimer's disease (AD), Parkinson's disease (PD), stroke or vascular ischemia, MS, etc.) is unknown, there is substantial evidence that oxidative stress is a critical factor in common in all of these diseases (Emerit, Edeas, & Bricaire, 2004).

The whole nervous system is rich in metals, particularly, the brain is a specialized organ that accumulates iron ions and it is especially susceptible to oxidative damage since it has a high metabolic activity and a high content of unsaturated fatty acid (Georgieff, 2007). The high level of brain iron may be essential, particularly during development, but it is also means the presence of brain injury, which may release iron ions, can lead to oxidative stress via the iron-catalyzed formation of ROS.

## 2.1 Alzheimer Disease

Harman has been estimated that the human lifespan could last for 5–10 years if free radicals are found both in the controlled environment and in food. The amount of fatty acids containing cell membranes is the preferential point of impact of free radicals. Excessive formation of free radicals is controlled biologically. In AD free radicals attack the membrane, free fatty acids, and proteins; ion channels disturb and increase membrane permeability. Alter the position, formation, and function of proteins causes changes in enzyme activity and cell cytoarchitecture. It demonstrated an increase in the proliferative phase phosphomonoesters and phosphodiesters in degenerative phase. This correlates to a deficiency of the enzyme phosphodiesterase. One of the major systems of protection against free radicals is represented by the enzyme SOD, which catalyzes the dismutation of superoxide radicals, hazardous oxygen, and hydrogen peroxide. SOD produces excess thus paradoxically more free radicals than it consumes. SOD values decrease during normal brain aging. Increased SOD values in trisomic is explained considering that the gene encoding this enzyme is provided by the chromosome. It has very increased values in cultured human fibroblasts from individuals carrying a type dementia Alzheimer. Recent advances in genetic association studies of chromosome with this process group appear to support this hypothesis (Pohanka, 2013).

The term Alzheimer's disease "broadly" corresponds to Reisberg (Reisberg et al., 1987), a clinical syndrome characterized by the progressive and irreversible impairment of cognitive functions, comprising neurofibrillary degeneration, granulovacuolar degeneration, and neuritic plaques. These findings can clinically manifest before or after 65 ages. Although the unification of the two major aspects of the disease appears more or less justified in the anatomical plane, a distinction must prevail. There are obvious semiotic differences according to age at onset and neuropsychological box differs markedly from one extreme to another. The prevalence of the process in populations of more than 65 years old varies according to studies between 1% and 5.8%. Numerous pathogenic hypotheses related to this problem, as the primary genetic alteration of neurotransmission, the viral, toxic, vascular and metabolic origin, autoimmune, and free radicals, among others (Querfurth & LaFerla, 2010).

Oxidative balance in the brain cells is important in the understanding of Alzheimer's. In examining brains with this disease is a major oxidative damage associated with both as neuritic plaques, neurofibrillary tangles, as well as pyramidal neurons of normal appearance (Fig. 4). Oxidative stress occurs in

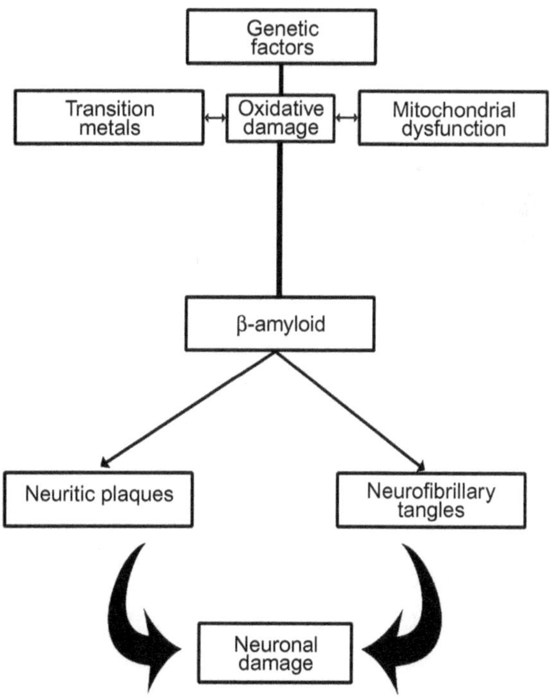


**Fig. 4** Factors related to neuronal damage in Alzheimer's disease. The precise cause involved in the development of Alzheimer's disease is unknown, however, have associated genetic, environmental factors (oxidative damage), transition metals, and mitochondrial dysfunction in to development of neuritic plaques and neurofibrillary tangles in pyramidal neurons damage in the early stages of the disease.

the early stages of Alzheimer's, but the mechanisms by which the redox balance is disturbed; today the most part is unknown. There are a number of possible causes that may play an important role in the production of free radicals: mitochondrial dysfunction, high levels of β-amyloid peptide, the transition metal accumulation, and genetic factors such as apolipoprotein E and presenilin. Understanding these mechanisms lead to a better understanding of the pathogenesis of AD and new therapeutic approaches (Gella & Durany, 2009).

## 2.2 Alzheimer, Mitochondria, and ROS

Analysis of brain tissue from autopsy from deceased patients with AD suggests a correlation between mitochondrial alterations and high levels of the amyloid precursor protein (APP). Some authors have reported decreased activity of complex IV (cytochrome oxidase, or COX) in the brain mitochondrial and platelets, although other studies have found no significant differences in the activity of COX (Swerdlow, Burns, & Khan, 2010). In platelets and lymphocytes of AD patients, the expression of mRNA encoding some of the subunits of COX decreased in the brain of AD patients on the other hand has detected an increase in mitochondrial DNA point mutations in the parietal cortex, hippocampus, and cerebellum. Previous analyzes in mitochondria, the start potential source of oxidative stress in AD, have given rise to controversy about the existence of mtDNA polymorphisms associated with EA (Mancuso, Orsucci, Siciliano, & Murri, 2008). Ultrastructural analysis also indicated the presence of mtDNA in lysosomes of neurons in AD. This study suggests that phagocytosis of mitochondria can result in the output of metals that may facilitate cytoplasmic redox processes. These redox reactions can take place in the endoplasmic reticulum, a badly damaged neurons in AD area. Metals such as iron or copper catalyze oxidation processes that have been found in AD, metal-binding redox reactions involved in nucleic acids result in the appearance of a single nucleic acid chain degeneration sensitive S1 nuclease.

## 2.3 AD: Lipid Peroxidation and ROS

Most studies exhibited an increase in markers of LPO in the brain of AD patients, including malonyldialdehyde (MDA)–thiobarbituric acid reactive substances (TBARS), 4-hydroxy-2-nonenal (HNE), and some isoprostanes (Castellani, Rolston, & Smith, 2011), particularly in the temporal cortex and hippocampus. Recently described an increase in markers of lipid peroxidation in the olfactory epithelium, in which there have been no data suggestive oxidation of proteins and DNA. Plasma membranes senile dystrophic neurites appear more sensitive to β-amyloid deposition and lipid peroxidation processes that the membranes of normal cells or other neurites of neuritic plaques. The mitochondrial membrane fluidity is decreased in the cerebral cortex of AD patients, data that appear to be due to increased lipid peroxidation processes; the derived membrane phospholipids and phosphatidyl inositol fostatidiletanolamin decrease in the hippocampus and the first well in the parietal cortex.

The lipofosfatidilcolina/phosphatidylcholine ratio is reduced in the cerebrospinal fluid (CSF). A potentially interesting is the fact that some products of lipid peroxidation, as the MDA and HNE, are able to modify the structure of the INA apolipoprotein E3 (apoE3) and alter their metabolism in cell cultures.

## 2.4 AD: Nitric Oxide and ROS

The levels of nitrates and nitrites found decreased in the frontal cortex of AD patients compared to young controls, and are normal in CSF and plasma of these compared with controls of similar age and sex. In neurofibrillary tangles of AD patients has been detected nitrotyrosine, but not in brains of controls without such structures, which implies the formation of NO and peroxynitrite free radical in the pathogenesis of AD associated to oxidative stress (Fig. 5). Nitrotyrosine was detected in neurons, astrocytes, and blood vessels in AD patients and not in controls, also with abnormal expression of neuronal NO synthase (NOS) in cortical pyramidal cells. Dimetilargininasa levels, involved in the regulation of NOS activity, increase in neurons of AD patients subjected to oxidative stress and decrease or are normal in the CSF of patients with AD. NO production by human macrophages is stimulated seems therefore apoE as β-amyloid, which also increases the production of NO by astrocytes. Finally, it described the presence of a 3-nitrotyrosine immunoreactivity in the olfactory epithelium of patients with AD, which is not observed in controls (Togo, Katsuse, & Iseki, 2004).

Acute and chronic application fragment 25–35 and 1–42 of amyloid β to cell cultures of the cerebral cortex and processes of tau protein glycation in paired helical filaments induce oxidative stress. 1–42 and 1–40 fragments of human β-amyloid protein are able to generate by themselves an output of

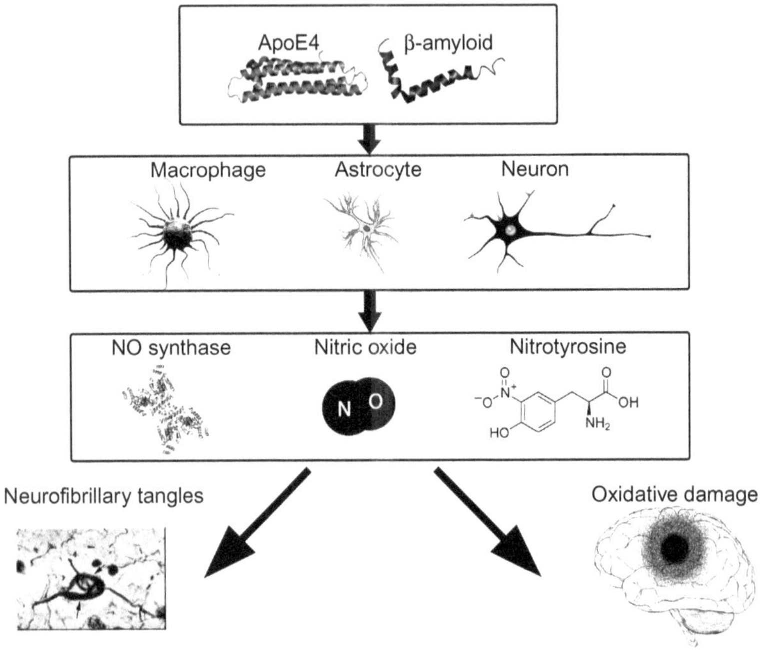


**Fig. 5** AD: Nitric oxide (NO) and reactive oxygen species (ROS). Several studies have shown the induction of ROS and increased nitric oxide production by neural cells: neurons and astrocytes in addition to immune cells such as macrophages; these alterations are possibly influenced by factors such as the expression of apoE and amyloid beta peptide harmful condition also an increase in the concentration of nitric oxide synthase and high levels of nitrotyrosine. These mechanisms have been linked to the development of neurofibrillary tangles and the generation of oxidative stress in the brain of AD patients.

hydrogen peroxide by a mechanism that involves the reduction of $Fe^{3+}$ or $Cu_2$ by Fenton reaction. Fragment 25–35 of β-amyloid protein is also capable of inducing mitochondrial oxidative damage DNA and induce apoptotic neuronal death, which is mediated by hydrogen peroxide, HNE, and caspase-3 and which can be prevented by vitamin E and *N*-acetylcysteine. In transgenic mouse models of AD with increased expression of APP and in other experimental models, it has been observed increased oxidative damage, including lipid peroxidation which precedes the formation of β-amyloid (neuritic plate), nitrotyrosine production, and the expression of C/Zn-SOD and heme-oxygenase-1 around β-amyloid deposits. Also it described increased oxidative stress in transgenic mice with mutations in presenilin-1 (Butterfield, Swomley, & Sultana, 2013).

Pretreatment with some antioxidants such as idebenone and α-tocopherol appears to prevent the development of learning and memory disorders caused by β-amyloid (1–42) in rats. Vitamin E also appears to prevent oxidative damage to proteins induced by β-amyloid 25–35. Some studies showed no increase in cell viability in culture treatment with vitamin E, contrary to what happens with *N*-acetylcysteine, dithiothreitol, and cyclosporin, which do have a protective effect. However, other studies have shown a protective effect of vitamin E both as inhibitors of caspase-3 (Yatin, Varadarajan, & Butterfield, 2000).

## 3. PD AND ROS

PD is a progressive neurodegenerative disorder characterized by severe motor symptoms. Histopathology reveals a pronounced loss of dopaminergic neurons in the *substantia nigra*, resulting in depletion of dopamine in the *striatum*, the structure receiving the projections of those neurons (Forno, 1996). Other neuronal systems, including some catecholaminergic nuclei are also affected in PD, but with less severity.

One of the most surprising aspects of neurodegenerative diseases (including the PD) is the selective vulnerability of damaged neuronal population. Thus, although the α-synuclein, for example, is expressed in large areas of the central nervous system (CNS), neurodegeneration is mainly restricted to the substantia nigra. Dopaminergic neurons are particularly exposed to oxidative stress due to the metabolism of dopamine that produces a series of molecules that are potentially toxic if they are not adequately removed. Dopamine can autooxidize at physiological pH forming toxic quinones species (dopamine-*o*-quinone, aminochrome and indole-quinone, superoxide radicals, and hydrogen peroxide). It can also be enzymatically deaminated by monoaminooxidase (MAO) in 3,4-dihydroxyphenylacetic nontoxic metabolite acid (DOPAC) and hydrogen peroxide, and by other oxidative processes. Thus, the metabolism of dopamine generates high concentrations of ROS, which can activate and induce apoptotic cell death cascades. ROS accumulation is toxic *per se* (Maker, Weiss, Silides, & Cohen, 1981) and generates oxidative stress as a result of depletion of cellular antioxidants (Vitamin E and reduced glutathione), increasing the membrane lipid peroxidation, DNA damage, and oxidation alteration of protein folding. Besides the general oxidative damage, there is evidence that the interaction between α-synuclein and dopamine metabolites determines the preferential neurodegeneration of dopaminergic neurons (Junn & Mouradian, 2002).

Along with a number of possible changes, abnormal protein aggregates could also elicits a chronic inflammatory reaction, synaptic changes, and neuronal death. In fact, mounting findings suggest the existence of a chronic inflammation process include the presence of microglial activation and astrocytosis in the brain of these patients, particularly in the vicinity of protein aggregates. In addition, the compounds released from damaged neurons can induce microglial release of neurotoxic factors aggravating neurodegeneration. Those compounds include neuromelanin, a strong iron chelator. The neuromelanin–iron complex activates microglia in vitro, causing the release of neurotoxic compounds such as tumor necrosis factor-alpha (TNF-α), interleukin-6 (IL-6), and nitric oxide (NO). It has been described increased concentration of total iron in the substantia nigra in PD, although the underlying mechanism is not known. Iron also contributes to increasing the generation of ROS, oxidative stress, and increased protein aggregation, including α-synuclein aggregation. Furthermore, the rapid aggregation of α-synuclein in turn can induce the formation of ROS (Lotharius & Brundin, 2002).

Lastly, dopamine stabilizes the form of protofibrilar α-synuclein, which would be toxic. Thus, in the oxidative environment of a dopaminergic neuron, α-synuclein is involved in generating a vicious circle which leads to neuronal death. In the CNS, the catecholamines are an important source of free radicals, such step is catalyzed by MAO during metabolic breakdown of DA, serotonin, and noradrenaline (NA) produces peroxide ($H_2O_2$). DA is a biogenic monoamine produced at substantia nigra neurons that project to the caudate and putamen, which only under certain conditions cross the blood–brain barrier. This neurotransmitter is synthesized from L-tirosina through the sequential action of tyrosine hydroxylase (TH) and L-amino acid decarboxylase (AADC). The conversion of L-tyrosine to hydroxylated L-DOPA is achieved from the TH enzyme and the required cofactor tetrahidrobiopteridinas. HT is the reaction-limiting step in the synthesis of DA and this soluble enzyme is located in the dopaminergic nerve terminals. DA metabolism under conditions of homeostasis is itself oxidative. MAO oxidizes DA to 3,4-dihydroxyphenylacetic (DOPAC), generating $H_2O_2$ in the presence of ferrous ion, which is relatively abundant in the basal ganglia, then hydroxyl free radicals is generated. However, in a redox balance, free radicals are offset by the brain antioxidant systems. When the antioxidant protective mechanisms are inadequate and fail to counteract the increase of free radicals, a state of oxidative stress, which induces damage and death in dopaminergic neurons occurs. Furthermore, it has been reported that DA can form reactive metabolites through a secondary reaction pathway, which compounds are also directly toxic to the cell. Due to the unstable nature of the phenolic ring of DA, it can be oxidized to quinones reactive molecules called DAQ's, which once formed can react with cysteine and produce cysteinyl residues; these structures are capable of inhibiting the function of many proteins, and decreased ATP production by blocking the mitochondrial respiratory chain (Fig. 6). However, part of the reactive species generated is neutralized primarily by neuromelanin, which is a polymer with iron chelating center. Paradoxically neuromelanin it formed from DAQ's (The DA-*p*-quinone imine). In addition to this, DA metabolism tends to collaterally form other toxic species that can damage dopaminergic neurons nonspecifically (LaVoie, Ostaszewski, Weihofen, Schlossmacher, & Selkoe, 2005).

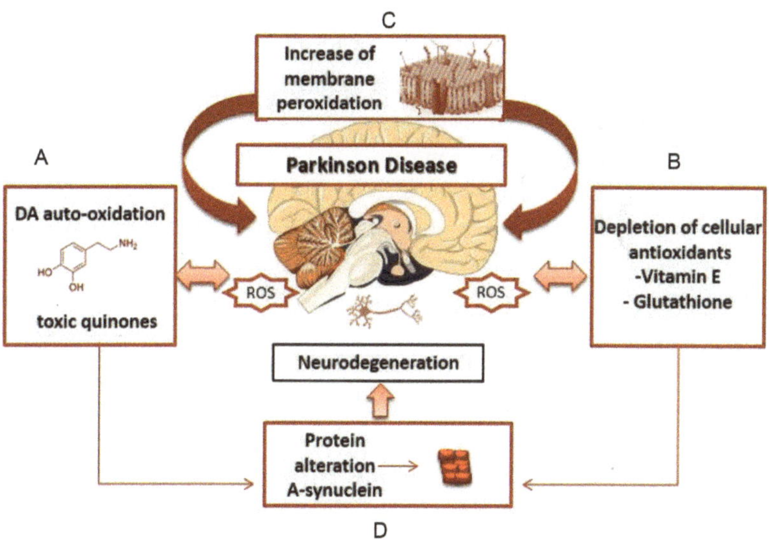


**Fig. 6** Oxidative stress and Parkinson's disease. Different factors contribute to high concentration of reactive oxygen species (ROS) producing oxidative damage that leads to neurodegeneration, destruction of dopaminergic neurons, and cell death including (A) high susceptibility of dopamine (DA) autooxidation producing toxic compounds such as quinones, (B) reduced of endogenous antioxidant activity, (C) high susceptibility of membrane lipoperoxidation, (D) PD-associated proteins alteration, abnormal protein aggregates could also cause a chronic inflammatory reaction and induce synaptic changes.

## 4. CEREBRAL ISCHEMIA AND ROS

The human brain lacks energy stores, so the CNS cells require a continuous blood flow and oxygen supply necessary to maintain their metabolic activity; this makes the brain especially vulnerable to ischemia. With cerebral ischemia a series of events that can lead to cell death if ischemia is of sufficient intensity and prolonged time are triggered. Initially, an energy depletion occurs and molecules that excitotoxicity and oxidative stress in the ischemic tissue (ischemic cascade) are released; at the same time starts a local inflammatory response, longer in time, which can amplify the ischemic brain damage (neuroinflammatory cascade) (Rodríguez-Yáñez & Castillo, 2008). These changes will not be viewed in isolation but in the broader context of interrelationship between the different neural elements. The cells of the nervous system interact with the extracellular matrix in order to maintain adequate functionalism. Neurons, astrocyte, and endothelial cells forming microvessels represent a "neurovascular unit" because there is a close relationship between them, but also involved other cell types (Del Zoppo, 2010). The "neurovascular unit" is a concept that aims to integrate the changes occurring in the brain tissue during ischemia, such as alteration of the blood–brain barrier due to the activation of matrix metalloproteases and the effects turn this disruption causes the elements of the neurovascular unit. During cerebral ischemia production of energy by mitochondrial ATP synthase abruptly ceases. Then the $Na+/K+$-ATPase function fails and a profound loss of ionic gradients and the depolarization of neurons and astrocytes. Membrane depolarization and changes in the concentration gradients of $Na+$ and $K+$ across the plasma membrane result in activation of voltage-gated calcium channels. This leads to excessive release of excitatory glutamate to the extracellular compartment (Fig. 7; Del Zoppo, 2010).

### 4.1 The Massive Influx of Calcium Into the Cell

Among other effects, the anoxic depolarization determines a flow of extracellular calcium into the cell but also the mobilization of calcium of the endoplasmic reticulum to the cytosol. Besides lactic acidosis caused by anaerobic metabolism of glucose displaces calcium binding to intracellular proteins, further increasing the concentration of intracellular free calcium (García-Sancho, 2014). The increase in intracellular calcium concentration has proved instrumental in the process of cellular damage it induces the following phenomena:

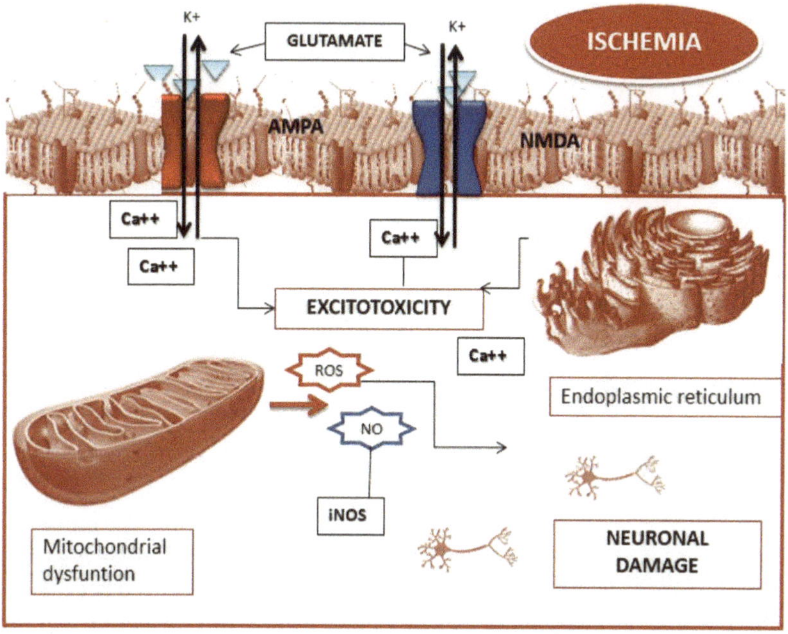


**Fig. 7** Oxidative stress in cerebral ischemia. With cerebral ischemia, excitotoxicity and oxidative stress in the ischemic tissue are triggered; anoxic depolarization determines a flow of extracellular calcium into the cell but also the mobilization of calcium found in the endoplasmic reticulum to the cytosol, the release of neurotransmitters such as glutamate acts mainly by AMPA and NMDA receptors. Glutamate also induces the formation of free radicals and reactive oxygen species (ROS) due to mitochondrial dysfunction, microglial activation induce the activity of the inducible isoform of the enzyme nitric oxide synthase (iNOS) increasing the concentration of nitric oxide (NO), all involved in neuronal damage that occurs during ischemia that can lead to cell death.

- The release of neurotransmitters such as glutamate, responsible for much of the phenomena of excitotoxicity, with synthesis and release of free radicals in ischemic brain tissue.
- Inhibits the production of energy (ATP).
- Active enzymes involved in the degradation of proteins, nucleic acids, and phospholipids.

## 4.2 Glutamate Excitotoxicity

Depolarization that occurs during cerebral ischemia itself induces neurotransmitter release. Of all the neurotransmitters released during the ischemic cascade glutamate is playing a greater role for their toxicity on neurons.

Glutamate acts on the one hand, on their mainly AMPA and NMDA receptors displacing magnesium type in rest position, it acts to block the passage of other ions; thus, a new path which allows passage of calcium ions into the cell is opened. Glutamate also induces the formation of free radicals and ROS. Free radicals such as nitric oxide (NO) and ROS have, under normal conditions, physiological functions; in fact in the brain, it does not act as a second messenger and causes arteriolar vasodilatation when it acts on the blood vessels. NO is synthesized from L-arginine by the action of the enzyme NOS, of which three known isoforms: form neuronal (nNOS), endothelial form (eNOS), and inducible form (iNOS). NO has a very short half life and under normal conditions, small quantities produced are metabolized by enzyme mechanisms (SOD, CAT, and GSH-Px) and nonenzymatic mechanisms such as vitamins C and E and glutathione; so there is a balance between the radicals that are produced and which are removed (García-Sancho, 2014). During cerebral ischemia abnormally high concentrations of these oxidants, among which are the superoxide anion, hydroxyl radical, nitric oxide, and peroxynitrite anion produced; all involved in neuronal damage that occurs during ischemia; while the $O_2$ is considered the most important oxidizing agent, it causes brain damage and reactivity directly with NO, ONOO – when generated. The activity of the neuronal isoforms (nNOS) and endothelial (eNOS) of NOS, localized in neurons and endothelial cells, respectively, is regulated by calcium. Its activation during cerebral ischemia occurs primarily in the initial phases. The inducible isoform (iNOS) has been identified in different cell types of the nervous system as astroglia, microglia, neurons, smooth muscle cells and vascular endothelium, and in the infiltrated neutrophils in ischemic brain tissue (Pérez-Asensio et al., 2005). This isoform is induced by inflammatory mediators and is independent of calcium, and causes, after a delay, significant increases in the concentration of NO in the ischemic area that contribute importantly to the progression of brain damage.

## 4.3 The Harmful Effects of Free Radicals Released in the Ischemic Brain

- NO produced in large quantities reacts with $O_2^-$, leading to peroxynitrite anion (ONOO –), which is highly toxic due to their high oxidizing ability, among other effects occurring lipid peroxidation, nitration of tyrosine groups, oxidation and nitrosylation groups DNA breakage, and sulfhydryl (Rodríguez-González, Hurtado, Sobrino, & Castillo, 2007).

- NO also inhibits antioxidant enzymes such as glutathione peroxidase and cytochrome c oxidase found in mitochondria, the respiratory chain altering. NO and $O_2$ compete for binding to cytochrome oxidase. High concentrations of NO can displace $O_2^-$, which in turn may react with NO to form peroxynitrite anion (ONOO −). Inhibition of mitochondrial respiratory chain causes the production of ATP is reduced (Fernandez-Velasco, Ruiz-Hurtado, Hurtado, Moro, & Delgado, 2007).
- NO induces the fusion of synaptic vesicles to the membrane with release of neurotransmitters such as glutamate into the synaptic cleft, further contributing to excitotoxic damage.

It seems that free radicals have a beneficial effect when they act in isolation at the vascular level, as they have a vasodilator and antiagregante effect (Cuenca-López et al., 2010).

During cerebral ischemia glutamatergic stimulation is maintained over time as ischemic depolarization prevents the glutamate transporter, bound to the cell membrane, to function properly, internalizing glutamate into synaptic vesicles; thus, the concentration of glutamate in the synaptic cleft is greater excitotoxic stimulus and maintains high concentrations of intracellular calcium and increased formation of free radicals and ROS, with their deleterious effect on cells by oxidative stress.

Astrocytes are fundamental in controlling the action of glutamate, because inside the glutamate is converted into glutamine by the action of the enzyme glutamine synthetase. Thus, glutamine may be used again by neurons for the synthesis of glutamate and GABA. During cerebral ischemia, as a result of power failure and dysfunction of the ion channels, cell edema involving first astrocytes, so glutamate uptake by these cells is reduced occurs (Castellanos et al., 2006).

## 4.4 Free Radicals Derived From Phospholipase A2 Activity

Phospholipase A2, which exists various isoforms, is an enzyme that is overactive during cerebral ischemia due to the influx of calcium into the cell. Their action causes an accumulation of fatty acids such as arachidonic acid and docosahaxaenoic acid that can decouple oxidative phosphorylation and alter the permeability of the cell membrane and its ion channels. Furthermore, overstimulation of phospholipase A2 leads to increased degradation products membrane phospholipids. Some accumulated lysophospholipids are readily convertible to platelet activating factor, a promoting effect on platelet aggregation, and inducers of the inflammatory response by adhesion and aggregation of leukocytes (Nanda et al., 2007).

## 5. MULTIPLE SCLEROSIS

MS is a chronic inflammatory disease, demyelinating CNS, considered as a multifactorial disease in which participate genetic, immunological, and environmental factors involved in the autoimmune process and the own pathological changes of the disease. It has been proposed to certain viruses carriers autoantigens generate molecular mimicry with myelin proteins and lead to a loss of tolerance against these, which results in the destruction of myelin mediated by activated T cells in the white matter of the brain, sometimes it is extending into the gray matter, resulting in defects in the conduction of nerve impulses, resulting in a wide range of symptoms as the affected site of the brain or spinal cord. Depending on the areas of myelin destruction motor or sensory symptoms were present, with balance disorders or vision. Between an "outbreak" or relapse (appearance of new symptoms or worsening of existing ones) may take days or years. Demyelinating lesions (plaques) spread through the CNS and differ both size and location. The onset of symptoms and response to treatment depends on each patient. MS is the most common cause of neurological disability in young adults. It is characterized by perivenous infiltration of lymphocytes and macrophages in the brain parenchyma with disseminated lesions typical. According to the clinical course, MS is classified in relapsing-remitting, primary progressive, and secondary progressive. His clinical picture can vary from a benign self-limiting disorder to severe and highly disabling disease (Olson, Ercolini, & Miller, 2005).

### 5.1 Oxidative Stress and MS

Chronic inflammation increases ROS production so antioxidant defenses are overcome leading to oxidative stress (Miller, Mrowicka, Zołyński, & Kedziora, 2009). The SNC is particularly vulnerable to oxidative damage, since by its active mitochondrial metabolism generates high levels of intracellular superoxide. Moreover, oligodendrocytes have low levels of antioxidant enzymes, high iron content, and a high ratio of lipid to protein in myelin so it is a preferred target of ROS (45). Studies in our laboratory together with other reports have corroborated an increase in markers of oxidative stress in MS. In this respect, we found a significant increase in serum levels of nitric oxide catabolites (nitrite–nitrate), products of lipid peroxidation (malondialdehyde and 4-hidroxialkenals), and glutathione peroxidase activity in patients with remitting-recurrent MS compared to healthy subjects (Ortiz et al., 2009).

Oxidative stress levels are directly related to the progression of MS. There are reports suggesting that the loss of myelin nerve sheath is possible because the immune system also participates with defects in mitochondria, in the generation of free radicals of nitrogen and oxygen. Macrophages and monocytes release mediators of oxidative stress in the cells and myelin directly attack by the trimolecular complex as mentioned earlier. Indeed, ROS have been implicated as mediators of axonal demyelination in MS and experimental autoimmune encephalomyelitis (EAE) (Constantinescu, Farooqi, O'Brien, & Gran, 2011). Furthermore, the direct examination of MS plaques revealed increased activity of free radicals, together with the decreased levels of important antioxidants such as glutathione and alpha-tocopherol (Van Horssen, Witte, Schreibelt, & de Vries, 2011).

The main proposed mechanisms to explain the interaction of ROS and the immune system that contribute to neurological damage in MS are

- low levels of antioxidants which promote increased activity of the lipoxygenase enzyme, favoring the production of leukotrienes and increasing the immunoinflammatory processes in the cerebral cortex
- excess reactive oxygen and nitrogen species starts increased activity of T cells via the arachidonic acid cascade and produce direct damage to myelin
- oxidative stress provides important information about the underlying CNS tissue damage in MS by markers hydroxyl radical activity of lipid peroxides level of reduced glutathione, the activity of SOD and GSH-Px (Fig. 8) (Lutsky, Zemskov, & Razinkin, 2014).

Other reports implicating ROS in demyelination in MS constitute paraclinical studies have shown an increase in RNA in serum and CSF lymphocytes of MS patients, correlating them with pathological studies. Peroxynitrite (ONOO −), reactive nitrogen species, is closely associated with acute but rarely seen in the inactive chronic inflammatory lesions plates (Yatin et al., 2000).

Damage to axons mediated by the combination of ROS produced by mitochondrial dysfunction and immunological participation: Nitric oxide decreases mitochondrial energy metabolism and inhibits the respiratory chain, which causes:

**(1)** Decreased activity of ATPase $Na+/K+$ and altered glutamate transporters dependent $Na+$;

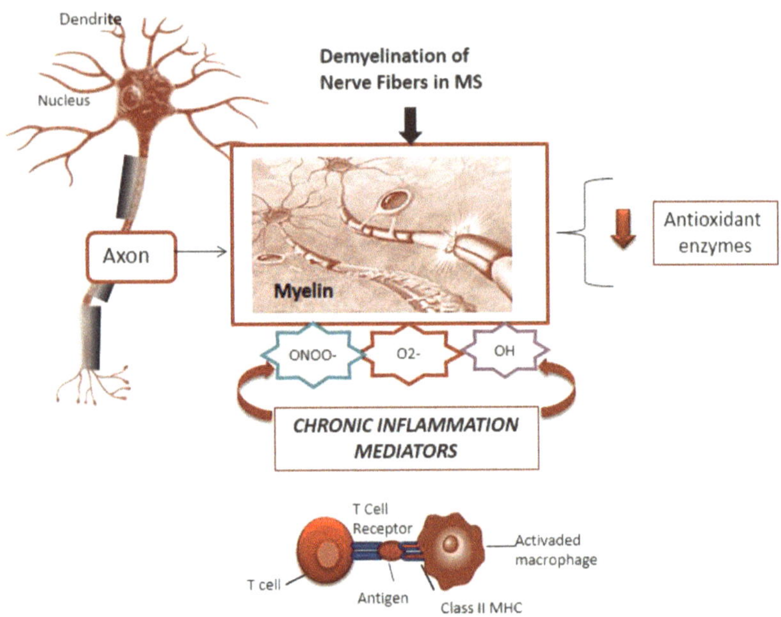


**Fig. 8** Oxidative stress, demyelination, and axonal damage in MS. Oxidative stress plays a major role in the pathogenesis of multiple sclerosis (MS), free radicals are generated in excess primarily by macrophages and have been implicated as mediators of demyelination and axonal damage. Chronic inflammation increases ROS production so antioxidant defenses are overcome, resulting in axonal demyelination, which leads to the slowing or blockade of conduction.

(2) Overexpression of glutamate receptors;
(3) Oligodendroglial excitotoxicity;
(4) Massive influx of extracellular $Ca^{2+}$;
(5) Activation of proteases;
(6) Alteration of the cytoskeleton and the axonal flow.

Of these mechanisms, it produces nitric oxide, and nitrosative stress produces glutamate excitotoxicity. Nitric oxide is a highly toxic free radical itself at high concentrations produces conduction block, especially in demyelinated axons and stimulates apoptosis. When the oxide NO reacts with superoxide produces peroxynitrite prooxidant high capacity. Glutamate in turn causes neurodegeneration through the AMPA and NMDA receptors in oligodendrocytes and astrocytes. The role of mediators has been shown both in experimental models and experimental autoimmune encephalitis occur to protection against experimental disease after administration of antagonists of such receptors. Under physiologic conditions, nitric oxide is produced from L-arginine by nitric oxide synthase constitutive (cNOS) and mediate a variety of important biological functions such as immunoregulation of inflammatory reactions, the downregulation of TNFa production, MHC II expression on macrophages, inducing apoptosis in CD4 cells, the physiological regulation of the mitochondrial respiratory chain, as well as inhibition of antigen presentation, leukocyte adhesion, and migration. However, during inflammatory reactions, exposure of macrophages to IFNY and TNF results in the activation of the inducible form of NOS (iNOS), an enzyme that produces, for longer periods of time, an amount up to 10 times nitric oxide and superoxide • the cNOS. Increasing these two radicals facilitates the formation of peroxynitrite radical. This is achieved under both free radicals interact at a rate whose speed depends only on the diffusion and thus is much faster than the rate at which the SOD captures the $O_2$. Only cells capable of generating a high flow of nitric oxide may have the potential to cause nitrosative stress. The role of nitric oxide in MS is, therefore, complex; in fact, it is very likely that its derivatives, especially ONOO −, are definitely more toxic than the same NO. Nitric oxide and glutamate also play a role in mediating chronic axonal damage (Yuste, Tarragon, Campuzano, & Ros-Bernal, 2015).

## 5.2 ROS, Cytokines, and Axonal Damage in MS

Possible mechanisms of axonal damage include cytotoxicity mediators, TNF, MMPs, ROS, antibodies, and excitotoxicity. It documented increased glutamate and aspartate excitotoxicity in MS patients. Glutamate is increased in patients with MS both active lesions and in normal-appearing white matter. It is noteworthy that the mature oligodendrocytes and astrocytes are highly sensitive to glutamate due to the expression of AMPA and NMDA receptors.

The PBM is blank oligodendrocyte CD8 α/β, associated with classical cytotoxic MHC class I. The myelin sheath can be damaged by cytokines, autoantibodies, ROS, proteolytic enzymes, and phagocytosis. Increased ROS by activated microglia during the immune response generates a state of increased lipid peroxidation, and oligodendrocyte cell is the cell more susceptible to damage by ROS. Myelin degradation may be the result of lipid peroxidation-mediated peroxides, but the role of these specific toxic factors in the pathogenesis of MS remained obscure until very recently.

At the end of the day, we exhibited two realities: we design a system that proved to be the most efficient way to produce energy called ATP; but this as all (at least in this world) has errors and drawbacks, the formation of ROS. Moreover also we designed a self-recognition system (immune system) that generates ROS to protect us, but their actions are also in many cases exacerbated and lead to collateral damage (Yuste et al., 2015).

The CNS remains highly vulnerable place; it has a special barrier (blood–brain barrier), and play with own rules (biochemical-immunological), and makes us look even too young to try to explain many unresolved problems; not only in the normal biochemical and physiological explanations. In the case harmful of ROS: hero and villain only the interdisciplinary and systematic research will give us the answers that we even asked to this marvelous organ (brain), which apparently is the only one who studying himself.

## REFERENCES

Absi, E., Ayala, A., Machado, A., & Parrado, J. (2000). Protective effect of melatonin against the 1-methyl-4-phenylpyridinium-induced inhibition of complex I of the mitochondrial respiratory chain. *Journal of Pineal Research*, *29*(1), 40–47. http://dx.doi.org/10.1034/j.1600-079X.2000.290106.x.

Barja, G., & Herrero, A. (2000). Oxidative damage to mitochondrial DNA is inversely related to maximum life span in the heart and brain of mammals. *FASEB Journal: Official Publication of the Federation of American Societies for Experimental Biology*, *14*(2), 312–318. http://dx.doi.org/0892-6638/00/0014-0312.

Butterfield, D. A., Swomley, A. M., & Sultana, R. (2013). Amyloid β-peptide (1–42)-induced oxidative stress in Alzheimer disease: Importance in disease pathogenesis and progression. *Antioxidants & Redox Signaling*, *19*(8), 823–835. http://dx.doi.org/10.1089/ars.2012.5027.

Castellani, R. J., Rolston, R. K., & Smith, M. A. (2011). Alzheimer disease. *Disease-a-Month*, *56*(9), 484–546. http://dx.doi.org/10.1016/j.disamonth.2010.06.001.

Castellanos, M., Sobrino, T., & Castillo, J. (2006). Evolving paradigms for neuroprotection: Molecular identification of ischemic penumbra. *Cerebrovascular Diseases (Basel, Switzerland)*, *21*(Suppl. 2), 71–79. http://dx.doi.org/10.1159/000091706.

Constantinescu, C. S., Farooqi, N., O'Brien, K., & Gran, B. (2011). Experimental autoimmune encephalomyelitis (EAE) as a model for multiple sclerosis (MS). *British Journal of Pharmacology*, *164*(4), 1079–1106. http://dx.doi.org/10.1111/j.1476-5381.2011.01302.x.

Cuenca-López, M. D., Brea, D., Galindo, M. F., Antón-Martínez, D., Sanz, M. J., Agulla, J., … Jordán, J. (2010). Inflammatory response during ischaemic processes: Adhesion molecules and immunomodulation. *Revista de Neurologia*, *51*(1), 30–40.

Davies, K. J. (1995). Oxidative stress: The paradox of aerobic life. *Biochemical Society Symposium*, *61*, 1–31. http://dx.doi.org/10.1042/bss0610001.

Del Zoppo, G. J. (2010). The neurovascular unit in the setting of stroke. *Journal of Internal Medicine*, *267*(2), 156–171. http://dx.doi.org/10.1111/j.1365-2796.2009.02199.x.

Emerit, J., Edeas, M., & Bricaire, F. (2004). Neurodegenerative diseases and oxidative stress. *Biomedicine & Pharmacotherapy = Biomédecine & Pharmacothérapie*, *58*(1), 39–46. http://dx.doi.org/10.1016/j.biopha.2003.11.004.

Fernandez-Velasco, M., Ruiz-Hurtado, G., Hurtado, O., Moro, M. A., & Delgado, C. (2007). TNF-alpha downregulates transient outward potassium current in rat ventricular myocytes through iNOS overexpression and oxidant species generation. *American Journal of Physiology. Heart and Circulatory Physiology, 293*(1), H238–H245. http://dx.doi.org/10.1152/ajpheart.01122.2006.

Forno, L. S. (1996). Neuropathology of Parkinson's disease. *Journal of Neuropathology and Experimental Neurology, 55*(3), 259–272. http://dx.doi.org/10.1097/00005072-199603000-00001.

García-Sancho, J. (2014). The coupling of plasma membrane calcium entry to calcium uptake by endoplasmic reticulum and mitochondria. *The Journal of Physiology, 592*(2), 261–268. http://dx.doi.org/10.1113/jphysiol.2013.255661.

Gella, A., & Durany, N. (2009). Oxidative stress in Alzheimer disease. *Cell Adhesion & Migration, 3*(1), 88–93. http://dx.doi.org/10.4161/cam.3.1.7402.

Georgieff, M. K. (2007). Nutrition and the developing brain. *The American Journal of Clinical Nutrition, 85*(2), 614S–620S. http://dx.doi.org/10.1038/266751a0.

Halliwell, B. (2006). Oxidative stress and neurodegeneration: Where are we now? *Journal of Neurochemistry, 97*(6), 1634–1658. http://dx.doi.org/10.1111/j.1471-4159.2006.03907.x.

Halliwell, B., & Gutteridge, J. M. (1995). The definition and measurement of antioxidants in biological systems. *Free Radical Biology and Medicine, 18*(1), 125–126. http://dx.doi.org/10.1016/0891-5849(95)91457-3.

Jan, A., Azam, M., Siddiqui, K., Ali, A., Choi, I., & Haq, Q. (2015). Heavy metals and human health: Mechanistic insight into toxicity and counter defense system of antioxidants. *International Journal of Molecular Sciences, 16*(12), 29592–29630. http://dx.doi.org/10.3390/ijms161226183.

Junn, E., & Mouradian, M. M. (2002). Human α-Synuclein over-expression increases intracellular reactive oxygen species levels and susceptibility to dopamine. *Neuroscience Letters, 320*(3), 146–150. http://dx.doi.org/10.1016/S0304-3940(02)00016-2.

Juránek, I., Nikitovic, D., Kouretas, D., Hayes, A. W., & Tsatsakis, A. M. (2013). Biological importance of reactive oxygen species in relation to difficulties of treating pathologies involving oxidative stress by exogenous antioxidants. *Food and Chemical Toxicology, 61*, 240–247. http://dx.doi.org/10.1016/j.fct.2013.08.074.

Kang, D., Narabayashi, H., Sata, T., & Takeshige, K. (1983). Kinetics of superoxide formation by respiratory chain NADH-dehydrogenase of bovine heart mitochondria. *Journal of Biochemistry, 94*(4), 1301–1306.

Kardeh, S., Ashkani-Esfahani, S., & Alizadeh, A. M. (2014). Paradoxical action of reactive oxygen species in creation and therapy of cancer. *European Journal of Pharmacology, 735*(1), 150–168. http://dx.doi.org/10.1016/j.ejphar.2014.04.023.

LaVoie, M. J., Ostaszewski, B. L., Weihofen, A., Schlossmacher, M. G., & Selkoe, D. J. (2005). Dopamine covalently modifies and functionally inactivates parkin. *Nature Medicine, 11*(11), 1214–1221. http://dx.doi.org/10.1038/nm1314.

Lobo, V., Patil, A., Phatak, A., & Chandra, N. (2010). Free radicals, antioxidants and functional foods: Impact on human health. *Pharmacognosy Reviews, 4*(8), 118–126. http://dx.doi.org/10.4103/0973-7847.70902.

Lotharius, J., & Brundin, P. (2002). Pathogenesis of Parkinson's disease: Dopamine, vesicles and alpha-synuclein. *Nature Reviews. Neuroscience, 3*(12), 932–942. http://dx.doi.org/10.1038/nrn983.

Lutsky, M. A., Zemskov, A. M., & Razinkin, K. A. (2014). [Biochemical markers of oxidative stress in different forms and phases of multiple sclerosis]. *Zhurnal Nevrologii I Psikhiatrii Imeni S.S Korsakova, 114*(11), 74–77.

Majmundar, A. J., Wong, W. J., & Simon, M. C. (2010). Hypoxia-inducible factors and the response to hypoxic stress. *Molecular Cell, 40*(2), 294–309. http://dx.doi.org/10.1016/j.molcel.2010.09.022.

Maker, H. S., Weiss, C., Silides, D. J., & Cohen, G. (1981). Coupling of dopamine oxidation (monoamine oxidase activity) to glutathione oxidation via the generation of hydrogen peroxide in rat brain homogenates. *Journal of Neurochemistry, 36*(2), 589–593. http://dx.doi.org/10.1111/j.1471-4159.1981.tb01631.x.

Mancuso, M., Orsucci, D., Siciliano, G., & Murri, L. (2008). Mitochondria, mitochondrial DNA and Alzheimer's disease. What comes first? *Current Alzheimer Research, 5*(5), 457–468. http://dx.doi.org/10.2174/156720508785908946.

McCord, J. M. (2000). The evolution of free radicals and oxidative stress. *American Journal of Medicine, 108*(8), 652–659. http://dx.doi.org/10.1016/S0002-9343(00)00412-5.

Miller, E., Mrowicka, M., Zołyński, K., & Kedziora, J. (2009). Oxidative stress in multiple sclerosis. *Polski Merkuriusz Lekarski: Organ Polskiego Towarzystwa Lekarskiego, 27*(162), 499–502.

Nanda, B., Nataraju, A., Rajesh, R., Rangappa, K., Shekar, M., & Vishwanath, B. (2007). PLA2 mediated arachidonate free radicals: PLA2 inhibition and neutralization of free radicals by anti-oxidants—A New role as anti-inflammatory molecule. *Current Topics in Medicinal Chemistry*, 7(8), 765–777. http://dx.doi.org/10.2174/156802607780487623.

Olson, J. K., Ercolini, A. M., & Miller, S. D. (2005). A virus-induced molecular mimicry model of multiple sclerosis. *Current Topics in Microbiology and Immunology, 296*(2), 39–53. http://dx.doi.org/10.1172/JCI200113032.

Ortiz, G. G., Macías-Islas, M. A., Pacheco-Moisés, F. P., Cruz-Ramos, J. A., Sustersik, S., Barba, E. A., & Aguayo, A. (2009). Oxidative stress is increased in serum from Mexican patients with relapsing-remitting multiple sclerosis. *Disease Markers, 26*(1), 35–39. http://dx.doi.org/10.3233/DMA-2009-0602.

Pérez-Asensio, F. J., Hurtado, O., Burguete, M. C., Moro, M. A., Salom, J. B., Lizasoain, I., … Lorenzo, P. (2005). Inhibition of iNOS activity by 1400 W decreases glutamate release and ameliorates stroke outcome after experimental ischemia. *Neurobiology of Disease, 18*(2), 375–384. http://dx.doi.org/10.1016/j.nbd.2004.10.018.

Pham-Huy, L. A., He, H., & Pham-Huy, C. (2008). Free radicals, antioxidants in disease and health. *International Journal of Biomedical Science, 4*(2), 89–96.

Pohanka, M. (2013). Alzheimer's disease and oxidative stress: A review. *Current Medicinal Chemistry, 21*(3), 356–364. http://dx.doi.org/10.2174/09298673113206660258.

Poljsak, B., & Milisav, I. (2013). Aging, oxidative stress and antioxidants. In J. A. Morales-González (Ed.), *Oxidative stress and chronic degenerative diseases–A role for antioxidants* (pp. 331–353). Croatia: InTech.

Querfurth, H. W., & LaFerla, F. M. (2010). Alzheimer's disease. *New England Journal of Medicine, 362*(4), 329–344. http://dx.doi.org/10.1056/NEJMra0909142.

Rahman, K. (2007). Studies on free radicals, antioxidants, and co-factors. *Clinical Interventions in Aging, 2*(2), 219–236.

Reisberg, B., Borenstein, J., Salob, S. P., Ferris, S. H., Franssen, E., & Georgotas, A. (1987). Behavioral symptoms in Alzheimer's disease: Phenomenology and treatment. *The Journal of Clinical Psychiatry, 48*(Suppl.), 9–15.

Rodríguez-González, R., Hurtado, O., Sobrino, T., & Castillo, J. (2007). Neuroplasticity and cellular therapy in cerebral infarction. *Cerebrovascular Diseases (Basel, Switzerland), 24*(Suppl. 1), 167–180. http://dx.doi.org/10.1159/000107393.

Rodríguez-Yáñez, M., & Castillo, J. (2008). Role of inflammatory markers in brain ischemia. *Current Opinion in Neurology, 21*(3), 353–357. http://dx.doi.org/10.1097/WCO.0b013e3282ffafbf.

Schultz, A. M., Chao, S. M., & Mcginnis, J. M. (2009). *Integrative medicine and the health of the public: A summary of the february 2009 Summit*. Institute of Medicine, USA: The National Academies Press. http://dx.doi.org/10.17226/12668.

Swerdlow, R., Burns, J., & Khan, S. (2010). The AD mitochondrial cascade hypothesis. *Journal of Alzheimer's Disease, 20*(Suppl. 2), S265–S279. http://dx.doi.org/10.3233/JAD-2010-100339.

Togo, T., Katsuse, O., & Iseki, E. (2004). Nitric oxide pathways in Alzheimer's disease and other neurodegenerative dementias. *Neurological Research, 26*(5), 563–566. http://dx.doi.org/10.1179/016164104225016236.

Van Horssen, J., Witte, M. E., Schreibelt, G., & de Vries, H. E. (2011). Radical changes in multiple sclerosis pathogenesis. *Biochimica et Biophysica Acta-Molecular Basis of Disease, 1812*(2), 141–150. http://dx.doi.org/10.1016/j.bbadis.2010.06.011.

Yatin, S. M., Varadarajan, S., & Butterfield, D. A. (2000). Vitamin E prevents Alzheimer's amyloid beta-peptide (1-42)-induced neuronal protein oxidation and reactive oxygen species production. *Journal of Alzheimer's Disease, 2*(2), 123–131.

Yuste, J. E., Tarragon, E., Campuzano, C. M., & Ros-Bernal, F. (2015). Implications of glial nitric oxide in neurodegenerative diseases. *Frontiers in Cellular Neuroscience, 9*, 322. http://dx.doi.org/10.3389/fncel.2015.00322.

CHAPTER TWO

# Neuroinflammation in Alzheimer's Disease: The Preventive and Therapeutic Potential of Polyphenolic Nutraceuticals

**Yousef Sawikr**[*,1], **Nagendra Sastry Yarla**[†,1], **Ilaria Peluso**[‡], **Mohammad Amjad Kamal**[§,¶], **Gjumrakch Aliev**[||,#,**,2], **Anupam Bishayee**[††,3]

[*]Faculty of Medicine, University of Benghazi, Benghazi, Libya
[†]Institute of Science, GITAM University, Visakhapatnam, Andhra Pradesh, India
[‡]Centre for Food and Nutrition, Council for Agricultural Research and Economics, Rome, Italy
[§]Metabolomics and Enzymology Unit, Fundamental and Applied Biology Group, King Fahd Medical Research Center, King Abdulaziz University, Jeddah, Saudi Arabia
[¶]Enzymoics and Novel Global Community Educational Foundation, Hebersham, NSW, Australia
[||]"GALLY" International Biomedical Research Consulting LLC, San Antonio, TX, United States
[#]School of Health Sciences and Healthcare Administration, University of Atlanta, Johns Creek, GA, United States
[**]Institute of Physiologically Active Compounds Russian Academy of Sciences, Chernogolovka, Russia
[††]College of Pharmacy, Larkin Health Sciences Institute, Miami, FL, United States
[2]Corresponding author: e-mail address: aliev03@gmail.com
[3]Corresponding author: e-mail addresses: abishayee@ULarkin.org, abishayee@gmail.com

## Contents



## Abstract

Brain inflammation, characterized by increased microglia and astrocyte activation, increases during aging and is a key feature of neurodegenerative diseases, such as Alzheimer's disease (AD). In AD, neuronal death and synaptic impairment, induced by amyloid-β (Aβ) peptide, are at least in part mediated by microglia and astrocyte activation. Glial activation results in the sustained production of proinflammatory cytokines

[1] These authors contributed equally.

*Advances in Protein Chemistry and Structural Biology*, Volume 108
ISSN 1876-1623
http://dx.doi.org/10.1016/bs.apcsb.2017.02.001

and reactive oxygen species, giving rise to a chronic inflammatory process. Astrocytes are the most abundant glial cells in the central nervous system and are involved in the neuroinflammation. Astrocytes can be activated by numerous factors, including free saturated fatty acids, pathogens, lipopolysaccharide, and oxidative stress. Activation of astrocytes produces inflammatory cytokines and the enzyme cyclooxygenase-2, enhancing the production of Aβ. Furthermore, the role of the receptor for advanced glycation end products/nuclear factor-κB (NF-κB) axis in neuroinflammation is in line with the nonenzymatic glycosylation theory of aging, suggesting a central role of the advanced glycation end products in the age-related cognitive and a possible role of nutraceuticals in the prevention of neuroinflammation and AD. However, modulation of P-glycoprotein, rather than antioxidant and anti-inflammatory effects, could be the major mechanism of polyphenolic compounds, including flavonoids. Curcumin, resvertrol, piperine, and other polyphenols have been explored as novel therapeutic and preventive agents for AD. The aim of this review is to critically analyze and discuss the mechanisms involved in neuroinflammation and the possible role of nutraceuticals in the prevention and therapy of AD by targeting neuroinflammation.

## 1. INTRODUCTION

Alzheimer's disease (AD) is a chronic irreversible neurodegenerative disease, with a progressive loss of basal forebrain cholinergic neurons and clinical characteristics of memory loss, dementia, and impairment in memory, accounting for 60%–70% of dementia cases in the elderly (Hebert, Scherr, Bienias, Bennett, & Evans, 2003). AD affects 20–30 million individuals worldwide (Selkoe, 2005). The prevalence of AD increases with age, affecting approximately 1%–3% of the population around 60 years of life, 3%–12% of the population between 70 and 80 years, and up to 25%–35% of the population older than 85 years (Walsh & Selkoe, 2004). The earliest sign of AD is an impairment of the recent memory function and of the attention followed by the failure of the language skills, abstract thinking, and judgment, as well as visualspatial orientation. AD gradually progresses to severe dementia and stupor. These defects are accompanied by alterations of personality (Nestor, Scheltens, & Hodges, 2006). Histopathological features of AD are the senile plaques (SPs), composed of the amyloid-β (Aβ) plagues and neurofibrillary tangles (NFTs). The latter are intracellular inclusions of hyperphosphorylated tau protein (Ries & Sastre, 2016). Aβ is a peptide of 42 amino acid residues produced by cleavage of transmembrane amyloid precursor proteins (APP) by the secretases β and γ (Haass & Selkoe, 1993). Aβ can directly induce neuronal cytotoxicity (Sidoryk-Wegrzynowicz, Wegrzynowicz, Lee, Bowman, & Aschner,

2011). Neuronal death and synaptic impairment, induced by Aβ in AD, are at least partially mediated by the activation of astrocyte and microglia. Glial activation results in the sustained production of proinflammatory cytokines and reactive oxygen species (ROS), giving rise to a chronic inflammatory process. Antioxidant and anti-inflammatory activities of pytochemicals have been reported by several investigators throughout the world. In this background, these nutraceuticals have been explored as preventive and therapeutic agents for AD. We aim to review these mechanisms that suggest a possible role of nutraceuticals in the prevention of neuroinflammation and AD.

## 2. NEUROINFLAMMATION IN AD (ALZHEIMER'S DISEASE)

Neuroinflammation is mainly carried out by activated microglia and reactive astrocytes. In the normal brain, microglia does not produce proinflammatory molecules or ROS. However, in pathological situations or after traumatic insults microglia become activated and release proinflammatory molecules (Yucesoy et al., 2006).

Elevated brain concentrations of inflammatory cytokines have been associated with AD (Zilka, Ferencik, & Hulin, 2006). The interleukin-1α (IL-1α), IL-β, IL-6, and tumor necrosis factor-α (TNF-α) are the most extensively studied cytokines in relation with the onset or progression of AD together with the enzyme cyclooxygenase 2 (COX-2) (Ferencik, Novak, Rovensky, & Rybar, 2001; Singhal, Jaehne, Corrigan, Toben, & Baune, 2014).

The production of cytokines in the central nervous system (CNS), partly mediated by Aβ, is accompained by ROS production by activated microglia (Kawaguchi, 2008; López, Jara, & Valenzuela, 2005; Rogers et al., 1996). In particular, IL-1 has a role in the development of AD. In the CNS, IL-1 can be released from astrocytes, microglia, and neurons (Medeiros & LaFerla, 2013). Elevated IL-1 was found in the serum, cerebrospinal fluid, and brain of patients with AD or other dementia (Déniz-Naranjo et al., 2008; Li, Zhao, & Gao, 2011). In astrocytes of the cortex and hippocampus, IL-1β level was dramatically increased by Aβ (Wyss-Coray, 2006). On the other hand, it seems that IL-1 induces astrocytes and neurons to produce Aβ, leading to the deposition of amyloid fibrils (Zilka et al., 2006). The role of IL-1β in the neuronal degeneration and astrogliosis has been suggested by clinical and experimental studies (Griffin, Liu, Li, Mrak, & Barger, 2006; Streit,

2006; Webster, Hansen, & Adame, 2006). IL-1β-induced astrogliosis of the Aβ-treated astrocytes by binding its receptors (Cruz, Tseng, Goldman, Shih, & Tsai, 2003). In particular, IL-1β secreted by astrocytes stimulated the production of APP and neurotoxic Aβ by neurons (Griffin et al., 2006; Wyss-Coray, 2006).

IL-1β activates also other cell types, such as astrocytes and microglia, to induce the cytokine release (e.g., IL-1β, IL-6, and IL-18) and the inducible nitric oxide synthase activity, leading to neurotoxicity (Liu & Chan, 2014). The IL-1β-induced signals are cell type specific. In glial cells, IL-1β induces the NF-κB signaling, increasing the cytokine production. In neurons, IL-1β activates the mitogen-activated protein kinase (MAPK)-p38 signaling cascade, increasing the secreted APP fragment cleaved by the β-secretase of APP (BACE1) and the formation of Aβ (Solito & Sastre, 2012). In fact, IL-1β is overexpressed at the neuroinflammatory sites and induces the MAPK-p38 pathway in the AD brain (Floyd & Hensley, 2000; Sheng et al., 2001).

In particular, IL-1β can induce the phosphorylation of the tau protein and the formation of NFTs through the MAPK-p38 pathway (Barreto, Gonzalez, Cardiológicas, & Taquini, 2008). The IL-1β-induced activation of MAPK-p38 in neurons is involved in the hyperphosphorylation of tau protein, the major component of the NFTs in AD brain (Sheng et al., 2001; Streit, 2006; Webster et al., 2006). Tau stabilizes growing axons and is necessary for the development and the growth of neurites. However, in AD, tau becomes hyperphosphorylated and appears in paired helical filaments, dystrophic neurites, and NFTs (Braak, Braak, & Strothjohann, 1994). This mechanism suggests a loss of axonal integrity and a decline in connectivity and synapses, both related with dementia in AD (Braak et al., 1994). The induction of tau phosphorylation by IL-1 in vitro (Li, Liu, Barger, & Griffin, 2003) and in vivo (Sheng et al., 2001) indicates that IL-1 contributes to the reorganization of the cytoskeleton and result in the loss of synapses. In fact in AD that a loss of synaptophysin is observed in tangle-bearing neurons has been observed (Callahan & Coleman, 1995) and the activated microglia correlate with the neurofibrillary pathology (DiPatre & Gelman, 1997; Sheng, Mrak, & Griffin, 1997).

On the other hand, IL-6, synthetised by microglia, astrocytes, neuronal, and endothelial cells, increases the inflammatory response initiated by IL-1β (Lee, Liu, Dickson, Brosnan, & Berman, 1993) and high levels of IL-6 mRNA have been found in the entorhinal cortex and in the superior temporal gyrus of AD patients (Ge & Lahiri, 2002).

## 3. ASTROGLIOSIS IN AD

Astrocytes constitute about 20%–50% of the human brain volume (Saido, 2003) and have many functions, including the energy reserves, the regulation of extracellular ions, as well as the clearance and metabolism of neurotransmitters (Dong & Benveniste, 2001). Their connections with neurons include also the modulation of oxidative stress. Each single gray matter astrocyte has been estimated to envelope as many as 100,000 synapses (Parpura & Haydon, 2009). Turmoil of the many neurosupportive astrocyte occupations have harmful outcome for the CNS (Wyss-Coray & Rogers, 2012). It has been shown that the neurotransmitter glutamate is released in neuroinflammatory conditions and to some degree under normal circumstances, which on the long-term is proved to be toxic to neurons. The neuroprotective action of astrocytes has also been attributed to their capacity to take up the neurotransmitter glutamate, convert it to glutamine, and recycle it to neurons (Vesce, Rossi, Brambilla, & Volterra, 2007). Aβ decreases the uptake of glutamate, increases oxidative stress, and activates MAPK pathways (Agostinho, Cunha, & Oliveira, 2010). Despite astrocytes are important for Aβ clearance and degradation (Li et al., 2011), for providing support to neurons, and for forming a protective barrier between Aβ deposits and neurons (Rossner & Lange-Dohna, 2005), they could also be a source for Aβ, due to their overexpression of BACE1 in response to chronic stress (Rossner & Lange-Dohna, 2005). Therefore, in addition to neurons, the mainly significant maker of Aβ, astrocytes can create low amount of Aβ (Li et al., 2011).

Migration of astrocytes to amyloid plaques is promoted by the chemokines CCL2 and CCL3, which are released by activated microglial cells (Weiner & Frenkel, 2006). Astrocytes that are recruited to Aβ plaques both mediate neurotoxicity and participate in the clearance of Aβ. In brain divisions from patients with AD, activated astrocytes and activated microglial cells, contain Aβ fragments (Mandrekar-Colucci & Landreth, 2010). Mouse astrocytes plated on amyloid-rich brain sections from *APP*-transgenic mice reduce the amyloid levels in these sections (Wyss-Coray et al., 2003). Furthermore, like microglia, astrocytes take action to pathology with changes in their antigenicity, morphology, and function, and these reactive states can have potentially beneficial and destructive consequences (Kahlson & Colodner, 2015).

It has been shown that astrocytes themselves actively contribute to the inflammatory response (Sidoryk-Wegrzynowicz & Aschner, 2013). Astrocytes respond to CNS insults through a process named reactive astrogliosis, an early pathological feature of AD and can represent a response to the accumulation of amyloid and or to the increasing number of degenerating neurons (Sofroniew, 2009). Astrocytes can be stimulated by oxidative stress, free saturated fatty acids, pathogens, and lipopolysaccharide (Liu, Martin, & Chan, 2013). The activation of astrocytes is longer than that of microglia (Yan, Xiao, Chen, & Cai, 2013). Contrary to quiescent astrocytes, reactive astrocytes can produce cytokines, such as TNF-α, interferon-γ (IFN-γ), and interleukins (Liu et al., 2013). Cytokines, such as IFN-β, TNF-α, and/or IL-1β, induce the generation of Aβ in primary human astrocytes and astrocytoma cells (Saido, 2003).

High levels of proinflammatory cytokines, such as IL-1β, IL6, and TNF-α, have been detected in the brain of AD subjects (Veerhuis, 2011). Proinflammatory molecules produced by the reactive astrocytes can elevate the expression of secretases in neurons, enhancing the production of Aβ and activating microglia to produce inflammatory factors (Li et al., 2011).

Astrogliosis is characterized also by high level of the astrocyte marker glial fibrillary acidic protein (GFAP) (Kamphuis et al., 2012). The latter occurs around amyloid deposits both in the brain parenchyma and in the cerebral microvasculature (Kamphuis et al., 2012). Senile plaques are associated with GFAP-positive-activated astrocytes (Nagele, Wegiel, & Venkataraman, 2004). In various neuropathological states, the increased expression of GFAP corresponds to the severity of astroglial activation (Axelsson, Malmeström, & Nilsson, 2011; Kashon et al., 2004; Notturno, Capasso, DeLauretis, Carpo, & Uncini, 2009; Pelinka et al., 2004; Simpson et al., 2010).

GFAP and protein S-100B are within the markers of astrogliosis in neuropathology in AD and others cognitive diseases (Jesse et al., 2009; Mollenhauer et al., 2005). S100B is expressed in varying abundance by cells of central and peripheral nervous, such as glia, astrocytes, growing oligodendrocytes, neuronal progenitor cells, pituicytes, and ependymocytes (Yang et al., 1995). Astrocytic S100B seems to be the major factor in the development of dystrophic neurites, which are concentrated around amyloid plaques (Cairns, Chadwick, Luthert, & Lantos, 1992; Mrak & Griffin, 2005). S100B release is driven by the developmental stage of the astrocytes (Van Eldik & Zimmer, 1987) and by the metabolic stress (e.g., oxygen, serum, or glucose deprivation) (Gerlach et al., 2006). S100B can be released in response to glutamate (Ciccarelli et al., 1999), TNF-α (Edwards & Robinson, 2006), IL-1β (de Souza et al., 2009), and Aβ (Pena, Brecher, & Marshak, 1995).

The receptor for advanced glycation end products (RAGE), interacts with Aβ and can also mediate neurotoxicity due to elevated levels of S100B (Leclerc & Heizmann, 2011; Sathe et al., 2012; Sedaghat & Notopoulos, 2008). S100B induces the NF-κB transcriptional activity in microglia in a RAGE-dependent manner, pointing to further RAGE-mediated effects in AD (Adami, Bianchi, Pula, & Donato, 2004).

## 4. ADVANCED GLYCATION END PRODUCTS IN AD

The multiligand receptor (for Aβ, S100B) RAGE is first identified as the receptor for advanced glycation end products (AGE) (Leclerc & Heizmann, 2011). The nonenzymatic glycosylation theory of aging suggested that the AGE-mediated crosslinking of proteins contributes to the age-related decline in the function of cells in normal aging (Monnier & Cerami, 1981). Humans are exposed to AGE produced in the body, in particular in subjects with unusual glucose metabolism, and to AGE ingested in foods (Peppa, Uribarri, & Vlassara, 2008; Semba, Nicklett, & Ferrucci, 2010). In fact, AGE make up a complex group of compounds formed endogenously through the aging progression and under circumstances of hyperglycemia and oxidative stress but also lifestyle factor for instance diet and cigarette smoke could be exogenous sources of AGE (Palimeri, Palioura, & Diamanti-Kandarakis, 2015). It is well known that an excess of dietary carbohydrates, chiefly fructose, and an insufficiency in cholesterol, may lead to AD (Seneff, Wainwright, & Mascitelli, 2011).

Due to their stability, the proteins that comprise the long-lived intracellular (NFTs and Hirano bodies) and extracellular protein deposits (senile plaques) could be the best substrates for glycation. This process occurs over a long time, even in condition of normal levels of glucose, resulting in the development of AGE, establishing a link among the appearance of RAGE and the pathophysiological changes in AD (Takeda, Sato, & Morishita, 2014). One of the suggested mechanisms of AGE-induced injury is the development of ROS, such as superoxide and hydrogen peroxide (Guglielmotto, Giliberto, Tamagno, & Tabaton, 2010). The activation of microglial RAGE by many ligands, such as AGE and Aβ, induces the release of ROS and cytokines (Schmidt, Braun, & Narlawar, 2005). These effects are due to the activation of NF-κB (Onyango, Tuttle, & Bennett, 2005) and it has been suggested that the RAGE/NF-κB axis could be a potential therapeutic target in AD (Bortolotto & Grilli, 2016; Maczurek, Shanmugam, & Münch, 2008).

Recent studies suggest that consumption of dietary AGE may encourage inflammation and oxidative stress (Clarke, Dordevic, Tan, Ryan, & Coughlan, 2016; Kellow & Coughlan, 2015; Uribarri et al., 2015) and short-term trials in humans indicate that a low AGE diet reduces oxidative stress and inflammatory markers (Palimeri et al., 2015; Peppa et al., 2008; Van Puyvelde, Mets, Njemini, Beyer, & Bautmans, 2014). In this context, dietary nutraceuticals possess inhibitory effects on the formation of AGE (Khangholi, Majid, Berwary, Ahmad, & Aziz, 2016).

## 5. POLYPHENOLIC NUTRACEUTICALS TARGETING NEUROINFLAMMATION FOR PREVENTION AND THERAPY OF AD

Phytochemicals including polyphenolic compounds that are widely distributed in the plant kingdom, possess antioxidant and anti-inflammatory activities (Biesalski, 2007), potentially preserving cognitive function during aging (Thangthaeng, Poulose, Miller, & Shukitt-Hale, 2016). Phytochemicals possess antioxidant activities, which are useful to act as natural neuroprotective agents with low adverse effects against AD (Ataie, Shadifar, & Ataee, 2016).

In this background, the potential therapeutic role of phytochemicals including various nutraceutical polyphenols in AD has been investigated by several investigators throughout the world. Therapeutic role of antioxidant vitamin C (Heo, Hyon-Lee, & Lee, 2013) and vitamin E (Farina, Isaac, Clark, Rusted, & Tabet, 2012; La Fata, Weber, & Mohajeri, 2014) in AD has been investigated and patients whose oxidative stress markers were lowered by vitamin E showed no significant difference in the percentage change in mini-mental state examination (MMSE) score (Farina et al., 2012). Hence, the therapeutic role of antioxidant vitamin C (Heo et al., 2013) and vitamin E (Farina et al., 2012; La Fata et al., 2014) in AD is controversial and needed detailed investigations.

Antioxidant and anti-inflammatory activities of phytochemicals, such as curcumin, catechins, licopene, resveratrol, piperine, and anthocyanins, have been reported using in vitro and in animal models (Darvesh, Carroll, Bishayee, Geldenhuys, & Van der Schyf, 2010; Samadi et al., 2015; Yarla et al., 2016).

Neuroinflammation in AD is associated with a significant rise in the inflammatory cytokines, such as IFN-γ, IL-6, and TNF-α (Belkhelfa et al., 2014; Gubandru et al., 2013), and altered levels of systemic markers of oxidative stress (Altunoglu et al., 2015; Gubandru et al., 2013; Moslemnezhad, Mahjoub, & Moghadasi, 2016; Schrag et al., 2013; Wang, Shinto, Connor, & Quinn, 2008; Yuan et al., 2016) (Fig. 1). Table 1 describes the studies reporting markers of oxidative stress and inflammatory cytokines in humans after chronic administration with nutraceuticals containing curcumin (*Curcuma longa* extract), anthocyanins

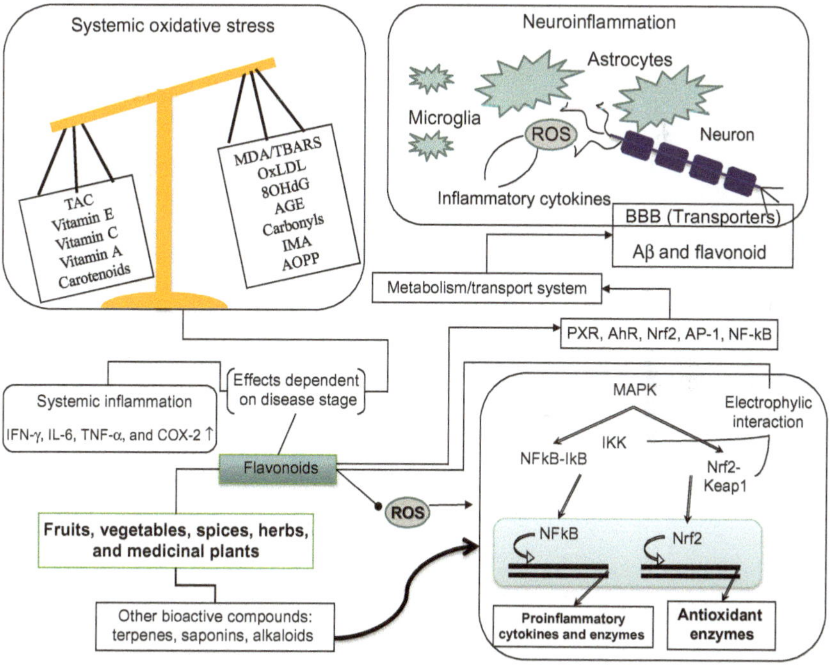


**Fig. 1** Effects of nutraceuticals on inflammation and oxidative stress in AD. AD is associated with systemic inflammation and oxidative stress. Despite the antioxidant activity of nutraceuticals at molecular level, the modulation of the inflammatory and antioxidant pathways is due to electrophylic interactions. Therefore, they act as prooxidants rather than antioxidants. On the other hand, flavonoids modulate the transporters of the blood–brain barrier (BBB), potentially affecting the Aβ peptides deposition. *8-OHdG*: 8-hydroxy-2′-deoxyguanosine; *AGE*: advanced glycation end products; *AhR*: aryl hydrocarbon receptor; *AOPP*: advanced oxidation protein products; *AP-1*: activation protein-1; *IKK*: I-kappa kinases; *IL*: interleukin; *IMA*: ischemia-modified albumin; *MAPK*: mitogen-activated protein kinase; *MDA*: malondialdehyde; *NF-κB*: nuclear factor-kappa B; *Nrf2*: nuclear factor-erythroid 2-related factor 2; *PXR*: pregnane X receptor; *TAC*: total antioxidant capacity; *TBARS*: thiobarbituric acid reactive substances; *TNF*: tumor necrosis factor.

**Table 1** Antioxidant and Anti-inflammatory Effects of Nutraceuticals in Human Interventions

| Study | Nutraceutical | Treatment Duration | Subjects | Oxidative Stress Markers | Inflammatory Cytokines |
|---|---|---|---|---|---|
| Nieman, Cialdella-Kam, Knab, and Shanely (2012) | *Curcuma longa* extract | 4 weeks | Overweight/obese women | ↔ Isoprostanes | ↔ TNF-α, IL-6 |
| Srivastava, Saksena, Khattri, Kumar, and Dagur (2016) | | 120 days | Osteoarthritis | ↓ MDA | ↓ IL-1β |
| Zern et al. (2005) | Grape powder | 4 weeks | Postmenopausal women | ↓ Isoprostanes | ↓ TNF-α |
| Barona et al. (2012) | | 4 weeks | Dyslipidemic | ↔ Isoprostanes | ↔ TNF-α, IL-6 |
| Bogdanski et al. (2012) | Green tea extract | 3 months | Obese | ↑ TAC | ↓ TNF-α |
| Hsu et al. (2007) | | 7 months | Hemodialysis | ↓ Peroxides | ↓ TNF-α |
| Markovits, Ben Amotz, and Levy (2009) | Lycopene | 4 weeks | Obese | ↔ Dienes | ↔ TNF-α, IL-6 |
| Wood et al. (2012) | | 14 weeks | Asthmatic | ↔ Isoprostanes | ↔ TNF-α, IL-6 |
| Hosseini, Saedisomeolia, Wood, Yaseri, and Tavasoli (2016) | Pomegranate extract | 30 days | Overweight/obese | ↓ MDA | ↓ IL-6 |
| Wu et al. (2015) | | 6 weeks | Hemodialysis | ↔ TAC and oxLDL | ↔ IL-6 |

↔: unchanged; ↓: decrease; ↑: increase; *IL*: interleukin; *MDA*: malondialdehyde; *oxLDL*: oxidized low-density lipoprotein; *TAC*: total antioxidant capacity; *TNF*: tumor necrosis factor.

(pomegranate extract), catechins (green tea extract), licopene (tomato derived), and resveratrol (grape powder).

Tomato-derived lycopene (Lyc-*o*-Mato) supplement did not affect neither peroxidation markers nor plasma cytokines concentration in obese and asthmatic subjects (Table 1). On the contrary, supplements containing polyphenols gave more promising results. Green tea extract improved both markers of oxidation and inflammation in obese subjects and in patients on hemodialysis (Table 1). *Curcuma longa* extract decreased both the inflammatory cytokine IL-1β and the peroxidation marker MDA in osteoarthritis patients, but it had no effect on overweight/obese subjects (Table 1).

In some human intervention trials, improvement of cognitive function has been reported after intake of cocoa flavanols, grape juice, blueberry, curcumin, *Ginkgo biloba*, and green tea catechins (Arab, Mahjoub, Hajian-Tilaki, & Moghadasi, 2016; Krikorian et al., 2012, 2010; Matias, Buosi, & Gomes, 2016; Solfrizzi & Panza, 2015; Venigalla, Sonego, Gyengesi, Sharman, & Münch, 2016; Williams & Spencer, 2012). In particular, in AD patients, MMSE and TAC increased, whereas MDA, 8-OHdG, and carbonyl decreased significantly, after 2 months of consumption of 2 g/day of green tea pills (Arab et al., 2016), suggesting a link between systemic oxidative stress and cognitive function.

Suggested mechanisms of neuroprotection of polyphenolic nutraceuticals include the inhibition of the formation of misfolded protein aggregates (Aβ and hyperphosphorylated tau protein) (Ho & Pasinetti, 2010; Williams & Spencer, 2012), of the production inflammatory cytokines (IL-1β and TNF-α) (Matias et al., 2016), by the inhibition of NF-κB and MAPK pathways, in microglia and astrocytes (Spencer, Vafeiadou, Williams, & Vauzour, 2012; Venigalla et al., 2016; Williams & Spencer, 2012); and the activation of the nuclear factor-erythroid 2-related factor 2 (Nrf2)/antioxidant responsive elements (ARE), which leads to the expression of several enzymes with antioxidant and detoxification capacities (Moosavi, Hosseini, Saso, & Firuzi, 2016).

NF-κB inhibition or Nrf2 activation has been reported also for other phytochemicals contained in spices, herbs, and medicinal plants, such as terpenoids (ginkgolide A, ginkgolide B, and bilobalide from *Ginkgo biloba*; Nada & Shah, 2012), saponins (saikosaponins from *Bupleurum falcatum*; Park et al., 2015), akebia saponin D from the herbal medicine *Dipsacus asper Wall* (Yu et al., 2012), and alkaloids (tetrandrine from the Chinese herb radix *Stephania tetrandra*; (He, Qiu, Li, et al., 2011; He, Qiu, Zhang, et al., 2011), SCM-198 from *Herbaleonuri* (Hong, Shi, Zhu, Wu, & Zhu, 2014), berberine from Chinese herb *Rhizoma coptidis* (Jia et al., 2012). A concerted modulation of the Nrf2/ARE and NF-κB has been suggested in inflammation and carcinogenesis (Hu, Saw, Yu, & Kong, 2010; Fig. 1).

Two studies (Rubio-Perez, Albaladejo, Zafrilla, Vidal-Guevara, & Morillas-Ruiz, 2016; Rubio-Perez & Morillas-Ruiz, 2013) evaluated the effect of the consumption (8 weeks) antioxidant beverage (AB), containing extracts of green tea leaves and apple, on inflammatory cytokines and biomarkers of oxidative stress in AD patients. The AB induced, at 4 months of intake, a decrease in serum levels of the proinflammatory IFN-γ and TNF-α in AD patients in initial phase (ADI), whereas significant increases in these cytokines were observed in AD patients in moderate phase (ADM) (Rubio-Perez & Morillas-Ruiz, 2013). IL-1α increased significantly at 8 months of the study in ADM taking the placebo beverage (PB), but not in ADI and controls and the AB did not produce a significant modification in IL-1α levels in any of the three groups of subjects (Rubio-Perez & Morillas-Ruiz, 2013). On the other hand, AB consumption for 8 weeks did not affect 8-OHdG, oxLDL, and isoprostanes, prevented the decrease in TAS in ADM, but not in ADI (Rubio-Perez et al., 2016). Therefore, the effect on inflammation and oxidative stress depends on both disease stage and treatment duration.

In this context, it must be take into account that the mechanism suggested for Nrf2 and/or NF-κB modulation by nutraceuticals is the interaction of electrophiles with cysteine residues of Keap-1 and IκB and/or Iκ kinases (IKK) (Cichocki, Blumczyńska, & Baer-Dubowska, 2010; Copple et al., 2014; Han, Hashimoto, & Fukushima, 2016; Ishii et al., 2009; Nair, Li, & Kong, 2007; Sasaki, Tozawa, Sugamoto, Matsushita, & Satoh, 2013; Sirota, Gibson, & Kohen, 2015; Son et al., 2010; Wang et al., 2014; Wu et al., 2010). Therefore, they act as prooxidants rather than antioxidants (Fig. 1).

However, due to the extensive metabolic activity during digestion, leading to different metabolites endowed with different bioactive ingredients from parental compounds (Del Rio et al., 2013). Some preclinical studies suggest that metabolites are more active than the original ones in reducing ROS production (Fang et al., 2003; Merfort, Heilmann, Weiss, Pietta, & Gardana, 1996; Suri et al., 2008) or inflammatory cytokines secretion (Monagas et al., 2009), suggesting that flavonoids may act as prodrugs (Peluso & Palmery, 2015). However, naringenin metabolites modulated in a different manner macrophage gene expression and in directions that were not always consistent with anti-inflammatory and antioxidant effects (Dall'asta et al., 2013). In particular, naringenin-4-O-glucuronide was able to increase TNF-α and Nrf2 expression, whereas naringenin-7-*O*-glucuronide reduced Nrf2 expression (Dall'asta et al., 2013).

On the other hand, it has been pointed out that it is important to remember that the brain is separated from the periphery by the blood–brain barrier (BBB), which prevents molecules from leaving or entering the CNS and that what is observed in the periphery regarding oxidative stress markers and antioxidants might not be reflected in the brain (Persson, Popescu, & Cedazo-Minguez, 2014). In this context, despite their low bioavailability, flavonoid or their metabolites have been detected in rodent brain after oral administration, suggesting that they are able to penetrate BBB (Matias et al., 2016; Williams & Spencer, 2012) depending on their interactions with specific efflux transporters expressed in the BBB, such as the P-glycoprotein (P-gp) (Williams & Spencer, 2012) (Fig. 1).

An increasing number of studies suggest that ATP-binding cassette (ABC) (i.e., multidrug resistance, MDR) transporters, including ABCB1 (P-gp), ABCG2 (breast cancer resistant protein, BCRP), ABCC1 (multidrug resistance protein 1, MRP1), and the cholesterol transporter ABCA1 in the pathogenesis of AD and Aβ peptides deposition inside the brain (Abuznait & Kaddoumi, 2012; Kuhnke et al., 2007). In particular, P-gp function is decreased in patients with AD (Deo et al., 2014; van Assema, Lubberink, Bauer, et al., 2012) and ABCB1 polymorphisms have been involved in the risk for AD and in the progression of disease (Fehér, Juhász, Pákáski, Kálmán, & Janka, 2014; van Assema, Lubberink, Rizzu, et al., 2012). From that, ABCB1 gene expression profiling of peripheral blood leukocytes (Chen et al., 2011) and ABC transporter-targeting agents (Pahnke, Langer, & Krohn, 2014) have been suggested, respectively, as biomarker and treatment for AD. Flavonoids and their derivatives may exert their effect through competitive inhibition of the MDR after acute consumption, as well as by the induction, after long-term consumption, of phase I, phase II, and phase III drug metabolism/transport systems, through the activation protein-1 (AP-1), the NF-κB, the Nrf2, the aryl hydrocarbon receptor (AhR), and the pregnane X receptor (PXR) pathways (Peluso & Palmery, 2015). The latter is a major transcriptional regulator of P-gp (Jain, Rathod, Prajapati, Nandekar, & Sangamwar, 2014). Therefore, pharmacological rather than antioxidant mechanisms could be involved in the effect of flavonoids in AD (Fig. 1).

Head et al. (2012) formulated a cocktail mixture of a turmeric extract containing 95% curcuminoids, green tea extract with 50% epigallocatechingallate, a black pepper extract with 95% piperine, and others.

The study suggests that this nutraceutical cocktail may be useful for treatment of aging and AD.

Piperine, a chief constituent of *Piper nigrum*, has been reported to possess anti-inflammatory and antioxidant activities (Fig. 2). Piperine has been reported neuroprotective potential in AD. Hence, piperine can be preventive and therapeutic agent for AD. However, oral bioavailability is low due to its hydrophobicity and internal metabolism. However, several nanoparticle-mediated delivery systems have been developed to enhance bioavailability of piperine for prevention and treatment of AD (Elnaggar, Etman, Abdelmonsif, & Abdallah, 2015).

Curcumin is found in rhizome of *Curcuma longa* and several reports demonstrated that it inhibits inflammation-associated AD. A recent study by Liu et al. (2016) demonstrated that curcumin attenuated Aβ-induced neuroinflammation via activation of PPARγ in rats with AD. Curcumin also reduced the activation of microglia and astrocytes, as well as proinflammatory cytokine production and inhibited NF-κB-associated signaling pathway. This study suggesting that the neuroprotective effects of curcumin on AD are attributable to the suppression of neuroinflammation. However, poor bioavailability and internal metabolization of curcumin are the major

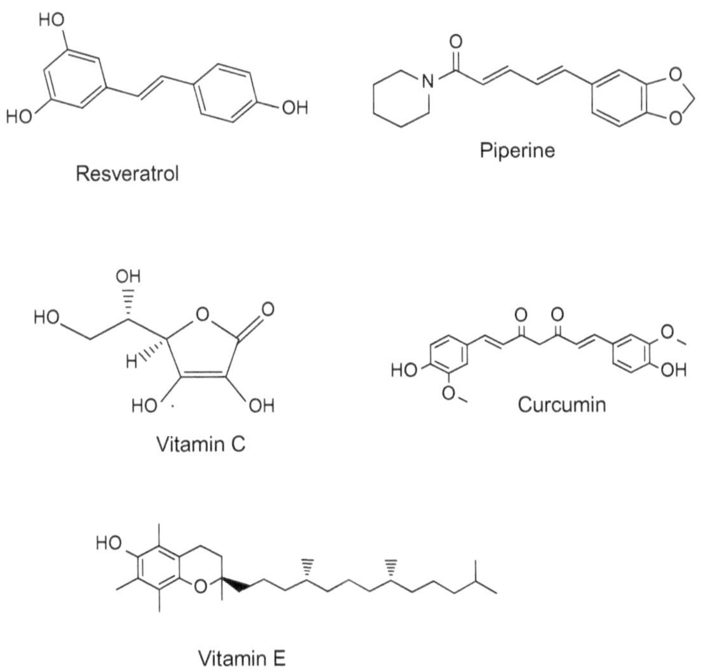


**Fig. 2** Bioactive phytochemicals targeting neuroinflammation for prevention and therapy of Alzhiemer's disease.

problems in its clinical applications (Begum et al., 2008). Several nanoparticles loaded with curcumin were synthesized and exhibiting improved therapeutic potential in AD (Tiwari et al., 2014).

Resveratrol administration prevented AD in rats by antioxidation. Resveratrol inhibited Aβ1-42-mediated neuroinflammation by downregulating the expression of NF-κB and protects the integrity of the BBB in rats with AD (Zhao et al., 2015). A recent clinical study by Moussa et al. (2017) demonstrated that resveratrol modulates neuroinflammation-associated mediators and CSF Aβ40 in patients with AD (ClinicalTrials.gov NCT01504854). Previous studies suggested that resveratrol can be chemopreventive and therapeutic agent for AD and warranted detailed clinical trials.

## 6. CONCLUSIONS AND FUTURE DIRECTIONS

Astrocytes play an important role both in normal function of the mammalian nervous system and in neurodegenerative diseases, coordinating many of the initial and subsequent responses of astrocytes to injury. Astroglial cells regulate synaptic transmission and plasticity, protect neurons against toxic compounds, and support metabolically to ensure their optimal functioning. On the other hand, astrocytes are involved in the initiation and in the progression of AD.

Therapeutic potential of antioxidant vitamins and polyphenolic nutraceuticals in AD has been well investigated by several research groups worldwide. However, the benefit of antioxidant vitamin C and vitamin E treatment for AD is still under considerable discussion and debate. Some studies suggest polyphenols act as novel preventive and curative agents for AD by inhibiting molecular pathways associated with neuroinflammation. Curcumin, resvertrol, piperine, and other polyphenols have been explored as novel therapeutic and preventive agents for AD. However, poor bioavailability and internal metabolism of these natural products are major problems in their clinical applications. In this scenario, further detailed studies including clinical are needed to develop aforementioned natural products to develop them as novel preventive and therapeutic agents for AD.

## REFERENCES

Abuznait, A. H., & Kaddoumi, A. (2012). Role of ABC transporters in the pathogenesis of Alzheimer's disease. *ACS Chemical Neuroscience*, *3*, 820–831.

Adami, C., Bianchi, R., Pula, G., & Donato, R. (2004). S100B-stimulated NO production by BV-2 microglia is independent of RAGE transducing activity but dependent on RAGE extracellular domain. *Biochimica et Biophysica Acta*, *1742*, 169–177.

Agostinho, P., Cunha, R. A., & Oliveira, C. (2010). Neuroinflammation, oxidative stress and the pathogenesis of Alzheimer's disease. *Current Pharmaceutical Design*, *16*, 2766–2778.

Altunoglu, E., Guntas, G., Erdenen, F., Akkaya, E., Topac, I., Irmak, H., et al. (2015). Ischemia-modified albumin and advanced oxidation protein products as potential biomarkers of protein oxidation in Alzheimer's disease. *Geriatric and Gerontology International*, *15*, 872–880.

Arab, H., Mahjoub, S., Hajian-Tilaki, K., & Moghadasi, M. (2016). The effect of green tea consumption on oxidative stress markers and cognitive function in patients with Alzheimer's disease: A prospective intervention study. *Caspian Journal of Internal Medicine*, *7*, 188–194.

Ataie, A., Shadifar, M., & Ataee, R. (2016). Polyphenolic antioxidants and neuronal regeneration. *Basic and Clinical Neuroscience*, 7(2), 81–90.

Axelsson, M., Malmeström, C., & Nilsson, S. (2011). Glial fibrillary acidic protein: A potential biomarker for progression in multiple sclerosis. *Journal of Neurology*, *258*, 882–888.

Barona, J., Blesso, C. N., Andersen, C. J., Park, Y., Lee, J., & Fernandez, M. L. (2012). Grape consumption increases anti-inflammatory markers and upregulates peripheral nitric oxide synthase in the absence of dyslipidemias in men with metabolic syndrome. *Nutrients*, *4*, 1945–1957.

Barreto, G. E., Gonzalez, J., Cardiológicas, I. D. I., & Taquini, P. A. (2008). Role of astrocytes in neurodegenerative diseases. In R. C.-C. Chang (Ed.), *Neurodegenerative diseases—Processes, prevention, protection and monitoring* (pp. 258–272, 570 pages). InTech. ISBN 978-953-307-485-6, http://dx.doi.org/10.5772/1252. http://www.intechopen.com/books/neurodegenerative-diseases-processes-prevention-protection-and-monitoring.

Begum, A. N., Jones, M. R., Lim, G. P., Morihara, T., Kim, P., Heath, D. D., et al. (2008). Curcumin structure-function, bioavailability, and efficacy in models of neuroinflammation and Alzheimer's disease. *The Journal of Pharmacology and Experimental Therapeutics*, *326*(1), 196–208. http://dx.doi.org/10.1124/jpet.108.137455.

Belkhelfa, M., Rafa, H., Medjeber, O., Arroul-Lammali, A., Behairi, N., Abada-Bendib, M., et al. (2014). IFN-γ and TNF-α are involved during Alzheimer disease progression and correlate with nitric oxide production: A study in Algerian patients. *Journal of Interferon & Cytokine Research*, *34*, 839–847.

Biesalski, H. K. (2007). Polyphenols and inflammation: Basic interactions. *Current Opinion in Clinical Nutrition & Metabolic Care*, *10*, 724–728.

Bogdanski, P., Suliburska, J., Szulinska, M., Stepien, M., Pupek-Musialik, D., & Jablecka, A. (2012). Green tea extract reduces blood pressure, inflammatory biomarkers, and oxidative stress and improves parameters associated with insulin resistance in obese, hypertensive patients. *Nutrition Research*, *32*, 421–427.

Bortolotto, V., & Grilli, M. (2016). Every cloud has a silver lining: Proneurogenic effects of Aβ oligomers and HMGB-1 via activation of the RAGE-NF-κB axis. *CNS & Neurological Disorders Drug Targets*, *15*, 1–14. ISSN 1871-5273/1996-3181. http://dx.doi.org/10.2174/1871527315666160803153459.

Braak, H., Braak, E., & Strothjohann, M. (1994). Abnormally phosphorylated tau protein related to the formation of neurofibrillary tangles and neuropil threads in the cerebral cortex of sheep and goat. *Neuroscience Letters*, *171*, 1–4.

Cairns, N. J., Chadwick, A., Luthert, P. J., & Lantos, P. L. (1992). Astrocytosis, βA4-protein deposition and paired helical filament formation in Alzheimer's disease. *Journal of the Neurological Sciences*, *112*, 68–75.

Callahan, L. M., & Coleman, P. D. (1995). Neurons bearing neurofibrillary tangles are responsible for selected synaptic deficits in Alzheimer's disease. *Neurobiology of Aging*, *16*, 311–314.

Chen, K. D., Chang, P. T., Ping, Y. H., Lee, H. C., Yeh, C. W., & Wang, P. N. (2011). Gene expression profiling of peripheral blood leukocytes identifies and validates ABCB1 as a novel biomarker for Alzheimer's disease. *Neurobiology of Disease, 43*, 698–705.

Ciccarelli, R., Di Iorio, P., Bruno, V., Battaglia, G., D'Alimonte, I., D'Onofrio, M., et al. (1999). Activation of A1 adenosine or mGlu3 metabotropic glutamate receptors enhances the release of nerve growth factor and S-100 protein from cultured astrocytes. *GLIA, 27*, 275–281.

Cichocki, M., Blumczyńska, J., & Baer-Dubowska, W. (2010). Naturally occurring phenolic acids inhibit 12-O-tetradecanoylphorbol-13-acetate induced NF-kappaB, iNOS and COX-2 activation in mouse epidermis. *Toxicology, 268*, 118–124.

Clarke, R. E., Dordevic, A. L., Tan, S. M., Ryan, L., & Coughlan, M. T. (2016). Dietary advanced glycation end products and risk factors for chronic disease: A systematic review of randomised controlled trials. *Nutrients, 8*(3), 125. http://dx.doi.org/10.3390/nu8030125.

Copple, I. M., Shelton, L. M., Walsh, J., Kratschmar, D. V., Lister, A., Odermatt, A., et al. (2014). Chemical tuning enhances both potency toward nrf2 and in vitro therapeutic index of triterpenoids. *Toxicological Sciences, 140*, 462–469.

Cruz, J. C., Tseng, H.-C., Goldman, J. A., Shih, H., & Tsai, L.-H. (2003). Aberrant Cdk5 activation by p25 triggers pathological events leading to neurodegeneration and neurofibrillary tangles. *Neuron, 40*, 471–483.

Dall'asta, M., Derlindati, E., Curella, V., Mena, P., Calani, L., Ray, S., et al. (2013). Effects of naringenin and its phase II metabolites on in vitro human macrophage gene expression. *International Journal of Food Sciences and Nutrition, 64*, 843–849.

Darvesh, A. S., Carroll, R. T., Bishayee, A., Geldenhuys, W. J., & Van der Schyf, C. J. (2010). Oxidative stress and Alzheimer's disease: Dietary polyphenols as potential therapeutic agents. *Expert Review of Neurotherapeutics, 10*, 729–745.

de Souza, D. F., Leite, M. C., Quincozes-Santos, A., Nardin, P., Tortorelli, L. S., Rigo, M. M., et al. (2009). S100B secretion is stimulated by IL-1beta in glial cultures and hippocampal slices of rats: Likely involvement of MAPK pathway. *Journal of Neuroimmunology, 206*, 52–57.

Del Rio, D., Rodriguez-Mateos, A., Spencer, J. P., Tognolini, M., Borges, G., & Crozier, A. (2013). Dietary (poly)phenolics in human health: Structures, bioavailability, and evidence of protective effects against chronic diseases. *Antioxidants & Redox Signaling, 18*, 1818–1892.

Déniz-Naranjo, M. C., Muñoz-Fernandez, C., Alemany-Rodríguez, M. J., Pérez-Vieitez, M. C., Aladro-Benito, Y., Irurita-Latasa, J., et al. (2008). Cytokine IL-1 beta but not IL-1 alpha promoter polymorphism is associated with Alzheimer disease in a population from the Canary islands, Spain. *European Journal of Neurology, 15*, 1080–1084.

Deo, A. K., Borson, S., Link, J. M., Domino, K., Eary, J. F., Ke, B., et al. (2014). Activity of P-glycoprotein, a β-amyloid transporter at the blood-brain barrier, is compromised in patients with mild Alzheimer disease. *Journal of Nuclear Medicine: Official Publication, Society of Nuclear Medicine, 55*, 1106–1111.

DiPatre, P. L., & Gelman, B. B. (1997). Microglial cell activation in aging and Alzheimer disease: Partial linkage with neurofibrillary tangle burden in the hippocampus. *Journal of Neuropathology and Experimental Neurology, 56*, 143–149.

Dong, Y., & Benveniste, E. N. (2001). Immune function of astrocytes. *Glia, 36*, 180–190.

Edwards, M. M., & Robinson, S. R. (2006). TNF alpha affects the expression of GFAP and S100B: Implications for Alzheimer's disease. *Journal of Neural Transmission, 113*, 1709–1715.

Elnaggar, Y. S., Etman, S. M., Abdelmonsif, D. A., & Abdallah, O. Y. (2015). Intranasal piperine-loaded chitosan nanoparticles as brain-targeted therapy in Alzheimer's disease:

Optimization, biological efficacy, and potential toxicity. *Journal of Pharmaceutical Sciences, 104*(10), 3544–3556.
Fang, S. H., Hou, Y. C., Chang, W. C., Hsiu, S. L., Chao, P. D., & Chiang, B. L. (2003). Morin sulfates/glucuronides exert anti-inflammatory activity on activated macrophages and decreased the incidence of septic shock. *Life Sciences, 74*, 743–756.
Farina, N., Isaac, M. G., Clark, A. R., Rusted, J., & Tabet, N. (2012). Vitamin E for Alzheimer's dementia and mild cognitive impairment. *The Cochrane Database of Systematic Reviews, 11*. CD002854. http://dx.doi.org/10.1002/14651858.
Fehér, Á., Juhász, A., Pákáski, M., Kálmán, J., & Janka, Z. (2014). ABCB1 C3435T polymorphism influences the risk for Alzheimer's disease. *Journal of Molecular Neuroscience, 54*, 826–829.
Ferencik, M., Novak, M., Rovensky, J., & Rybar, I. (2001). Alzheimer's disease, inflammation and non-steroidal anti-inflammatory drugs. *Bratislavské Lekárske Listy, 102*(3), 123–132.
Floyd, R. A., & Hensley, K. (2000). Nitrone inhibition of age-associated oxidative damage. *Annals of the New York Academy of Sciences, 899*, 222–237.
Ge, Y.-W., & Lahiri, D. K. (2002). Regulation of promoter activity of the APP gene by cytokines and growth factors. *Annals of the New York Academy of Sciences, 973*, 463–467.
Gerlach, R., Demel, G., König, H. G., Gross, U., Prehn, J. H. M., Raabe, A., et al. (2006). Active secretion of S100B from astrocytes during metabolic stress. *Neuroscience, 141*, 1697–1701.
Griffin, W. S. T., Liu, L., Li, Y., Mrak, R. E., & Barger, S. W. (2006). Interleukin-1 mediates Alzheimer and Lewy body pathologies. *Journal of Neuroinflammation, 3*(1), 5.
Gubandru, M., Margina, D., Tsitsimpikou, C., Goutzourelas, N., Tsarouhas, K., Ilie, M., et al. (2013). Alzheimer's disease treated patients showed different patterns for oxidative stress and inflammation markers. *Food and Chemical Toxicology, 61*, 209–214.
Guglielmotto, M., Giliberto, L., Tamagno, E., & Tabaton, M. (2010). Oxidative stress mediates the pathogenic effect of different Alzheimer's disease risk factors. *Frontiers in Aging Neuroscience, 2*, 3.
Haass, C., & Selkoe, D. J. (1993). Cellular processing of beta-amyloid precursor protein and the genesis of amyloid beta-peptide. *Cell, 75*(6), 1039–1042.
Han, K. H., Hashimoto, N., & Fukushima, M. (2016). Relationships among alcoholic liver disease, antioxidants, and antioxidant enzymes. *World Journal of Gastroenterology, 22*, 37–49.
He, F. Q., Qiu, B. Y., Li, T. K., Xie, Q., Cui, D. J., Huang, X. L., et al. (2011a). Tetrandrine suppresses amyloid-β-induced inflammatory cytokines by inhibiting NF-κB pathway in murine BV2 microglial cells. *International Immunopharmacology, 11*, 1220–1225.
He, F. Q., Qiu, B. Y., Zhang, X. H., Li, T. K., Xie, Q., Cui, D. J., et al. (2011b). Tetrandrine attenuates spatial memory impairment and hippocampal neuroinflammation via inhibiting NF-κB activation in a rat model of Alzheimer's disease induced by amyloid-β(1-42). *Brain Research, 1384*, 89–96.
Head, E., Murphey, H. L., Dowling, A. L., McCarty, K. L., Bethel, S. R., Nitz, J. A., et al. (2012). A combination cocktail improves spatial attention in a canine model of human aging and Alzheimer's disease. *Journal of Alzheimer's Disease, 32*(4), 1029–1042. http://dx.doi.org/10.3233/JAD-2012-120937.
Hebert, L. E., Scherr, P. A., Bienias, J. L., Bennett, D. A., & Evans, D. A. (2003). Alzheimer disease in the US population. *Archives of Neurology, 60*, 1119–1122.
Heo, J. H., Hyon-Lee, & Lee, K. M. (2013). The possible role of antioxidant vitamin C in Alzheimer's disease treatment and prevention. *American Journal of Alzheimer's Disease and Other Dementias, 28*, 120–125.
Ho, L., & Pasinetti, G. M. (2010). Polyphenolic compounds for treating neurodegenerative disorders involving protein misfolding. *Expert Review of Proteomics, 7*, 579–589.

Hong, Z. Y., Shi, X. R., Zhu, K., Wu, T. T., & Zhu, Y. Z. (2014). SCM-198 inhibits microglial overactivation and attenuates Aβ(1-40)-induced cognitive impairments in rats via JNK and NF-κB pathways. *Journal of Neuroinflammation, 11*, 147. http://dx.doi.org/10.1186/s12974-014-0147-x.

Hosseini, B., Saedisomeolia, A., Wood, L. G., Yaseri, M., & Tavasoli, S. (2016). Effects of pomegranate extract supplementation on inflammation in overweight and obese individuals: A randomized controlled clinical trial. *Complementary Therapies in Clinical Practice, 22*, 44–50.

Hsu, S. P., Wu, M. S., Yang, C. C., Huang, K. C., Liou, S. Y., Hsu, S. M., et al. (2007). Chronic green tea extract supplementation reduces hemodialysis-enhanced production of hydrogen peroxide and hypochlorous acid, atherosclerotic factors, and proinflammatory cytokines. *American Journal of Clinical Nutrition, 86*, 1539–1547.

Hu, R., Saw, C. L., Yu, R., & Kong, A. N. (2010). Regulation of NF-E2-related factor 2 signaling for cancer chemoprevention: Antioxidant coupled with antiinflammatory. *Antioxidants & Redox Signaling, 13*, 1679–1698.

Ishii, T., Ishikawa, M., Miyoshi, N., Yasunaga, M., Akagawa, M., Uchida, K., et al. (2009). Catechol type polyphenol is a potential modifier of protein sulfhydryls: Development and application of a new probe for understanding the dietary polyphenol actions. *Chemical Research in Toxicology, 22*, 1689–1698.

Jain, S., Rathod, V., Prajapati, R., Nandekar, P. P., & Sangamwar, A. T. (2014). Pregnane X receptor and P-glycoprotein: A connexion for Alzheimer's disease management. *Molecular Diversity, 18*, 895–909.

Jesse, S., Steinacker, P., Cepek, L., von Arnim, C. A., Tumani, H., Lehnert, S., et al. (2009). Glial fibrillary acidic protein and protein S-100B: Different concentration pattern of glial proteins in cerebrospinal fluid of patients with Alzheimer's disease and Creutzfeldt-Jakob disease. *Journal of Alzheimer's Disease, 17*, 541–551.

Jia, L., Liu, J., Song, Z., Pan, X., Chen, L., Cui, X., et al. (2012). Berberine suppresses amyloid-beta-induced inflammatory response in microglia by inhibiting nuclear factor-kappaB and mitogen-activated protein kinase signalling pathways. *Journal of Pharmacy and Pharmacology, 64*, 1510–1521.

Kahlson, M. A., & Colodner, K. J. (2015). Glial Tau pathology in tauopathies: Functional consequences. *Journal of Experimental Neuroscience, 9*(Suppl. 2), 43–50.

Kamphuis, W., Mamber, C., Moeton, M., Kooijman, L., Sluijs, J. A., Jansen, A. H. P., et al. (2012). GFAP isoforms in adult mouse brain with a focus on neurogenic astrocytes and reactive astrogliosis in mouse models of Alzheimer disease. *PloS One*, 7(8). e42823.

Kashon, M. L., Ross, G. W., O'Callaghan, J. P., Miller, D. B., Petrovitch, H., Burchfiel, C. M., et al. (2004). Associations of cortical astrogliosis with cognitive performance and dementia status. *Journal of Alzheimer's Disease, 6*, 595–604.

Kawaguchi, Y. (2008). IL-lα gene expression and protein production by fibroblasts from patients with systemic sclerosis. *Clinical & Experimental Immunology, 97*, 445–450.

Kellow, N. J., & Coughlan, M. T. (2015). Effect of diet-derived advanced glycation end products on inflammation. *Nutrition Reviews, 73*, 737–759.

Khangholi, S., Majid, F. A., Berwary, N. J., Ahmad, F., & Aziz, R. B. (2016). The mechanisms of inhibition of advanced glycation End products formation through polyphenols in hyperglycemic condition. *Planta Medica, 82*, 32–45.

Krikorian, R., Boespflug, E. L., Fleck, D. E., Stein, A. L., Wightman, J. D., Shidler, M. D., et al. (2012). Concord grape juice supplementation and neurocognitive function in human aging. *Journal of Agricultural and Food Chemistry, 60*, 5736–5742.

Krikorian, R., Shidler, M. D., Nash, T. A., Kalt, W., Vinqvist-Tymchuk, M. R., Shukitt-Hale, B., et al. (2010). Blueberry supplementation improves memory in older adults. *Journal of Agricultural and Food Chemistry, 58*, 3996–4000.

Kuhnke, D., Jedlitschky, G., Grube, M., Krohn, M., Jucker, M., Mosyagin, I., et al. (2007). MDR1-P-glycoprotein (ABCB1) mediates transport of Alzheimer's amyloid-beta peptides—Implications for the mechanisms of abeta clearance at the blood-brain barrier. *Brain Pathology*, *17*, 347–353.

La Fata, G., Weber, P., & Mohajeri, M. H. (2014). Effects of vitamin E on cognitive performance during ageing and in Alzheimer's disease. *Nutrients*, *6*, 5453–5472.

Leclerc, E., & Heizmann, C. W. (2011). The importance of Ca2+/Zn2+ signaling S100 proteins and RAGE in translational medicine. *Frontiers in Bioscience (Scholar Edition)*, *3*, 1232–1262.

Lee, S. C., Liu, W., Dickson, D. W., Brosnan, C. F., & Berman, J. W. (1993). Cytokine production by human fetal microglia and astrocytes. Differential induction by lipopolysaccharide and IL-1 beta. *Journal of Immunology*, *150*, 2659–2667.

Li, Y., Liu, L., Barger, S. W., & Griffin, W. S. T. (2003). Interleukin-1 mediates pathological effects of microglia on tau phosphorylation and on synaptophysin synthesis in cortical neurons through a p38-MAPK pathway. *Journal of Neuroscience*, *23*, 1605–1611.

Li, C., Zhao, R., & Gao, K. (2011). Astrocytes: Implications for neuroinflammatory pathogenesis of Alzheimer's disease. *Current Alzheimer Research*, *8*, 67–80.

Liu, L., & Chan, C. (2014). The role of inflammasome in Alzheimer's disease. *Ageing Research Reviews*, *15*, 6–15.

Liu, Z. J., Li, Z. H., Liu, L., Tang, W. X., Wang, Y., Dong, M. R., et al. (2016). Curcumin attenuates beta-amyloid-induced neuroinflammation via activation of peroxisome proliferator-activated receptor-gamma function in a rat model of Alzheimer's disease. *Frontiers in Pharmacology*, *7*, 261.

Liu, L., Martin, R., & Chan, C. (2013). Palmitate-activated astrocytes via serine palmitoyltransferase increase BACE1 in primary neurons by sphingomyelinases. *Neurobiology of Aging*, *34*, 540–550.

López, N. J., Jara, L., & Valenzuela, C. Y. (2005). Association of interleukin-1 polymorphisms with periodontal disease. *The Journal of Periodontology*, *76*, 234–243.

Maczurek, A., Shanmugam, K., & Münch, G. (2008). Inflammation and the redox-sensitive AGE-RAGE pathway as a therapeutic target in Alzheimer's disease. *Annals of the New York Academy of Sciences*, *1126*, 147–151.

Mandrekar-Colucci, S., & Landreth, G. E. (2010). Microglia and inflammation in Alzheimer's disease. *CNS & Neurological Disorders Drug Targets*, *9*, 156–167.

Markovits, N., Ben Amotz, A., & Levy, Y. (2009). The effect of tomato-derived lycopene on low carotenoids and enhanced systemic inflammation and oxidation in severe obesity. *The Israel Medical Association Journal*, *11*, 598–601.

Matias, I., Buosi, A. S., & Gomes, F. C. (2016). Functions of flavonoids in the central nervous system: Astrocytes as targets for natural compounds. *Neurochemistry International*, *95*, 85–91.

Medeiros, R., & LaFerla, F. M. (2013). Astrocytes: Conductors of the Alzheimer disease neuroinflammatory symphony. *Experimental Neurology*, *239*, 133–138.

Merfort, I., Heilmann, J., Weiss, M., Pietta, P., & Gardana, C. (1996). Radical scavenger activity of three flavonoid metabolites studied by inhibition of chemiluminescence in human PMNs. *Planta Medica*, *62*, 289–292.

Mollenhauer, B., Bibl, M., Trenkwalder, C., Stiens, G., Cepek, L., Steinacker, P., et al. (2005). Follow-up investigations in cerebrospinal fluid of patients with dementia with Lewy bodies and Alzheimer's disease. *Journal of Neural Transmission*, *112*, 933–948.

Monagas, M., Khan, N., Andrés-Lacueva, C., Urpí-Sardá, M., Vázquez-Agell, M., Lamuela-Raventós, R. M., et al. (2009). Dihydroxylated phenolic acids derived from microbial metabolism reduce lipopolysaccharide-stimulated cytokine secretion by human peripheral blood mononuclear cells. *British Journal of Nutrition*, *102*, 201–206.

Monnier, V., & Cerami, A. (1981). Nonenzymatic browning in vivo: Possible process for aging of long-lived proteins. *Science*, *211*, 491–493.

Moosavi, F., Hosseini, R., Saso, L., & Firuzi, O. (2016). Modulation of neurotrophic signaling pathways by polyphenols. *Drug Design, Development and Therapy*, *10*, 23–42.

Moslemnezhad, A., Mahjoub, S., & Moghadasi, M. (2016). Altered plasma marker of oxidative DNA damage and total antioxidant capacity in patients with Alzheimer's disease. *Caspian Journal of Internal Medicine*, 7, 88–92.

Moussa, C., Hebron, M., Huang, X., Ahn, J., Rissman, R. A., Aisen, P. S., et al. (2017). Resveratrol regulates neuro-inflammation and induces adaptive immunity in Alzheimer's disease. *Journal of Neuroinflammation*, *14*, 1.

Mrak, R. E., & Griffin, W. S. T. (2005). Glia and their cytokines in progression of neurodegeneration. *Neurobiology of Aging*, *26*, 349–354.

Nada, S. E., & Shah, Z. A. (2012). Preconditioning with Ginkgo biloba (EGb 761®) provides neuroprotection through HO1 and CRMP2. *Neurobiology of Disease*, *46*, 180–189.

Nagele, R., Wegiel, J., & Venkataraman, V. (2004). Contribution of glial cells to the development of amyloid plaques in Alzheimer's disease. *Neurobiology of Aging*, *25*, 663–674.

Nair, S., Li, W., & Kong, A. N. (2007). Natural dietary anti-cancer chemopreventive compounds: Redox-mediated differential signaling mechanisms in cytoprotection of normal cells versus cytotoxicity in tumor cells. *Acta Pharmacologica Sinica*, *28*, 459–472.

Nestor, P. J., Scheltens, P., & Hodges, J. (2006). Advances in the early detection of Alzheimer's disease. *Nature Medicine*, *12*, s34–s41.

Nieman, D. C., Cialdella-Kam, L., Knab, A. M., & Shanely, R. A. (2012). Influence of red pepper spice and turmeric on inflammation and oxidative stress biomarkers in overweight females: A metabolomics approach. *Plant Foods for Human Nutrition*, *67*, 415–421.

Notturno, F., Capasso, M., DeLauretis, A., Carpo, M., & Uncini, A. (2009). Glial fibrillary acidic protein as a marker of axonal damage in chronic neuropathies. *Muscle & Nerve*, *40*, 50–54.

Onyango, I. G., Tuttle, J. B., & Bennett, J. P. (2005). Altered intracellular signaling and reduced viability of Alzheimer's disease neuronal cybrids is reproduced by beta-amyloid peptide acting through receptor for advanced glycation end products (RAGE). *Molecular and Cellular Neurosciences*, *29*, 333–343.

Pahnke, J., Langer, O., & Krohn, M. (2014). Alzheimer's and ABC transporters—New opportunities for diagnostics and treatment. *Neurobiology of Disease*, *72 Pt. A*, 54–60.

Palimeri, S., Palioura, E., & Diamanti-Kandarakis, E. (2015). Current perspectives on the health risks associated with the consumption of advanced glycation end products: Recommendations for dietary management. *Diabetes, Metabolic Syndrome and Obesity: Targets and Therapy*, *8*, 415–426.

Park, W. H., Kang, S., Piao, Y., Pak, C. J., Oh, M. S., Kim, J., et al. (2015). Ethanol extract of bupleurum falcatum and saikosaponins inhibit neuroinflammation via inhibition of NF-κB. *Journal of Ethnopharmacology*, *174*, 37–44.

Parpura, V., & Haydon, P. G. (Eds.), (2009). *Astrocytes in (Patho)Physiology of the Nervous System*. Boston, MA: Springer US. http://doi.org/10.1007/978-0-387-79492-1.

Pelinka, L. E., Kroepfl, A., Leixnering, M., Buchinger, W., Raabe, A., & Redl, H. (2004). GFAP versus S100B in serum after traumatic brain injury: Relationship to brain damage and outcome. *Journal of Neurotrauma*, *21*, 1553–1561.

Peluso, I., & Palmery, M. (2015). Flavonoids at the pharma-nutrition interface: Is a therapeutic index in demand? *Biomedicine & Pharmacotherapy*, *71*, 102–107.

Pena, L. A., Brecher, C. W., & Marshak, D. R. (1995). beta-Amyloid regulates gene expression of glial trophic substance S100 beta in C6 glioma and primary astrocyte cultures. *Brain Research. Molecular Brain Research*, *34*, 118–126.

Peppa, M., Uribarri, J., & Vlassara, H. (2008). Aging and glycoxidant stress. *Hormones*, *7*, 123–132.

Persson, T., Popescu, B. O., & Cedazo-Minguez, A. (2014). Oxidative stress in Alzheimer's disease: Why did antioxidant therapy fail? *Oxidative Medicine and Cellular Longevity*, *2014*, 11. Article ID 427318. http://dx.doi.org/10.1155/2014/427318.

Ries, M., & Sastre, M. (2016). Mechanisms of Aβ clearance and degradation by glial cells. *Frontiers in Aging Neuroscience*, *8*, 160.

Rogers, J., Webster, S., Lue, L.-F., Brachova, L., Harold Civin, W., Emmerling, M., et al. (1996). Inflammation and Alzheimer's disease pathogenesis. *Neurobiology of Aging*, *17*, 681–686.

Rossner, S., & Lange-Dohna, C. (2005). Alzheimer's disease β secretase BACE1 is not a neuron specific enzyme. *Journal of Neurochemistry*, *92*, 226–234.

Rubio-Perez, J. M., Albaladejo, M. D., Zafrilla, P., Vidal-Guevara, M. L., & Morillas-Ruiz, J. M. (2016). Effects of an antioxidant beverage on biomarkers of oxidative stress in Alzheimer's patients. *European Journal of Nutrition*, *55*, 2105–2116.

Rubio-Perez, J. M., & Morillas-Ruiz, J. M. (2013). Serum cytokine profile in Alzheimer's disease patients after ingestion of an antioxidant beverage. *CNS & Neurological Disorders Drug Targets*, *12*, 1233–1241.

Saido, T. (2003). Metabolism and Alzheimer's disease. *Proceedings of the Japan Academy. Series B Physical and Biological Sciences*, 7, 321–339.

Samadi, A. K., Bilsland, A., Georgakilas, A. G., Amedei, A., Amin, A., Bishayee, A., et al. (2015). A multi-targeted approach to suppress tumor-promoting inflammation. *Seminars in Cancer Biology*, *35*, S151–S184.

Sasaki, S., Tozawa, T., Sugamoto, K., Matsushita, Y., & Satoh, T. (2013). A novel di terpene para-hydroquinone compound derived from cryptoquinone protects neuronal cells against oxidative stress and activates the Nrf2/ARE pathway. *Neuroscience Letters*, *548*, 132–136.

Sathe, K., Maetzler, W., Lang, J. D., Mounsey, R. B., Fleckenstein, C., Martin, H. L., et al. (2012). S100B is increased in Parkinson's disease and ablation protects against MPTP-induced toxicity through the RAGE and TNF-α pathway. *Brain*, *135*, 3336–3347.

Schmidt, B., Braun, H. A., & Narlawar, R. (2005). Drug development and PET-diagnostics for Alzheimer's disease. *Current Medicinal Chemistry*, *12*, 1677–1695.

Schrag, M., Mueller, C., Zabel, M., Crofton, A., Kirsch, W. M., Ghribi, O., et al. (2013). Oxidative stress in blood in Alzheimer's disease and mild cognitive impairment: A meta-analysis. *Neurobiology of Disease*, *59*, 100–110.

Sedaghat, F., & Notopoulos, A. (2008). S100 protein family and its application in clinical practice. *Hippokratia*, *12*, 198–204.

Selkoe, D. J. (2005). Defining molecular targets to prevent Alzheimer disease. *Archives of Neurology*, *62*, 192–195.

Semba, R. D., Nicklett, E. J., & Ferrucci, L. (2010). Does accumulation of advanced glycation end products contribute to the aging phenotype? *Journals of Gerontology. Series A, Biological Sciences and Medical Sciences*, *65*, 963–975.

Seneff, S., Wainwright, G., & Mascitelli, L. (2011). Nutrition and Alzheimer's disease: The detrimental role of a high carbohydrate diet. *European Journal of Internal Medicine*, *22*, 134–140.

Sheng, J. G., Jones, R. A., Zhou, X. Q., McGinness, J. M., Van Eldik, L. J., Mrak, R. E., et al. (2001). Interleukin-1 promotion of MAPK-p38 overexpression in experimental animals and in Alzheimer's disease: Potential significance for tau protein phosphorylation. *Neurochemistry International*, *39*, 341–348.

Sheng, J. G., Mrak, R. E., & Griffin, W. S. (1997). Glial-neuronal interactions in Alzheimer disease: Progressive association of IL-1alpha + microglia and S100beta + astrocytes with neurofibrillary tangle stages. *Journal of Neuropathology and Experimental Neurology*, *56*, 285–290.

Sidoryk-Wegrzynowicz, M., & Aschner, M. (2013). Role of astrocytes in manganese mediated neurotoxicity. *BMC Pharmacology & Toxicology*, *14*, 23.

Sidoryk-Wegrzynowicz, M., Wegrzynowicz, M., Lee, E., Bowman, A. B., & Aschner, M. (2011). Role of astrocytes in brain function and disease. *Toxicologic Pathology*, *39*, 115–123.

Simpson, J. E., Ince, P. G., Lace, G., Forster, G., Shaw, P. J., Matthews, F., et al. (2010). Astrocyte phenotype in relation to Alzheimer-type pathology in the ageing brain. *Neurobiology of Aging*, *31*, 578–590.

Singhal, G., Jaehne, E. J., Corrigan, F., Toben, C., & Baune, B. T. (2014). Inflammasomes in neuroinflammation and changes in brain function: A focused review. *Frontiers in Neuroscience*, *8*, 1–22.

Sirota, R., Gibson, D., & Kohen, R. (2015). The role of the catecholic and the electrophilic moieties of caffeic acid in Nrf2/Keap1 pathway activation in ovarian carcinoma cell lines. *Redox Biology*, *4*, 48–59.

Sofroniew, M. V. (2009). Molecular dissection of reactive astrogliosis and glial scar formation. *Trends in Neurosciences*, *32*, 638–647.

Solfrizzi, V., & Panza, F. (2015). Plant-based nutraceutical interventions against cognitive impairment and dementia: Meta-analytic evidence of efficacy of a standardized Gingko biloba extract. *Journal of Alzheimers Disease*, *43*, 605–611.

Solito, E., & Sastre, M. (2012). Microglia function in Alzheimer's disease. *Frontiers in Pharmacology*, *3*, 14. http://dx.doi.org/10.3389/fphar.2012.00014.

Son, P. S., Park, S. A., Na, H. K., Jue, D. M., Kim, S., & Surh, Y. J. (2010). Piceatannol, a catechol-type polyphenol, inhibits phorbol ester-induced NF-{kappa}B activation and cyclooxygenase-2 expression in human breast epithelial cells: Cysteine 179 of IKK{beta} as a potential target. *Carcinogenesis*, *31*, 1442–1449.

Spencer, J. P., Vafeiadou, K., Williams, R. J., & Vauzour, D. (2012). Neuroinflammation: Modulation by flavonoids and mechanisms of action. *Molecular Aspects of Medicine*, *33*, 83–97.

Srivastava, S., Saksena, A. K., Khattri, S., Kumar, S., & Dagur, R. S. (2016). *Curcuma longa* extract reduces inflammatory and oxidative stress biomarkers in osteoarthritis of knee: A four-month, double-blind, randomized, placebo-controlled trial. *Inflammopharmacology*, *24*, 377–388.

Streit, W. J. (2006). Microglial senescence: Does the brain's immune system have an expiration date? *Trends in Neurosciences*, *29*, 506–510.

Suri, S., Taylor, M. A., Verity, A., Tribolo, S., Needs, P. W., Kroon, P. A., et al. (2008). A comparative study of the effects of quercetin and its glucuronide and sulfate metabolites on human neutrophil function in vitro. *Biochemical Pharmacology*, *76*, 645–653.

Takeda, S., Sato, N., & Morishita, R. (2014). Systemic inflammation, blood-brain barrier vulnerability and cognitive/non-cognitive symptoms in Alzheimer disease: Relevance to pathogenesis and therapy. *Frontiers in Aging Neuroscience*, *6*, 171.

Thangthaeng, N., Poulose, S. M., Miller, M. G., & Shukitt-Hale, B. (2016). Preserving brain function in aging: The anti-glycative potential of berry fruit. *Neuromolecular Medicine*, *18*, 465–473.

Tiwari, S. K., Agarwal, S., Seth, B., Yadav, A., Nair, S., Bhatnagar, P., et al. (2014). Curcumin-loaded nanoparticles potently induce adult neurogenesis and reverse cognitive deficits in Alzheimer's disease model via canonical Wnt/β-catenin pathway. *ACS Nano*, *8*(1), 76–103. http://dx.doi.org/10.1021/nn405077y.

Uribarri, J., del Castillo, M. D., de la Maza, M. P., Filip, R., Gugliucci, A., Luevano-Contreras, C., et al. (2015). Dietary advanced glycation end products and their role in health and disease. *Advances in Nutrition*, *6*, 461–473.

van Assema, D. M., Lubberink, M., Bauer, M., van der Flier, W. M., Schuit, R. C., Windhorst, A. D., et al. (2012a). Blood-brain barrier P-glycoprotein function in Alzheimer's disease. *Brain*, *135*, 181–189.

van Assema, D. M., Lubberink, M., Rizzu, P., van Swieten, J. C., Schuit, R. C., Eriksson, J., et al. (2012b). Blood-brain barrier P-glycoprotein function in healthy subjects and Alzheimer's disease patients: Effect of polymorphisms in the ABCB1 gene. *EJNMMI Research*, *2*(1), 57. http://dx.doi.org/10.1186/2191-219X-2-57.
Van Eldik, L. J., & Zimmer, D. B. (1987). Secretion of S-100 from rat C6 glioma cells. *Brain Research*, *436*, 367–370.
Van Puyvelde, K., Mets, T., Njemini, R., Beyer, I., & Bautmans, I. (2014). Effect of advanced glycation end product intake on inflammation and aging: A systematic review. *Nutrition Reviews*, *72*, 638–650.
Veerhuis, R. (2011). Histological and direct evidence for the role of complement in the neuroinflammation of AD. *Current Alzheimer Research*, *8*, 34–58.
Venigalla, M., Sonego, S., Gyengesi, E., Sharman, M. J., & Münch, G. (2016). Novel promising therapeutics against chronic neuroinflammation and neurodegeneration in Alzheimer's disease. *Neurochemistry International*, *95*, 63–74.
Vesce, S., Rossi, D., Brambilla, L., & Volterra, A. (2007). Glutamate release from astrocytes in physiological conditions and in neurodegenerative disorders characterized by neuroinflammation. *International Review of Neurobiology*, *82*, 57–71.
Walsh, D. M., & Selkoe, D. J. (2004). Deciphering the molecular basis of memory failure in Alzheimer's disease. *Neuron*, *44*, 181–193.
Wang, L., Gao, S., Jiang, W., Luo, C., Xu, M., Bohlin, L., et al. (2014). Antioxidative dietary compounds modulate gene expression associated with apoptosis, DNA repair, inhibition of cell proliferation and migration. *International Journal of Molecular Sciences*, *15*, 16226–16245.
Wang, W., Shinto, L., Connor, W. E., & Quinn, J. F. (2008). Nutritional biomarkers in Alzheimer's disease: The association between carotenoids, n-3 fatty acids, and dementia severity. *Journal of Alzheimers Disease*, *13*, 31–38.
Webster, B., Hansen, L., & Adame, A. (2006). Astroglial activation of extracellular-regulated kinase in early stages of Alzheimer disease. *Journal of Neuropathology & Experimental Neurology*, *65*, 142–151.
Weiner, H. L., & Frenkel, D. (2006). Immunology and immunotherapy of Alzheimer's disease. *Nature Reviews. Immunology*, *6*, 404–416.
Williams, R. J., & Spencer, J. P. (2012). Flavonoids, cognition, and dementia: Actions, mechanisms, and potential therapeutic utility for Alzheimer disease. *Free Radical Biology and Medicine*, *52*, 35–45.
Wood, L. G., Garg, M. L., Smart, J. M., Scott, H. A., Barker, D., & Gibson, P. G. (2012). Manipulating antioxidant intake in asthma: A randomized controlled trial. *American Journal of Clinical Nutrition*, *96*, 534–543.
Wu, P. T., Fitschen, P. J., Kistler, B. M., Jeong, J. H., Chung, H. R., Aviram, M., et al. (2015). Effects of pomegranate extract supplementation on cardiovascular risk factors and physical function in hemodialysis patients. *Journal of Medicinal Food*, *18*, 941–949.
Wu, R. P., Hayashi, T., Cottam, H. B., Jin, G., Yao, S., Wu, C. C., et al. (2010). Nrf2 responses and the therapeutic selectivity of electrophilic compounds in chronic lymphocytic leukemia. *Proceedings of the National Academy of Sciences of the United States of America*, *107*, 7479–7484.
Wyss-Coray, T. (2006). Inflammation in Alzheimer disease: Driving force, bystander or beneficial response? *Nature Medicine*, *12*, 1005–1015.
Wyss-Coray, T., Loike, J. D., Brionne, T. C., Lu, E., Anankov, R., Yan, F., et al. (2003). Adult mouse astrocytes degrade amyloid-β in vitro and in situ. *Nature Medicine*, *9*(4), 453–457.
Wyss-Coray, T., & Rogers, J. (2012). Inflammation in Alzheimer disease-a brief review of the basic science and clinical literature. *Cold Spring Harbor Perspectives in Medicine*, *2*(1), a006346.

Yan, L., Xiao, M., Chen, R., & Cai, Z. (2013). Metabolic dysfunction of astrocyte: An initiating factor in beta-amyloid pathology? *Aging and Neurodegeneration*, *1*, 7–14.

Yang, Q., Hamberger, A., Hyden, H., Wang, S., Stigbrand, T., & Haglid, K. G. (1995). S-100β has a neuronal localisation in the rat hindbrain revealed by an antigen retrieval method. *Brain Research*, *696*, 49–61.

Yarla, N. S., Bishayee, A., Sethi, G., Reddanna, P., Kalle, A. M., Dhananjaya, B. L., et al. (2016). Targeting arachidonic acid pathway by natural products for cancer prevention and therapy. *Seminars in Cancer Biology*, *40–41*, 48–81.

Yu, X., Wang, L. N., Du, Q. M., Ma, L., Chen, L., You, R., et al. (2012). Akebia Saponin D attenuates amyloid β-induced cognitive deficits and inflammatory response in rats: Involvement of Akt/NF-κB pathway. *Behavioural Brain Research*, *235*, 200–209.

Yuan, L., Liu, J., Ma, W., Dong, L., Wang, W., Che, R., et al. (2016). Dietary pattern and antioxidants in plasma and erythrocyte in patients with mild cognitive impairment from China. *Nutrition*, *32*, 193–198.

Yucesoy, B., Peila, R., White, L. R., Wu, K. M., Johnson, V. J., Kashon, M. L., et al. (2006). Association of interleukin-1 gene polymorphisms with dementia in a community-based sample: The Honolulu-Asia aging study. *Neurobiology of Aging*, *27*, 211–217.

Zern, T. L., Wood, R. J., Greene, C., West, K. L., Liu, Y., Aggarwal, D., et al. (2005). Grape polyphenols exert a cardioprotective effect in pre- and postmenopausal women by lowering plasma lipids and reducing oxidative stress. *Journal of Nutrition*, *135*, 1911–1917.

Zhao, H. F., Li, N., Wang, Q., Cheng, X. J., Li, X. M., & Liu, T. T. (2015). Resveratrol decreases the insoluble Aβ1-42 level in hippocampus and protects the integrity of the blood-brain barrier in AD rats. *Neuroscience*, *310*, 641–649.

Zilka, N., Ferencik, M., & Hulin, I. (2006). Neuroinflammation in Alzheimer's disease: Protector or promoter? *Bratislavské Lekárske Listy*, *107*, 374–383.

CHAPTER THREE

# Inflammation in Epileptic Encephalopathies

**Oleksii Shandra*, Solomon L. Moshé*,†, Aristea S. Galanopoulou*,†,1**
*Laboratory of Developmental Epilepsy, Albert Einstein College of Medicine, Bronx, NY, United States
†Montefiore/Einstein Epilepsy Center, Montefiore Medical Center, Bronx, NY, United States
[1]Corresponding author: e-mail address: aristea.galanopoulou@einstein.yu.edu

## Contents



## Abstract

West syndrome (WS) is an infantile epileptic encephalopathy that manifests with infantile spasms (IS), hypsarrhythmia (in ~60% of infants), and poor neurodevelopmental outcomes. The etiologies of WS can be structural–metabolic pathologies (~60%), genetic (12%–15%), or of unknown origin. The current treatment options include hormonal treatment (adrenocorticotropic hormone and high-dose steroids) and the GABA aminotransferase inhibitor vigabatrin, while ketogenic diet can be given as add-on treatment in refractory IS. There is a need to identify new therapeutic targets and more effective treatments for WS.

*Advances in Protein Chemistry and Structural Biology*, Volume 108
ISSN 1876-1623
http://dx.doi.org/10.1016/bs.apcsb.2017.01.005

Theories about the role of inflammatory pathways in the pathogenesis and treatment of WS have emerged, being supported by both clinical and preclinical data from animal models of WS. Ongoing advances in genetics have revealed numerous genes involved in the pathogenesis of WS, including genes directly or indirectly involved in inflammation. Inflammatory pathways also interact with other signaling pathways implicated in WS, such as the neuroendocrine pathway. Furthermore, seizures may also activate proinflammatory pathways raising the possibility that inflammation can be a consequence of seizures and epileptogenic processes. With this targeted review, we plan to discuss the evidence pro and against the following key questions. Does activation of inflammatory pathways in the brain cause epilepsy in WS and does it contribute to the associated comorbidities and progression? Can activation of certain inflammatory pathways be a compensatory or protective event? Are there interactions between inflammation and the neuroendocrine system that contribute to the pathogenesis of WS? Does activation of brain inflammatory signaling pathways contribute to the transition of WS to Lennox–Gastaut syndrome? Are there any lead candidates or unexplored targets for future therapy development for WS targeting inflammation?

## 1. INTRODUCTION

West syndrome (WS) is an age-specific epileptic encephalopathy (EE), which typically occurs in infants and has poor epilepsy, neurodevelopmental prognosis, and high risk of early mortality (Dulac, 2001; Fukuyama, 2001; Galanopoulou, 2013; Galanopoulou & Moshe, 2015a, 2015b). WS manifests with at least two of the following features: (a) ictal events of flexion or extension spasms, called infantile spasms (IS) that usually appear in clusters, (b) interictal chaotic high amplitude and multifocal epileptic interictal background (hypsarrhythmia), and (c) intellectual or neurodevelopmental disabilities (Galanopoulou & Moshe, 2015a, 2015b).

A wide range of possible etiologies for WS have been described, ranging from structural or metabolic (including malformations, vascular, inflammatory or immune, hypoxia, etc.) to genetic, albeit in many infants the causes are yet unknown (Berg et al., 2010; Epi4K Consortium et al., 2013; Frost & Hrachovy, 2005; Osborne et al., 2010; Paciorkowski et al., 2011; Pellock et al., 2010). Mortality can be significant ranging between 9% and 35% (Autry, Trevathan, Van Naarden Braun, & Yeargin-Allsopp, 2010; Pellock et al., 2010; Riikonen, 2001; Trevathan, Murphy, & Yeargin-Allsopp, 1999), although higher rates have been reported with long-term follow-up (Silanpaa, Riikonen, Saarinen, & Schmidt, 2016). IS can progress into other types of seizures and epilepsies, which might have poor response to treatment. Epilepsy develops in 50%–70% of surviving children (Koo, Hwang, & Logan, 1993; Riikonen, 2001) and evolution to Lennox–Gastaut syndrome (LGS) has been reported in 15%–25% of these patients, although reports of up to 54% have been published (Galanopoulou, 2013; Hrachovy & Frost, 2003; Lombroso, 1983; Rantala & Putkonen, 1999; Trevathan et al., 1999). Cognitive and neurodevelopmental impairments may be significant and often leading to mental retardation (70%–90%) (Galanopoulou & Moshe, 2015a, 2015b; Koo et al., 1993; Riikonen, 2001; Sidenvall & Eeg-Olofsson, 1995; Trevathan et al., 1999). Further evolution to autism is seen in 15%, although infants with IS due to underlying structural or metabolic pathologies are at higher risk (Galanopoulou, 2013; Riikonen & Amnell, 1981; Saemundsen, Ludvigsson, & Rafnsson, 2008). For example, 57% of infants with IS due to tuberous sclerosis complex (TSC) had ASD (Hunt & Dennis, 1987).

The first-line treatment options for IS are hormonal therapies [adrenocorticotropic hormone (ACTH), glucocorticoid steroids] or the GABA aminotransferase inhibitor vigabatrin (Galanopoulou & Moshe, 2015a, 2015b; Go et al., 2012; Mackay et al., 2004; Pellock et al., 2010; Riikonen, 2014). Ketogenic diet has been proposed as an alternative adjunctive therapy for IS to be considered after failure of corticosteroids and/or vigabatrin and has shown some efficacy in refractory IS (Hong, Turner, Hamdy, & Kossoff, 2010; Pires et al., 2013). A smaller percentage of patients with IS may respond to other antiseizure drugs (valproate, topiramate, zonisamide) or vitamin B6 (Pellock et al., 2010; Riikonen, 2014). Considering the devastating prognosis of patients with WS, the identification of better therapies for WS has been the focus of both clinical and preclinical studies using the animal models. Inflammation has attracted interest as a potential target for new therapy development for WS due to the availability of clinically available drugs targeting the inflammatory signaling pathways and the emerging evidence for a potential role in the pathogenesis of WS.

In this review, we will focus on selected inflammatory pathways and molecular targets investigated in infants and animal models of WS, and discuss the evidence that inflammation could be a cause of WS, its associated comorbidities, progression, and/or evolution of WS into LGS, or simply a consequence. Finally, we will present an update on the progress in new inflammation-targeting therapies based on recent clinical and experimental evidence from animal models (multiple-hit rat model of IS).

## 2. INFLAMMATION AND NEUROINFLAMMATION: DEFINITIONS, TARGETS, AND ROLE IN EPILEPSY AND SEIZURES

Inflammation can be defined as the body's immune system response to various stimuli, involving the cascade of reactions and signals which lead to activation of innate immune cells (in particular neutrophils and macrophages) and adaptive immune response (in particular lymphocytes, which further remove the targeted pathogens). The innate immunity is a nonspecific defense mechanisms activated either immediately or within hours after antigen invasion. The adaptive or acquired immunity is a specialized process involving antigen presentation that aims to eliminate or prevent pathogen growth. Inflammation can be nonsterile or sterile, depending upon whether the trigger is a microorganism or a noninfectious pathogen/antigen. Both types of inflammation manifest with recruitment of neutrophils and macrophages and production of proinflammatory cytokines and chemokines (e.g., tumor necrosis factor alpha [TNF-α] and interleukins [IL]) (Chen & Nuñez, 2010).

Neuroinflammation is the innate immunological response within the nervous system, involving microglia and astrocytes (Graeber, Li, & Rodriguez, 2011). Today, brain inflammation or neuroinflammation has been associated with many central nervous system (CNS) pathologies (Dey, Kang, Qiu, Du, & Jiang, 2016). Such processes may have beneficial effects, protecting against exogenous insults or promoting healing, but under certain situations they can be pathogenic. Prior studies associated elevated levels of proinflammatory cytokines with seizures, the pathogenesis of epilepsy, and pathologies manifesting epilepsy (Table 1) (Vitaliti, Pavone, Mahmood, Nunnari, & Falsaperla, 2014). Recent studies have introduced the neurological sequelae, in which proinflammatory cytokines with proictogenic properties (such as IL-1β, high-mobility group box 1 [HMGB1], cyclooxygenase-2 [COX-2], prostaglandin E2 [PGE2], IL-6, TNF-α, and nicotinamide adenine dinucleotide phosphate-oxidase [NOX2]) play an important role in seizure generation and exacerbation (Dey et al., 2016; Vezzani, Auvin, Ravizza, & Aronica, 2012; Wu & Huang, 2015). IL-1β and HMGB1 induce the proinflammatory innate immunity IL-1 receptor type 1 (IL-1R1)/toll-like receptor 4 (TLR4) signaling (Ravizza, Kostoula, & Vezzani, 2013; Vezzani, Lang, & Aronica, 2016), leading to neuroinflammation and its perpetuation and thus

**Table 1** Association of Proinflammatory Cytokines With Seizures in Patients With Epilepsy and Seizures

| Condition | Specimen | Increase in | Decrease in | References |
|---|---|---|---|---|
| TLE | TL tissue | Beta-amyloid precursor, IL-1 | NR | Sheng, Boop, Mrak, and Griffin (1994) |
| FCD type IA, IIA, and IIB | FCD brain tissue | IL-17, IL-17r | NR | He et al. (2013) |
| Tonic–clonic seizures (24 h following seizure) | Serum/ CSF | IL-6, leukocyte counts, C-reactive protein | NR | Peltola et al. (2002) |
| Drug-resistant epilepsy | Serum/ CSF | IL-6 | IL-1ra, IL-1ra: IL-1β ratio | Lehtimaki et al. (2004) and Lehtimaki, Keranen, Palmio, and Peltola (2010) |

Abbreviations: *CSF*, cerebrospinal fluid; *FCD*, focal cortical dysplasia; *IL*, interleukin; *IL-1ra*, interleukin-1 receptor antagonist; *IL-17r*, IL-17 receptor; *NR*, not reported; *TL*, temporal lobe; *TLE*, temporal lobe epilepsy.

modulating seizure susceptibility, lowering seizure threshold, and epileptogenesis (Maroso et al., 2010; Riazi, Galic, & Pittman, 2010; Vezzani et al., 2016; Vitaliti et al., 2014).

Toll-like receptors (TLRs) are the key proteins in the mammalian innate immune response to both infectious and noninfectious CNS diseases, including epilepsy (Akira, Takeda, & Kaisho 2001; Akira, Uematsu, & Takeuchi, 2006; Kawai & Akira, 2010; Pernhorst et al., 2013; Wang, Lin, & Yang, 2011). Each TLR family member, with the exception of TLR3, signals through the MyD88-dependent pathway, initiated by the MyD88 adaptor protein, resulting in the activation of the nuclear factor kappa-light-chain-enhancer of activated B cells (NF-κB) (a mediator of acute inflammation (Lawrence, 2009)), mammalian target of rapamycin (mTOR) (pathway involved in regulation of multiple processes including inflammation (Galanopoulou, 2013)), and generation of the proinflammatory cytokines, such as IL-6 and TNF-α (Ravizza et al., 2013; Takeda & Akira, 2004; Wang et al., 2011). This evidence arose from surgically resected epileptogenic tissues of adult and pediatric patients with epilepsy showing the presence of inflammatory molecules in activated glial cells, neurons, and endothelial cells of the blood–brain barrier (BBB) and indicating the strong link between inflammation and seizures (Aronica et al., 2007; Balosso, Ravizza, Aronica, & Vezzani, 2013; Crespel et al., 2002; Iori et al., 2013; Maroso et al., 2010; Pernhorst et al., 2013; Ravizza et al., 2006, 2008; Vezzani, French, Bartfai, & Baram, 2011; Zurolo et al., 2011).

Additional studies have also shown that loss of BBB integrity may be an important link between neuroinflammation and epileptogenesis (Bar-Klein et al., 2014; Fabene et al., 2008; Friedman et al., 2014; Kim, Kang, Dustin, & McGavern, 2009; Ransohoff, 2009; Salar et al., 2016; Vitaliti et al., 2014). Overactivated mast cells, which can be potentially activated by stress hormones, such as corticotropin-releasing hormone (CRH), lead to BBB disruption (Esposito et al., 2002; Theoharides, 1990; Theoharides & Konstantinidou, 2007; Vitaliti et al., 2014). In addition, extravasation of albumin during vascular injury activates the transforming growth factor beta (TGF-β) receptor I signaling pathway and contributes to epileptogenesis (Bar-Klein et al., 2014). Losartan, an inhibitor of the angiotensin II type 1 receptor and blocker of TGF-β signaling, prevented epileptogenesis in a model of BBB disruption induced by deoxycholate (Bar-Klein et al., 2014).

There is a two-way interaction between inflammation and seizures. Inflammation can decrease the seizure threshold and cause more severe seizures and comorbidities (Mazarati, Maroso, Iori, Vezzani, & Carli, 2011; Vezzani et al., 2011; Vezzani, Aronica, Mazarati, & Pittman, 2013). On the other hand, seizures, independent of their trigger, might cause neuroinflammation in structures involved in the onset, control, and generalization of epileptic activity, and may worsen outcomes (Auvin, Cilio, & Vezzani, 2016; Dube, Vezzani, Behrens, Bartfai, & Baram, 2005; Dube et al., 2010; Marcon et al., 2009; Rizzi et al., 2003).

In epilepsy, several pathogenic mechanisms leading to epileptogenesis and its progression or contributing to pharmacoresistance have been associated with proinflammatory cytokines as predisposing factors (Vitaliti et al., 2014). For example, cytokines may reduce astrocytic glutamate reuptake and increase the extracellular glutamate concentration (Takaki et al., 2012; Ye & Sontheimer, 1996), leading to neuronal excitation. Increase in the expression of astrocytic glutamate transporters through treatment with ceftriaxone has been shown to reduce seizures in a mouse model of tuberous sclerosis (Zeng, Bero, Zhang, Holtzman, & Wong, 2010). Inflammation or its downstream signaling mediators may alter the neurotransmitter receptor composition or expression or function aggravating seizures, their comorbidities, or their pathological sequelae, including cell death, neurogenesis, angiogenesis, synaptic reorganization, and BBB disruption (Lubin, Ren, Xu, & Anderson, 2007; Mazarati et al., 2011; Rigau et al., 2007; Sankar, Auvin, Mazarati, & Shin, 2007a, 2007b; Vezzani et al., 2011, 2013; Yang et al., 2010).

## 3. DOES ACTIVATION OF THE INFLAMMATORY PATHWAYS IN THE BRAIN CAUSE EPILEPSY IN WS?

### 3.1 Clinical Evidence

In WS, there are numerous etiologies recorded, including at least 60 known genes affected, numerous chromosomal abnormalities, or other structural and metabolic etiologies (Galanopoulou & Moshe, 2015b). Few of the genes linked with WS are components of neuroinflammatory pathways, including complement regulatory factors (CD46), glucocorticoid signaling (glucocorticoid modulatory element-binding protein 2 GMEB2, glucocorticoid receptor NR3C1), mTOR pathway [tuberous sclerosis complex 1 and 2 (TSC1 and TSC2), STE20-related kinase adaptor alpha (STRADα), phosphatase and tensin homolog tumor suppressor (PTEN), phosphoinositide 3-kinase adapter protein (PIK3AP1)], and TNFα-induced protein 6 (TNFAIP6) (see Supplemental Table 1 in Galanopoulou & Moshe, 2015a, 2015b). Malformations of brain development and focal cortical dysplasias are also common pathologies in infants with WS, which have been linked with abnormal expression of components of the mTOR signaling: TSC1 or TSC2 in TSC, PTEN in hemimegalencephaly, STRADα in polyhydramnios, megalencephaly and symptomatic epilepsy syndrome (PMSE), and overexpression of phosphorylated S6 ribosomal protein (pS6) in focal cortical dysplasias type IIB (Baybis et al., 2004; Chu-Shore, Major, Camposano, Muzykewicz, & Thiele, 2010; Epi4K Consortium et al., 2013; Galanopoulou, Gorter, & Cepeda, 2012; Lim & Crino, 2013; Orlova et al., 2010; Pardo, Nabbout, & Galanopoulou, 2014). Among the acquired pathologies leading to IS, hypoxic–ischemic injury, CNS infections, perinatal strokes, metabolic disorders, or autoimmune conditions may also manifest inflammatory changes (Mota, Rezkallah-Iwasso, Peracoli, & Montelli, 1984; Steele et al., 2012 and reviewed in Pardo et al., 2014).

Clinical observations show the involvement of immune and inflammatory processes in WS as manifested by altered blood or CSF levels of various cytokines prior to treatment, although differences across studies are noted (see Table 2 for summary of published data). Among the studies that correlated the treatment response to the inflammatory indices before and after treatment, there is evidence that some of these changes are linked to treatment response (e.g., IL-1ra elevation or reduction of blood IL-1β or reduction in CD4+ or CD4+/CD8+ ratios) (Ohya et al., 2009; Shiihara et al., 2010; Yamanaka et al., 2010). However, conclusive studies are needed as well as confirmation by animal models.

An interesting link has been described in tuberous sclerosis (TSC), which is a genetic pathology which can manifest IS (Chudomelova et al., 2010; Chu-Shore et al., 2010; Curatolo, Seri, Verdecchia, & Bombardieri, 2001). The overactivated mTOR complex 1 (mTORC1) in TSC is associated with loss-of-function mutation in TSC1 and TSC2 genes (Crino, 2010; Raffo, Coppola, Ono, Briggs, & Galanopoulou, 2011). Inhibitors of mTOR pathway, such as rapamycin, may reverse some of the pathologic sequelae of the mTOR overactivation which lead to pathologies associated with epilepsies (e.g., dysplastic neurons, gliosis) and epilepsy development (Li et al., 2010; Parker et al., 2013; Raffo et al., 2011; Sharma et al., 2010; Talos, Kwiatkowski, Cordero, Black, & Jensen, 2008; Wang, Barbaro, & Baraban, 2006). Clinical studies specifically demonstrating the efficacy of mTOR inhibitors (e.g., everolimus, rapamycin) in WS patients with mTOR dysregulation are lacking, largely due to the difficulty in testing new therapies in very young patients, rendering the need for obtaining answers in animal models critical.

Interestingly, outcomes in WS patients were affected by the etiology, response to ACTH treatment, coexistence of other types of seizures, time lag in treatment initiation and response (primarily affecting the cognitive outcome), and persistent EEG abnormality (Koo et al., 1993; Pellock et al., 2010). Although the mechanism of action of ACTH may not be entirely antiinflammatory, studies have shown evidence of reduction of some of the elevated proinflammatory cytokines in WS patients treated with ACTH (Table 2). Shiihara et al. (2010) have proposed that impaired immune system may be more a consequence of the WS rather than a cause, although there is no clinical definitive evidence yet for this hypothesis. More studies are necessary to prove this hypothesis, focusing on comparison of the immunologic parameters between patients with WS (before therapy) and controls, as well as WS groups before and after treatment (including

**Table 2** Immunological Findings in Patients With WS

| References | Total Number of Patients With WS | Treatment | Principal Parameters Investigated | Specimen | Time of Specimen Collection | Total Number of Control Patients | Immunologic Markers: WS Before Treatment vs Controls | Immunologic Markers: WS After Treatment vs WS Before Treatment |
|---|---|---|---|---|---|---|---|---|
| Tekgul et al. (2006) | 12 (8—symptomatic; 4—cryptogenic) | None (newly diagnosed) | IL-6 | CSF | Before treatment | Patients with tonic–clonic seizures (CNS infection or trauma) | Reduced: IL-6 | NR |
| Dussaix, Lebon, Ponsot, Huault, and Tardieu (1985) | 12 | NR | IFN-α | CSF | NR | Patients with acute encephalitis or meningitis or other chronic neurological disorders | No IFN-α in CSF detected | NR |
| Liu, Wang, Wang, and Yang (2001) | 23 (13—symptomatic; 10—unknown) | ACTH (4–6 weeks), further CZP/VPA/TOP | IL-2, TNF-α, IFN-α | Blood | Before treatment | 15 (healthy controls) | Elevated: IL-2, TNF-α, IFN-α | NR |
| Montelli, Soares, and Peraçoli (2003) | 50 (including 17 post-WS[a]) | NR | Peripheral lymphocyte subsets; CD3+ cells, CD4+ cells, CD8+ cells, and CD4+/CD8+ ratio | Blood | Samples collected at infection and ACTH-free period (not reflected if prior or after therapy) | 20 (healthy controls) | Elevated: CD8+<br>Reduced: CD3+; CD4+; CD4/CD8 ratio<br>*(not reported whether the data were obtained before or after treatment)* | |

*Continued*

**Table 2** Immunological Findings in Patients With WS—cont'd

| References | Total Number of Patients With WS | Treatment | Principal Parameters Investigated | Specimen | Time of Specimen Collection | Total Number of Control Patients | Immunologic Markers | |
|---|---|---|---|---|---|---|---|---|
| | | | | | | | WS Before Treatment vs Controls | WS After Treatment vs WS Before Treatment |
| Ohya et al. (2009) | 18 (8—symptomatic; 10—unknown) | ACTH (Cortrosyn $Z^R$) Other treatments: VPA (16/18), vitamin B6 (8/18), CZP, etc. | WBC counts; lymphocyte counts; T cell, B cell, CD4+ T cell, CD8+ T cell counts; CD 4/8 ratio, and the levels of IgA, IgM, and IgG | Blood | Before treatment; immediately after, 1, 3, 6, and 12 months after treatment | NR | | Reduced by ACTH: CD4+; CD4/8 ratio; lymphocyte counts<br>No change: Ig |
| Shiihara et al. (2010) | 76 (44—symptomatic; 32—unknown) | ACTH | Peripheral lymphocyte subsets; IL-1β, IL-1ra, IL-5, IL-6, IL-12, IL-15; eotaxin; bFGF; IFN-γ-inducible protein-10; MIP-1β; CD3+ CD25+, CD19+, CD19+ CD95+, CD4+ cells, and CD4+/8+ ratio | Blood | Before and after treatment | 26 (healthy controls) | Elevated: IL-1ra; IL-5; IL-6; IL-15; eotaxin; bFGF; IFN-γ-IP-10<br>Reduced: CD3+ CD25+; CD19+; CD19+ CD95+ | Elevated: none<br>Reduced: CD4+; CD3+; CD4+/8+ratio; IL-1β; IL-12; MIP-1β |

| | | | | | | | | |
|---|---|---|---|---|---|---|---|---|
| Yamanaka et al. (2010) | 13 (9—symptomatic; 4—unknown) | ACTH Other treatments: vitamin B6, VPA, ZNS | IL-1β, IL-1ra | Blood and CSF | Before or after ACTH | NR | Before ACTH: no group differences noted | Elevated in responders: serum IL-1ra |
| Haginoya et al. (2009) | 24 (17—symptomatic; 7—unknown) | None (newly diagnosed) | IL-1β, IL-1ra, IL-6, TNF-α | CSF | Before ACTH treatment | 15 (healthy controls) and 16 (aseptic meningitis) | Reduced: IL-1ra No change: IL-1β; TNF-α; IL-6 | NR |

[a]Post-WS designate patients initially diagnosed WS later evolving to other epilepsy such as Lennox–Gastaut syndrome.

These findings indicate a possible role of the cytokines and immune status alteration in the pathogenesis of WS, although it is difficult to make a strong comparison due to variations in patient groups, specimens and parameters evaluated.

Abbreviations: *ACTH*, adrenocorticotropic hormone; *ASD*, antiseizure drugs; *bFGF*, basic fibroblast growth factor; *CSF*, cerebrospinal fluid; *CZP*, clonazepam; *IFN*, interferon; *IFN-γ-IP-10*, IFN-γ-inducible protein-10; *Ig*, immunoglobulin; *IL*, interleukin; *IL-1ra*, interleukin-1 receptor antagonist; *MIP*, macrophage inflammatory protein; *TNF*, tumor necrosis factor; *TOP*, topiramate; *VPA*, valproate; *WS*, West syndrome; *ZNS*, zonisamide.

follow-up periods). Additionally, some of the reported markers may have a dual action (either pro- or antiinflammatory), such as TNF-α (Zakharova & Ziegler, 2005). Difficulties related to number of patients with WS, different patient groups and controls (or absence of controls), as well as variations in principal parameters evaluated and source of specimen collection complicate the analysis of the previous studies, and thus, a need in new clinical studies is critical for better understanding of the underlying mechanisms of inflammation and their role in contribution to epilepsy development in WS.

## 3.2 Inflammation in Animal Models of IS: The Multiple-Hit Rat Model of IS

Several animal models of IS or WS or epileptic spasms have been developed (reviewed in Galanopoulou & Moshe, 2015a, 2015b, also see Dube et al., 2015; Pirone et al., 2016). The investigation of inflammatory pathways in IS has been done so far in the multiple-hit rat model, which will be described later.

In the multiple-hit rat model of WS, the intent was to generate a lesion that disrupts cortical–subcortical communication networks that were hypothesized to underlie the pathogenesis of IS (Lado & Moshe, 2002). Spasms in this model are induced in male Sprague–Dawley rats by acute intracerebral infusion of the cytotoxic agent doxorubicin and TLR-4 agonist lipopolysaccharide (LPS) on postnatal day 3 (PN3) followed by a single intraperitoneal injection of brain serotonin synthesis blocker *p*-chlorophenylalanine (PCPA) on PN5 (Scantlebury et al., 2010). PCPA was added due to old reports on abnormal serotonin metabolism in patients with IS (Langlais, Wardlow, & Yamamoto, 1991; Scantlebury et al., 2010; Silverstein & Johnston, 1984; Yamamoto, 1991). LPS is a prototypical inducer of inflammation in both periphery and in the brain, known to lower seizure threshold when added with proconvulsant compounds and worsen seizure sequelae (Galic et al., 2008; Sankar et al., 2007a, 2007b; Sayyah, Javad-Pour, & Ghazi-Khansari, 2003). Spasms in this model appear at PN4 and continue until PN13 with epileptic patterns, including electrodecremental responses and fast oscillations with an epileptic interictal background (Briggs, Mowrey, Hall, & Galanopoulou, 2014; Ono, Moshe, & Galanopoulou, 2011; Raffo et al., 2011; Scantlebury et al., 2010). The multiple-hit model manifests a structural lesion, mainly right hemispheric (described in Briggs et al., 2014; Jequier Gygax, Klein, White, Kim, & Galanopoulou, 2014; Scantlebury et al., 2010) and is considered as model of the more severe type of WS due to structural lesion which is refractory to current treatments (ACTH refractory, partially sensitive to vigabatrin) (Scantlebury et al., 2010). In addition to the epilepsy phenotype, the multiple-hit model also demonstrates transient impairment of neurodevelopmental milestones around the period of spasms, learning deficits (Barnes maze PN16–19), and impaired social interactions, consistent with the more severe prognosis of WS due to structural lesions. As expected, phenytoin has no effect on spasms in this model (Ono et al., 2011). This model has been extensively used for the screening of new potential therapies for IS, including a promising experimental drug CPP-115 (more potent vigabatrin analogue with lower risk for retinal toxicity) (Briggs et al., 2014) which was recently clinically tested in a case report with very similar efficacy (Doumlele, Conway, Hedlund, Tolete, & Devinsky, 2016). In addition, other new drugs showing efficacy in this model include a drug with broad-spectrum antiepileptic effect in experimental animal models of seizures (carisbamate) and an mTOR inhibitor rapamycin (Ono et al., 2011; Raffo et al., 2011).

Among the three drugs tested in this model, we found that doxorubicin and/or LPS are sufficient to induce IS. Histopathologically, we found pronounced inflammation in the cortical region around the infusions, including astrocytosis [glial fibrillary astrocytic protein (GFAP)-positive cells], microglia, and evidence of mTOR overactivation, as shown by increased pS6 immunoreactivity (Raffo et al., 2011). There are also increased levels of cytokines (IL-1β, TNFα) or downstream targets (e.g., NF-κB) in the peri-infusional cortical region (A. Galanopoulou, unpublished observations). Proof of concept that inflammation alone can induce spasms was provided by the right intracerebral LPS infusion, which can induce spasms with decremental responses in young rats (Ono et al., 2012).

Exploration of the therapeutic potential of various antiinflammatory drugs in the multiple-hit rat model produced few notable results. Following a pulse, high-dose rapamycin treatment started after spasms' onset between PN4–6, spasms stopped, and pups showed a partial improvement of learning in the Barnes maze after spasms stopped (PN16–19), suggesting partial disease modification (Raffo et al., 2011). Administration of an antiinflammatory, antioxidant drug that inhibits NF-κB and cytokine production as a single dose given after spasms manifested, celastrol, also acutely reduced behavioral and electroclinical spasms in this model (Shandra, Wang, Mowrey, Moshé, & Galanopoulou, 2015). Of interest, the NF-κB pathway can be inhibited by ACTH peptides and melanocortin receptors (thought to be activated by ACTH) (Catania, 2007; Moustafa et al., 2002). On the other hand, the proconvulsant hormone CRH can inhibit NF-κB in human melanocytes (Zbytek, Pfeffer, & Slominski, 2006), emphasizing the need to test the effect of modulators of such signaling pathways in vivo, since the mechanisms of action may be complex. However, the interesting results of celastrol in an ACTH-refractory model of IS raise the possibility that targeting a downstream mediator of ACTH with possibly a more specific inhibitor might have better therapeutic effect.

Although the multiple-hit and related inflammation-induced models of IS provide a good proof-of-concept evidence that inflammatory pathways may contribute or suffice to generate IS epileptogenesis, further across model validation may be useful to determine the exact role of inflammatory cascades in the phenotype and treatment of WS across etiologies and pathologies.

## 4. DOES ACTIVATION OF THE INFLAMMATORY PATHWAYS IN THE BRAIN CONTRIBUTE TO THE ASSOCIATED COMORBIDITIES AND PROGRESSION?

Between 1960 and 1988, 29 (10% of all patients) infants with IS admitted to the Children's Hospital of University of Helsinki have had infections as the primary etiologic factor, including cytomegalovirus (CMV), congenital rubella, herpes simplex, enterovirus, adenovirus, encephalitis, meningococcus, pneumococcus, and pertussis (Riikonen, 1993). The outcomes in patients with IS of infectious etiology appeared to be particularly poor compared with outcomes of the total number of patients evaluated in the study and included mental retardation (90%) and convulsions (62%) and abnormal EEG (generalized disturbance, diffuse polyspike, and slow wave discharges and focal abnormality) (89%) (Riikonen, 1993). Twenty-six (90% of the total) IS patients with infectious etiology (except two with acquired CMV infection and one with encephalitis with unknown etiology) developed intellectual disabilities, and 20 (69%) had cerebral palsy (Riikonen, 1993).

Emerging data from experimental settings report that exposure to inflammation during critical developmental period in rats leads to long-lasting brain excitability increase, reducing significantly the seizure threshold (Riazi et al., 2010). Further pharmacological targeting of the IL-1 receptor and TLR4 signaling pathways in rodents demonstrated tight connection between proinflammatory cytokines and epileptogenesis (Maroso et al., 2011a; Maroso et al., 2011b). Evidence from the multiple-hit model in rats indicates evolution of spasms into other types of seizures, together with learning and sociability deficits after PN9 and in the adulthood, emphasizing a possible important role of inflammation and focal lesion in the development of seizures and epilepsy in patients with WS (Galanopoulou, 2013; Scantlebury et al., 2010). The observation that antiinflammatory drugs, like rapamycin, may prevent some of the learning deficits in the multiple-hit model (Raffo et al., 2011) may also suggest that these pathways may be important determinants of the cognitive and neurodevelopmental outcomes. Whether concomitant inflammation may contribute to the pharmacoresistance of this model is also worth investigating, particularly in view of the beneficial effects of antiinflammatory treatments in this ACTH-refractory model of IS.

## 5. CAN ACTIVATION OF CERTAIN INFLAMMATORY PATHWAYS BE A COMPENSATORY OR PROTECTIVE EVENT?

Paradoxical spontaneous remission of IS has been reported after viral infections (exanthema subitum, rotavirus gastroenteritis, measles, chickenpox) or acute febrile illness (Hattori, 2001; Pintaudi et al., 2007), suggesting that neuroinflammation may have complex effects on networks involved in IS. The exact mechanisms for these effects are unknown. The role of the cytokines in the processes linked with epileptogenesis can be complex, including a possible protective or compensatory role. Dual effects of neuroinflammatory pathways have been demonstrated with COX-2 inhibitors in poststatus epilepticus injury, depending upon the context they are applied into (Jiang & Dingledine, 2013). As an example, increased production of the proinflammatory cytokines can increase the production of antiinflammatory cytokines, such as IL-10. IL-10 expression also protects neurons and glial brain cells mainly by inhibiting proapoptotic cytokines and by stimulating protective signaling reactions (Qian et al., 2006; Youn, Sung, & Lee, 2013). Alternatively, inflammation may affect neuroendocrine, stress-related, or other biological pathways which could alter the course of seizures. The full spectrum of inflammatory mediators in epileptic encephalopathies, like IS, needs to be more fully investigated.

## 6. ARE THERE INTERACTIONS BETWEEN INFLAMMATION AND THE NEUROENDOCRINE SYSTEM THAT CONTRIBUTE TO THE PATHOGENESIS OF WS?

Based on the clinical evidence on the efficacy of therapy with ACTH and corticosteroids, a hypothesis has been proposed about a specific role of the immune and neuroendocrine systems in the pathophysiology of WS, although the precise mechanisms of action of immune suppressants in WS remain unclear (Rao & Willis, 1987; Wilson, 1973). The CSF levels of ACTH were found to be lower in patients with IS, lending a biological reason to justify the response of IS to ACTH (Baram, Mitchell, Snead, Horton, & Saito, 1992). In the same study, cortisol and CRH levels were not statistically different in WS patients than in controls. The emerging stress hypothesis of IS postulated that exaggerated response to otherwise anticipated stressors might be the pathogenic trigger in subjects predisposed to IS (Baram, 1993). CRH was proposed to be the mediator of this stress-induced IS epileptogenesis and was indeed shown to have proconvulsant effects when injected intracerebrally (cerebral cortex or hippocampus) in PN5 and PN10 rats (Baram & Schultz, 1991). However, the observed motor seizures responded to phenytoin but not ACTH and were thought to be limbic, rather than IS. Follow-up studies by this group utilizing a model of early chronic stress induced by "fragmented and unpredictable nurturing behavior" reported spasm-like events in 48% of rats originating from the amygdala or corticohippocampal regions (Dube et al., 2015). Further characterization of this model is pending and the sensitivity of the spasms to drugs has not been tested yet.

The influence of stress has been explored in the NMDA model of spasms (Mares & Velisek, 1992) by exploring the impact of perinatal stress. Prenatal stress (forced restrain) or prenatal betamethasone exposure or perinatal adrenalectomy accelerates the onset and/or increases the severity of spasms acutely induced by NMDA in rats (Velisek, Jehle, Asche, & Veliskova, 2007; Wang, Zhang, Liang, Yang, & Zou, 2012; Yum, Chachua, Veliskova, & Velisek, 2012). Interestingly, however, prenatal betamethasone also increased the sensitivity of NMDA-induced spasms to ACTH, suggesting that ACTH may employ stress-related pathways in its effects on this model. The clinical and translational relevance of these observations needs further investigation in order to clarify what constitutes a pathological exaggerated response to anticipated everyday common stressors, and what elements (genetic, biological, epigenetic) shift stress responses from being adaptive to being pathogenic.

## 7. DOES ACTIVATION OF BRAIN INFLAMMATORY SIGNALING PATHWAYS CONTRIBUTE TO THE TRANSITION OF WS TO LGS?

LGS is an EE, typically appearing between the second and sixth year of life and is characterized by an electroclinical triad of generalized slow spike wave activity in the EEG, various epileptic seizures (particularly tonic [stiffening] and atonic [drop]), and intellectual disabilities (Beaumanoir, 1982; Markand, 2003). The severity of this syndrome in particular is that it is difficult to control seizures and there is a need in life-long treatment (Beaumanoir, 1982; Donat & Wright, 1991).

While there is a debate whether or not WS and LGS are clinical manifestations of the same underlying encephalopathic process and expressing different features depending on brain maturation levels, clinical reports indicate the estimate of 15%–54% evolution of WS to LGS (Beaumanoir, 1982; Galanopoulou, 2013; Hrachovy & Frost, 2003; Lombroso, 1983; Ohtahara, Ishida, & Oka, 1976; Rantala & Putkonen, 1999; Trevathan et al., 1999; You, Kim, & Kang, 2009). 70%–80% of the patients with LGS, in turn, have preceding history other than WS (Chevrie & Aicardi, 1972; Heiskala, 1997; Markand, 1977).

The putative antiepileptic mechanisms such as inhibition of inflammation and modification of mitochondrial metabolism, as well as their correction with hormonal therapy or ketogenic diet, respectively, have been suspected to play key roles in preventing EE evolution to LGS in patients with WS (Bough & Rho, 2007; Choi & Koh, 2008; You et al., 2009). However, there is no proof-of-concept evidence about the impact of these mechanisms on WS evolution to LGS. Animal models of LGS would be greatly useful in elucidating these questions.

## 8. ARE THERE ANY LEAD CANDIDATES OR UNEXPLORED TARGETS FOR FUTURE THERAPY DEVELOPMENT FOR WS TARGETING INFLAMMATION?

To date, ACTH and corticosteroids remain among the first-line treatment options for WS (Mackay et al., 2004; Yamamoto, Fukuda, Miyamoto, Murakami, & Kamiyama, 2007). Previous studies have suggested the immunosuppressive effect of ACTH and corticosteroids, together with antiinflammatory properties, as well as the direct inhibitory impact on CNS excitability and possible inhibition of endogenous proconvulsant factor synthesis (Baram & Hatalski, 1998; Brunson, Khan, Eghbal-Ahmadi, & Baram, 2001; Joels & Baram, 2009; Riikonen, 2000; Vezzani et al., 2011), although ACTH has been reported to be more efficient (Snead, 2001). In addition, therapy with immunomodulatory agents such as ACTH and corticosteroids or high-dose immunoglobulin has shown efficacy on spasms cessation in some patients with IS of infectious etiology; however, it was suggested that steroid therapy should be avoided in patients with a history of CMV and herpes simplex in the past (Ariizumi et al., 1983, 1987; Hattori, 2001; Riikonen, 1993; Wise, Rutledge, & Kuzniecky, 1996).

Insights from the multiple-hit rat model of IS further support evidence about focal lesion in the etiology of IS and provide proof-of-concept evidence that inflammatory processes are among the key triggers of IS of structural/metabolic origin (Raffo et al., 2011; Scantlebury et al., 2010) due to efficacy of the tested therapeutic compounds. An mTOR inhibitor rapamycin has been shown to be a potential disease-modifying therapeutic candidate, which promotes cessation of spasms and partial improvement of memory in multiple-hit model rats, although better analogues are necessary for testing due to transient weight loss as a side effect (Raffo et al., 2011). Our recent findings about NF-κB activation in the cortex of the multiple-hit model rats and successful spasm cessation with NF-κB inhibitor celastrol (Shandra et al., 2015) provide additional evidence on the effectiveness of antiinflammatory therapy on spasms.

Together, immune status alterations, activated inflammatory pathways, infectious etiology of IS, as well as efficacy of the antiinflammatory and immunoregulatory therapy on spasms indicate the importance of these processes in the pathogenesis of WS and the promising strategy and targets for potential new therapy development. However, a detailed study of related links between epilepsy and neuroinflammation in WS will contribute to the isolation of the key "nodes" of dysregulation, and thus determine the new targets for potential pharmacological corrections.

## 9. CONCLUDING REMARKS

Recent studies of epileptogenesis in infantile EE, such as WS, lead to findings of various processes involved in EE pathogenesis, including neuroinflammation, immune status alteration, and neuroendocrine system recruitment. Neuroimmunomodulation has been suggested to be among the key pathogenic mechanisms engaged in the epileptogenesis of WS.

Current interest in neuroinflammation research has provided certain knowledge of its pathways, targets, and mechanisms involved in epilepsy development, progression, and recurrence of seizures and development of pharmacorefractoriness. Nevertheless, there is an urge in continuous and extensive studies in neuroinflammation for better understanding of its role and implications in developing brain and EEs. Studies of animal models of early-life epilepsies have provided researchers with invaluable tools for studying specific pathogenic processes, testing known and novel therapeutic candidates for better understanding of mechanisms of action, and/or for identification of potential new therapies and molecular targets. Genetic advances have further advanced the research in neuroinflammatory genes and shown their relevance to several epilepsy syndromes. Animal models of early-life epilepsies and epileptic encephalopathies have started to generate an increasing amount of evidence for new therapy targets, candidate therapy targets, and putative mechanisms involved in epileptogenesis and comorbidogenesis of these early-life epilepsies. Already few of these candidate drugs tested in animal models have entered clinical trials with promising results, lending value to the hope that the horizon of new pediatric therapy discovery will change to include not only extrapolation trials but also novel pediatric preclinical therapy discovery using age-specific animal models of epileptic encephalopathies (Galanopoulou & Mowrey, 2016).

## ACKNOWLEDGMENTS

A.S.G. reports grants from NINDS-NS91170, NINDS-1U54NS100064, Department of Defense (W81XWH-13-1-0180), and the Infantile Spasms Initiative from CURE (Citizens United for Research in Epilepsy), and acknowledges also research funding from the Heffer Family and the Segal Family Foundations, and the Abbe Goldstein/Joshua Lurie and Laurie Marsh/Dan Levitz families. S.L.M. is the Charles Frost Chair in Neurosurgery and Neurology. He reports grants NINDS-NS020253, NINDS-NS43209, NINDS-NS45911, and NINDS-1U54NS100064 and a grant from the US Department of Defense (W81XWH-13-1-0180), CURE Infantile Spasms Initiative grant and donations from the Heffer Family and the Segal Family Foundations, and the Abbe Goldstein/Joshua Lurie and Laurie Marsh/Dan Levitz families.

## REFERENCES

Akira, S., et al. (2001). Toll-like receptors: Critical proteins linking innate and acquired immunity. *Nature Immunology*, *2*(8), 675–680.

Akira, S., et al. (2006). Pathogen recognition and innate immunity. *Cell*, *124*(4), 783–801.

Ariizumi, M., et al. (1983). High dose gammaglobulin for intractable childhood epilepsy. *The Lancet*, *322*(8342), 162–163. Originally published as Volume 2, Issue 8342.
Ariizumi, M., et al. (1987). Immunoglobulin therapy in the West syndrome. *Brain and Development*, *9*(4), 422–425.
Aronica, E., et al. (2007). Complement activation in experimental and human temporal lobe epilepsy. *Neurobiology of Disease*, *26*(3), 497–511.
Autry, A. R., et al. (2010). Increased risk of death among children with Lennox-Gastaut syndrome and infantile spasms. *Journal of Child Neurology*, *25*(4), 441–447.
Auvin, S., et al. (2016). Current understanding and neurobiology of epileptic encephalopathies. *Neurobiology of Disease*, *92*(Pt. A), 72–89.
Balosso, S., et al. (2013). The dual role of TNF-alpha and its receptors in seizures. *Experimental Neurology*, *247*, 267–271.
Baram, T. Z. (1993). Pathophysiology of massive infantile spasms: Perspective on the putative role of the brain adrenal axis. *Annals of Neurology*, *33*(3), 231–236.
Baram, T. Z., & Hatalski, C. G. (1998). Neuropeptide-mediated excitability: A key triggering mechanism for seizure generation in the developing brain. *Trends in Neurosciences*, *21*(11), 471–476.
Baram, T. Z., & Schultz, L. (1991). Corticotropin-releasing hormone is a rapid and potent convulsant in the infant rat. *Brain Research. Developmental Brain Research*, *61*(1), 97–101.
Baram, T. Z., et al. (1992). Brain-adrenal axis hormones are altered in the CSF of infants with massive infantile spasms. *Neurology*, *42*(6), 1171–1175.
Bar-Klein, G., et al. (2014). Losartan prevents acquired epilepsy via TGF-beta signaling suppression. *Annals of Neurology*, *75*(6), 864–875.
Baybis, M., et al. (2004). mTOR cascade activation distinguishes tubers from focal cortical dysplasia. *Annals of Neurology*, *56*(4), 478–487.
Beaumanoir, A. (1982). The Lennox-Gastaut syndrome: A personal study. *Electroencephalography and Clinical Neurophysiology. Supplement*, (35), 85–99.
Berg, A. T., et al. (2010). Revised terminology and concepts for organization of seizures and epilepsies: Report of the ILAE Commission on Classification and Terminology, 2005-2009. *Epilepsia*, *51*(4), 676–685.
Bough, K. J., & Rho, J. M. (2007). Anticonvulsant mechanisms of the ketogenic diet. *Epilepsia*, *48*(1), 43–58.
Briggs, S. W., et al. (2014). CPP-115, a vigabatrin analogue, decreases spasms in the multiple-hit rat model of infantile spasms. *Epilepsia*, *55*(1), 94–102.
Brunson, K. L., et al. (2001). Corticotropin (ACTH) acts directly on amygdala neurons to down-regulate corticotropin-releasing hormone gene expression. *Annals of Neurology*, *49*(3), 304–312.
Catania, A. (2007). The melanocortin system in leukocyte biology. *Journal of Leukocyte Biology*, *81*(2), 383–392.
Chen, G. Y., & Nuñez, G. (2010). Sterile inflammation: Sensing and reacting to damage. *Nature Reviews. Immunology*, *10*(12), 826–837.
Chevrie, J. J., & Aicardi, J. (1972). Childhood epileptic encephalopathy with slow spike-wave a statistical study of 80 cases. *Epilepsia*, *13*(2), 259–271.
Choi, J., & Koh, S. (2008). Role of brain inflammation in epileptogenesis. *Yonsei Medical Journal*, *49*(1), 1–18.
Chudomelova, L., et al. (2010). Modeling new therapies for infantile spasms. *Epilepsia*, *51*(Suppl. 3), 27–33.
Chu-Shore, C. J., et al. (2010). The natural history of epilepsy in tuberous sclerosis complex. *Epilepsia*, *51*(7), 1236–1241.
Crespel, A., et al. (2002). Inflammatory reactions in human medial temporal lobe epilepsy with hippocampal sclerosis. *Brain Research*, *952*(2), 159–169.

Crino, P. B. (2010). The pathophysiology of tuberous sclerosis complex. *Epilepsia, 51*(Suppl. 1), 27–29.

Curatolo, P., et al. (2001). Infantile spasms in tuberous sclerosis complex. *Brain & Development, 23*(7), 502–507.

Dey, A., et al. (2016). Anti-inflammatory small molecules to treat seizures and epilepsy: From bench to bedside. *Trends in Pharmacological Sciences, 37*(6), 463–484.

Donat, J. F., & Wright, F. S. (1991). Seizures in series: Similarities between seizures of the west and Lennox-Gastaut syndromes. *Epilepsia, 32*(4), 504–509.

Doumlele, K., et al. (2016). A case report on the efficacy of vigabatrin analogue (1S, 3S)-3-amino-4-difluoromethylenyl-1-cyclopentanoic acid (CPP-115) in a patient with infantile spasms. *Epilepsy & Behavior Case Reports, 6*, 67–69.

Dube, C., et al. (2005). Interleukin-1beta contributes to the generation of experimental febrile seizures. *Annals of Neurology, 57*(1), 152–155.

Dube, C. M., et al. (2010). Epileptogenesis provoked by prolonged experimental febrile seizures: Mechanisms and biomarkers. *The Journal of Neuroscience, 30*(22), 7484–7494.

Dube, C. M., et al. (2015). Hyper-excitability and epilepsy generated by chronic early-life stress. *Neurobiology of Stress, 2*, 10–19.

Dulac, O. (2001). What is West syndrome? *Brain & Development, 23*(7), 447–452.

Dussaix, E., et al. (1985). Intrathecal synthesis of different alpha-interferons in patients with various neurological diseases. *Acta Neurologica Scandinavica, 71*(6), 504–509.

Epi4K Consortium, et al. (2013). De novo mutations in epileptic encephalopathies. *Nature, 501*(7466), 217–221.

Esposito, P., et al. (2002). Corticotropin-releasing hormone and brain mast cells regulate blood-brain-barrier permeability induced by acute stress. *The Journal of Pharmacology and Experimental Therapeutics, 303*(3), 1061–1066.

Fabene, P. F., et al. (2008). A role for leukocyte-endothelial adhesion mechanisms in epilepsy. *Nature Medicine, 14*(12), 1377–1383.

Friedman, A., et al. (2014). Should losartan be administered following brain injury? *Expert Review of Neurotherapeutics, 14*(12), 1365–1375.

Frost, J. D., Jr., & Hrachovy, R. A. (2005). Pathogenesis of infantile spasms: A model based on developmental desynchronization. *Journal of Clinical Neurophysiology, 22*(1), 25–36.

Fukuyama, Y. (2001). A special note on terminology of West syndrome and infantile spasms. *Brain & Development, 23*(7), 442.

Galanopoulou, A. S. (2013). Basic mechanisms of catastrophic epilepsy—Overview from animal models. *Brain & Development, 35*(8), 748–756.

Galanopoulou, A. S., & Moshe, S. L. (2015a). Neonatal and infantile epilepsy: Acquired and genetic models. *Cold Spring Harbor Perspectives in Medicine, 6*(1), a022707.

Galanopoulou, A. S., & Moshe, S. L. (2015b). Pathogenesis and new candidate treatments for infantile spasms and early life epileptic encephalopathies: A view from preclinical studies. *Neurobiology of Disease, 79*, 135–149.

Galanopoulou, A. S., & Mowrey, W. (2016). Not all that glitters is gold: A guide to critical appraisal of animal drug trials in epilepsy. *Epilepsia Open, 1*(3–4), 86–101.

Galanopoulou, A. S., et al. (2012). Finding a better drug for epilepsy: The mTOR pathway as an antiepileptogenic target. *Epilepsia, 53*(7), 1119–1130.

Galic, M. A., et al. (2008). Postnatal inflammation increases seizure susceptibility in adult rats. *The Journal of Neuroscience: The Official Journal of the Society for Neuroscience, 28*(27), 6904–6913.

Go, C. Y., et al. (2012). Evidence-based guideline update: Medical treatment of infantile spasms. Report of the Guideline Development Subcommittee of the American Academy of Neurology and the Practice Committee of the Child Neurology Society. *Neurology, 78*(24), 1974–1980.

Graeber, M. B., et al. (2011). Role of microglia in CNS inflammation. *FEBS Letters*, *585*(23), 3798–3805.

Haginoya, K., et al. (2009). Reduced levels of interleukin-1 receptor antagonist in the cerebrospinal fluid in patients with West syndrome. *Epilepsy Research*, *85*(2–3), 314–317.

Hattori, H. (2001). Spontaneous remission of spasms in West syndrome—Implications of viral infection. *Brain and Development*, *23*(7), 705–707.

He, J. J., et al. (2013). The interleukin 17 system in cortical lesions in focal cortical dysplasias. *Journal of Neuropathology and Experimental Neurology*, *72*(2), 152–163.

Heiskala, H. (1997). Community-based study of Lennox-Gastaut syndrome. *Epilepsia*, *38*(5), 526–531.

Hong, A. M., et al. (2010). Infantile spasms treated with the ketogenic diet: Prospective single-center experience in 104 consecutive infants. *Epilepsia*, *51*(8), 1403–1407.

Hrachovy, R. A., & Frost, J. D. J. (2003). Infantile epileptic encephalopathy with hypsarrhythmia (infantile spasms/West syndrome). *Journal of Clinical Neurophysiology*, *20*(6), 408–425.

Hunt, A., & Dennis, J. (1987). Psychiatric disorder among children with tuberous sclerosis. *Developmental Medicine and Child Neurology*, *29*(2), 190–198.

Iori, V., et al. (2013). Receptor for advanced glycation endproducts is upregulated in temporal lobe epilepsy and contributes to experimental seizures. *Neurobiology of Disease*, *58*, 102–114.

Jequier Gygax, M., et al. (2014). Efficacy and tolerability of the galanin analog NAX 5055 in the multiple-hit rat model of symptomatic infantile spasms. *Epilepsy Research*, *108*(1), 98–108.

Jiang, J., & Dingledine, R. (2013). Prostaglandin receptor EP2 in the crosshairs of anti-inflammation, anti-cancer, and neuroprotection. *Trends in Pharmacological Sciences*, *34*(7), 413–423.

Joels, M., & Baram, T. Z. (2009). The neuro-symphony of stress. *Nature Reviews. Neuroscience*, *10*(6), 459–466.

Kawai, T., & Akira, S. (2010). The role of pattern-recognition receptors in innate immunity: Update on toll-like receptors. *Nature Immunology*, *11*(5), 373–384.

Kim, J. V., et al. (2009). Myelomonocytic cell recruitment causes fatal CNS vascular injury during acute viral meningitis. *Nature*, *457*(7226), 191–195.

Koo, B., et al. (1993). Infantile spasms: Outcome and prognostic factors of cryptogenic and symptomatic groups. *Neurology*, *43*(11), 2322–2327.

Lado, F. A., & Moshe, S. L. (2002). Role of subcortical structures in the pathogenesis of infantile spasms: What are possible subcortical mediators? *International Review of Neurobiology*, *49*, 115–140.

Langlais, P. J., et al. (1991). Changes in CSF neurotransmitters in infantile spasms. *Pediatric Neurology*, 7(6), 440–445.

Lawrence, T. (2009). The nuclear factor NF-κB pathway in inflammation. *Cold Spring Harbor Perspectives in Biology*, *1*(6), a001651.

Lehtimaki, K. A., et al. (2004). Regulation of IL-6 system in cerebrospinal fluid and serum compartments by seizures: The effect of seizure type and duration. *Journal of Neuroimmunology*, *152*(1–2), 121–125.

Lehtimaki, K. A., et al. (2010). Levels of IL-1beta and IL-1ra in cerebrospinal fluid of human patients after single and prolonged seizures. *Neuroimmunomodulation*, *17*(1), 19–22.

Li, N., et al. (2010). mTOR-dependent synapse formation underlies the rapid antidepressant effects of NMDA antagonists. *Science*, *329*(5994), 959–964.

Lim, K. C., & Crino, P. B. (2013). Focal malformations of cortical development: New vistas for molecular pathogenesis. *Neuroscience*, *252*, 262–276.

Liu, Z. S., et al. (2001). Serum cytokine levels are altered in patients with West syndrome. *Brain & Development*, *23*(7), 548–551.

Lombroso, C. T. (1983). A prospective study of infantile spasms: Clinical and therapeutic correlations. *Epilepsia, 24*(2), 135–158.

Lubin, F. D., et al. (2007). Nuclear factor-kappa B regulates seizure threshold and gene transcription following convulsant stimulation. *Journal of Neurochemistry, 103*(4), 1381–1395.

Mackay, M. T., et al. (2004). Practice parameter: Medical treatment of infantile spasms: Report of the American Academy of Neurology and the Child Neurology Society. *Neurology, 62*(10), 1668–1681.

Marcon, J., et al. (2009). Age-dependent vascular changes induced by status epilepticus in rat forebrain: Implications for epileptogenesis. *Neurobiology of Disease, 34*(1), 121–132.

Mares, P., & Velisek, L. (1992). N-methyl-D-aspartate (NMDA)-induced seizures in developing rats. *Brain Research. Developmental Brain Research, 65*(2), 185–189.

Markand, O. N. (1977). Slow spike-wave activity in EEG and associated clinical features: Often called 'Lennox' or "Lennox-Gastaut" syndrome. *Neurology, 27*(8), 746–757.

Markand, O. N. (2003). Lennox-Gastaut syndrome (childhood epileptic encephalopathy). *Journal of Clinical Neurophysiology, 20*(6), 426–441.

Maroso, M., et al. (2010). Toll-like receptor 4 and high-mobility group box-1 are involved in ictogenesis and can be targeted to reduce seizures. *Nature Medicine, 16*(4), 413–419.

Maroso, M., et al. (2011a). Interleukin-1beta biosynthesis inhibition reduces acute seizures and drug resistant chronic epileptic activity in mice. *Neurotherapeutics, 8*(2), 304–315.

Maroso, M., et al. (2011b). Interleukin-1 type 1 receptor/toll-like receptor signalling in epilepsy: The importance of IL-1beta and high-mobility group box 1. *Journal of Internal Medicine, 270*(4), 319–326.

Mazarati, A., et al. (2011). High-mobility group box-1 impairs memory in mice through both toll-like receptor 4 and receptor for advanced glycation end products. *Experimental Neurology, 232*(2), 143–148.

Montelli, T. C. B., et al. (2003). Immunologic aspects of West syndrome and evidence of plasma inhibitory effects on T cell function. *Arquivos de Neuro-Psiquiatria, 61*, 731–737.

Mota, N. G., et al. (1984). Demonstration of antibody and cellular immune response to brain extract in West and Lennox-Gastaut syndromes. *Arquivos de Neuro-Psiquiatria, 42*(2), 126–131.

Moustafa, M., et al. (2002). Inhibition of tumor necrosis factor-alpha stimulated NFkappaB/p65 in human keratinocytes by alpha-melanocyte stimulating hormone and adrenocorticotropic hormone peptides. *The Journal of Investigative Dermatology, 119*(6), 1244–1253.

Ohtahara, S., et al. (1976). On the specific age dependent epileptic syndrome: The early infantile epileptic encephalopathy with suppression burst (Japanese). *No to Hattatsu, 8*(4), 270–280.

Ohya, T., et al. (2009). A pilot study on the changes in immunity after ACTH therapy in patients with West syndrome. *Brain & Development, 31*(10), 739–743.

Ono, T., et al. (2011). Carisbamate acutely suppresses spasms in a rat model of symptomatic infantile spasms. *Epilepsia, 52*(9), 1678–1684.

Ono, T., et al. (2012). Intracortical injection of lipopolysaccharide induces epileptic spasms in neonatal rats: Another animal model of symptomatic infantile spasms. *Epilepsy Currents, 12 Suppl.* (1), Abstract 3.058.

Orlova, K. A., et al. (2010). STRADalpha deficiency results in aberrant mTORC1 signaling during corticogenesis in humans and mice. *The Journal of Clinical Investigation, 120*(5), 1591–1602.

Osborne, J. P., et al. (2010). The underlying etiology of infantile spasms (West syndrome): Information from the United Kingdom Infantile Spasms Study (UKISS) on contemporary causes and their classification. *Epilepsia, 51*(10), 2168–2174.

Paciorkowski, A. R., et al. (2011). Copy number variants and infantile spasms: Evidence for abnormalities in ventral forebrain development and pathways of synaptic function. *European Journal of Human Genetics, 19*(12), 1238–1245.

Pardo, C. A., et al. (2014). Mechanisms of epileptogenesis in pediatric epileptic syndromes: Rasmussen encephalitis, infantile spasms, and febrile infection-related epilepsy syndrome (FIRES). *Neurotherapeutics*, *11*(2), 297–310.

Parker, W. E., et al. (2013). Rapamycin prevents seizures after depletion of STRADA in a rare neurodevelopmental disorder. *Science Translational Medicine*, *5*(182), 182ra153.

Pellock, J. M., et al. (2010). Infantile spasms: A U.S. consensus report. *Epilepsia*, *51*(10), 2175–2189.

Peltola, J., et al. (2002). Indicators of inflammation after recent tonic-clonic epileptic seizures correlate with plasma interleukin-6 levels. *Seizure*, *11*(1), 44–46.

Pernhorst, K., et al. (2013). TLR4, ATF-3 and IL8 inflammation mediator expression correlates with seizure frequency in human epileptic brain tissue. *Seizure*, *22*(8), 675–678.

Pintaudi, M., et al. (2007). Can fever treat epileptic encephalopathies? *Epilepsy Research*, *77*(1), 44–61.

Pires, M. E., et al. (2013). Ketogenic diet for infantile spasms refractory to first-line treatments: An open prospective study. *Epilepsy Research*, *105*(1–2), 189–194.

Pirone, A., et al. (2016). APC conditional knock-out mouse is a model of infantile spasms with elevated neuronal beta-catenin levels, neonatal spasms, and chronic seizures. *Neurobiology of Disease*, *98*, 149–157.

Qian, L., et al. (2006). Interleukin-10 protects lipopolysaccharide-induced neurotoxicity in primary midbrain cultures by inhibiting the function of NADPH oxidase. *The Journal of Pharmacology and Experimental Therapeutics*, *319*(1), 44–52.

Raffo, E., et al. (2011). A pulse rapamycin therapy for infantile spasms and associated cognitive decline. *Neurobiology of Disease*, *43*(2), 322–329.

Ransohoff, R. M. (2009). Immunology: Barrier to electrical storms. *Nature*, *457*(7226), 155–156.

Rantala, H., & Putkonen, T. (1999). Occurrence, outcome, and prognostic factors of infantile spasms and Lennox-Gastaut syndrome. *Epilepsia*, *40*(3), 286–289.

Rao, J. K., & Willis, J. (1987). Hypothalamo-pituitary-adrenal function in infantile spasms: Effects of ACTH therapy. *Journal of Child Neurology*, *2*(3), 220–223.

Ravizza, T., et al. (2006). The IL-1beta system in epilepsy-associated malformations of cortical development. *Neurobiology of Disease*, *24*(1), 128–143.

Ravizza, T., et al. (2008). Innate and adaptive immunity during epileptogenesis and spontaneous seizures: Evidence from experimental models and human temporal lobe epilepsy. *Neurobiology of Disease*, *29*(1), 142–160.

Ravizza, T., et al. (2013). Immunity activation in brain cells in epilepsy: Mechanistic insights and pathological consequences. *Neuropediatrics*, *44*(6), 330–335.

Riazi, K., et al. (2010). Contributions of peripheral inflammation to seizure susceptibility: Cytokines and brain excitability. *Epilepsy Research*, *89*(1), 34–42.

Rigau, V., et al. (2007). Angiogenesis is associated with blood-brain barrier permeability in temporal lobe epilepsy. *Brain*, *130*(Pt. 7), 1942–1956.

Riikonen, R. (1993). Infantile spasms: Infectious disorders. *Neuropediatrics*, *24*(5), 274–280.

Riikonen, R. S. (2000). Steroids or vigabatrin in the treatment of infantile spasms? *Pediatric Neurology*, *23*(5), 403–408.

Riikonen, R. (2001). Long-term outcome of patients with West syndrome. *Brain & Development*, *23*(7), 683–687.

Riikonen, R. (2014). Recent advances in the pharmacotherapy of infantile spasms. *CNS Drugs*, *28*(4), 279–290.

Riikonen, R., & Amnell, G. (1981). Psychiatric disorders in children with earlier infantile spasms. *Developmental Medicine and Child Neurology*, *23*(6), 747–760.

Rizzi, M., et al. (2003). Glia activation and cytokine increase in rat hippocampus by kainic acid-induced status epilepticus during postnatal development. *Neurobiology of Disease*, *14*(3), 494–503.

Saemundsen, E., et al. (2008). Risk of autism spectrum disorders after infantile spasms: A population-based study nested in a cohort with seizures in the first year of life. *Epilepsia*, *49*(11), 1865–1870.
Salar, S., et al. (2016). Synaptic plasticity in area CA1 of rat hippocampal slices following intraventricular application of albumin. *Neurobiology of Disease*, *91*, 155–165.
Sankar, R., et al. (2007a). Inflammation contributes to seizure-induced hippocampal injury in the neonatal rat brain. *Acta Neurologica Scandinavica. Supplementum*, *186*, 16–20.
Sankar, R., et al. (2007b). Inflammation contributes to seizure-induced hippocampal injury in the neonatal rat brain. *Acta Neurologica Scandinavica*, *115*(4 Suppl.), 16–20.
Sayyah, M., et al. (2003). The bacterial endotoxin lipopolysaccharide enhances seizure susceptibility in mice: Involvement of proinflammatory factors: Nitric oxide and prostaglandins. *Neuroscience*, *122*(4), 1073–1080.
Scantlebury, M. H., et al. (2010). A model of symptomatic infantile spasms syndrome. *Neurobiology of Disease*, *37*(3), 604–612.
Shandra, O., et al. (2015). Effects of celastrol and edaravone in the multiple-hit rat model of infantile spasms. In *American epilepsy society annual meeting*, Philadelphia, PA: American Epilepsy Society. #2.048.
Sharma, A., et al. (2010). Dysregulation of mTOR signaling in fragile X syndrome. *The Journal of Neuroscience*, *30*(2), 694–702.
Sheng, J. G., et al. (1994). Increased neuronal beta-amyloid precursor protein expression in human temporal lobe epilepsy: Association with interleukin-1 alpha immunoreactivity. *Journal of Neurochemistry*, *63*(5), 1872–1879.
Shiihara, T., et al. (2010). Peripheral lymphocyte subset and serum cytokine profiles of patients with West syndrome. *Brain & Development*, *32*(9), 695–702.
Sidenvall, R., & Eeg-Olofsson, O. (1995). Epidemiology of infantile spasms in Sweden. *Epilepsia*, *36*(6), 572–574.
Silanpaa, M., et al. (2016). Long-term mortality of patients with West syndrome. *Epilepsia Open*, *1*(1–2), 61–66.
Silverstein, F., & Johnston, M. V. (1984). Cerebrospinal fluid monoamine metabolites in patients with infantile spasms. *Neurology*, *34*(1), 102–105.
Snead, O. C., 3rd. (2001). How does ACTH work against infantile spasms? Bedside to bench. *Annals of Neurology*, *49*(3), 288–289.
Steele, S. U., et al. (2012). Electroencephalographic and seizure manifestations in two patients with folate receptor autoimmune antibody-mediated primary cerebral folate deficiency. *Epilepsy & Behavior*, *24*(4), 507–512.
Takaki, J., et al. (2012). L-glutamate released from activated microglia downregulates astrocytic L-glutamate transporter expression in neuroinflammation: The 'collusion' hypothesis for increased extracellular L-glutamate concentration in neuroinflammation. *Journal of Neuroinflammation*, *9*, 275.
Takeda, K., & Akira, S. (2004). TLR signaling pathways. *Seminars in Immunology*, *16*(1), 3–9.
Talos, D. M., et al. (2008). Cell-specific alterations of glutamate receptor expression in tuberous sclerosis complex cortical tubers. *Annals of Neurology*, *63*(4), 454–465.
Tekgul, H., et al. (2006). Cerebrospinal fluid interleukin-6 levels in patients with West syndrome. *Brain & Development*, *28*(1), 19–23.
Theoharides, T. C. (1990). Mast cells: The immune gate to the brain. *Life Sciences*, *46*(9), 607–617.
Theoharides, T. C., & Konstantinidou, A. D. (2007). Corticotropin-releasing hormone and the blood-brain-barrier. *Frontiers in Bioscience*, *12*, 1615–1628.
Trevathan, E., et al. (1999). The descriptive epidemiology of infantile spasms among Atlanta children. *Epilepsia*, *40*(6), 748–751.
Velisek, L., et al. (2007). Model of infantile spasms induced by N-methyl-D-aspartic acid in prenatally impaired brain. *Annals of Neurology*, *61*(2), 109–119.

Vezzani, A., et al. (2011). The role of inflammation in epilepsy. *Nature Reviews. Neurology*, 7(1), 31–40.
Vezzani, A., et al. (2012). Glia-neuronal interactions in ictogenesis and epileptogenesis: Role of inflammatory mediators. In J. L. Noebels, M. Avoli, M. A. Rogawski, R. W. Olsen, & A. V. Delgado-Escueta (Eds.), *Jasper's basic mechanisms of the epilepsies*. Bethesda, MD: National Center for Biotechnology Information.
Vezzani, A., et al. (2013). Epilepsy and brain inflammation. *Experimental Neurology, 244*, 11–21.
Vezzani, A., et al. (2016). Immunity and inflammation in epilepsy. *Cold Spring Harbor Perspectives in Medicine, 6*(2), a022699.
Vitaliti, G., et al. (2014). Targeting inflammation as a therapeutic strategy for drug-resistant epilepsies: An update of new immune-modulating approaches. *Human Vaccines & Immunotherapeutics, 10*(4), 868–875.
Wang, Y., et al. (2006). A role for the mTOR pathway in surface expression of AMPA receptors. *Neuroscience Letters, 401*(1–2), 35–39.
Wang, Y. C., et al. (2011). Toll-like receptors in cerebral ischemic inflammatory injury. *Journal of Neuroinflammation, 8*, 134.
Wang, Y. J., et al. (2012). Effects of adrenal dysfunction and high-dose adrenocorticotropic hormone on NMDA-induced spasm seizures in young Wistar rats. *Epilepsy Research, 100*(1–2), 125–131.
Wilson, J. (1973). Proceedings: Neurological complications of DPT inoculation in infancy. *Archives of Disease in Childhood, 48*(10), 829–830.
Wise, M. S., et al. (1996). Rasmussen syndrome and long-term response to gamma globulin. *Pediatric Neurology, 14*(2), 149–152.
Wu, H.-M., & Huang, W.-Y. (2015). Role of NOX2 in seizure susceptibility of mice following systemic inflammation (P4. 251). *Neurology, 84*(14 Suppl.). P4. 251.
Yamamoto, H. (1991). Studies on CSF tryptophan metabolism in infantile spasms. *Pediatric Neurology*, 7(6), 411–414.
Yamamoto, H., et al. (2007). A new trial liposteroid (dexamethasone palmitate) therapy for intractable epileptic seizures in infancy. *Brain & Development, 29*(7), 421–424.
Yamanaka, G., et al. (2010). Increased level of serum interleukin-1 receptor antagonist subsequent to resolution of clinical symptoms in patients with West syndrome. *Journal of the Neurological Sciences, 298*(1–2), 106–109.
Yang, F., et al. (2010). Roles of astrocytes and microglia in seizure-induced aberrant neurogenesis in the hippocampus of adult rats. *Journal of Neuroscience Research, 88*(3), 519–529.
Ye, Z. C., & Sontheimer, H. (1996). Cytokine modulation of glial glutamate uptake: A possible involvement of nitric oxide. *Neuroreport*, 7(13), 2181–2185.
You, S. J., et al. (2009). Factors influencing the evolution of West syndrome to Lennox-Gastaut syndrome. *Pediatric Neurology, 41*(2), 111–113.
Youn, Y., et al. (2013). The role of cytokines in seizures: Interleukin (IL)-1β, IL-1Ra, IL-8, and IL-10. *Korean Journal of Pediatrics, 56*(7), 271–274.
Yum, M. S., et al. (2012). Prenatal stress promotes development of spasms in infant rats. *Epilepsia, 53*(3), e46–e49.
Zakharova, M., & Ziegler, H. K. (2005). Paradoxical anti-inflammatory actions of TNF-alpha: Inhibition of IL-12 and IL-23 via TNF receptor 1 in macrophages and dendritic cells. *Journal of Immunology, 175*(8), 5024–5033.
Zbytek, B., et al. (2006). CRH inhibits NF-kappa B signaling in human melanocytes. *Peptides, 27*(12), 3276–3283.
Zeng, L. H., et al. (2010). Modulation of astrocyte glutamate transporters decreases seizures in a mouse model of tuberous sclerosis complex. *Neurobiology of Disease, 37*(3), 764–771.
Zurolo, E., et al. (2011). Activation of toll-like receptor, RAGE and HMGB1 signalling in malformations of cortical development. *Brain, 134*(Pt. 4), 1015–1032.

CHAPTER FOUR

# Analyzing the Effect of V66M Mutation in BDNF in Causing Mood Disorders: A Computational Approach

**P. Sneha*, D. Thirumal Kumar*, Sugandhi Saini*, Kreeti Kajal*, R. Magesh†, R. Siva*,[1], C. George Priya Doss*,[1]**

*School of Biosciences and Technology, VIT University, Vellore, Tamil Nadu, India
†Faculty of Research and Bio Medical Sciences, Sri Ramachandra University, Chennai, Tamil Nadu, India
[1]Corresponding authors: e-mail addresses: siva.ramamoorthy@gmail.com; georgecp77@yahoo.co.in; georgepriyadoss@vit.ac.in

## Contents



## Abstract

Mental disorders or mood disorders are prevalent globally irrespective of region, race, and ethnic groups. Of the types of mood disorders, major depressive disorder (MDD) and bipolar disorder (BPD) are the most prevalent forms of psychiatric


ISSN 1876-1623
http://dx.doi.org/10.1016/bs.apcsb.2017.01.006

condition. A number of preclinical studies emphasize the essential role of brain-derived neurotrophic factor (BDNF) in the pathophysiology of mood disorders. Additionally, BDNF is the most common growth factor in the central nervous system along with their essential role during the neural development and the synaptic elasticity. A malfunctioning of this protein is associated with many types of mood disorders. The variant methionine replaces valine at 66th position is strongly related to BPD, and an individual with a homozygous condition of this allele is at a greater risk of developing MDD. There are very sparse reports suggesting the structural changes of the protein occurring upon the mutation. Consequently, in this study, we applied a computational pipeline to understand the effects caused by the mutation on the protein's structure and function. With the use of in silico tools and computational macroscopic methods, we identified a decrease in the alpha-helix nature, and an overall increase in the random coils that could have probably resulted in deformation of the protein.

## 1. INTRODUCTION

Mood disorder is a psychological condition with elevated moods such as depression and bipolar. The illnesses under mood disorders are categorized into few types based on the symptoms that include major depressive disorder (MDD), bipolar disorder (BPD), persistent depressive disorder, cyclothymia, and seasonal affective disorder. Of these the major and severe types are MDD and BPD. MDD is a psychological condition which alters the state of mind of an individual (Lee & Kim, 2010) and affects almost every age group. It mainly occurs due to psychological, biological, and social factors. The depression results due to the mutation in serotonin transporter (Arpawong et al., 2016; Goldman, Glei, Lin, & Weinstein, 2010), leading to the reduction in serotonin level, thereby lowering monoamine and norepinephrine levels (Marks et al., 2009). Reduction in neurotransmitter results in anxiety and other behavioral problems. Bipolar disorder is an episodic illness classically characterized by extreme fluctuations in mood including mania and depression. Mood swings and depressed states are components of bipolar disorder, on the other hand MDD is unipolar (Sklar et al., 2002). BPD is critical and may require immediate hospital care and the episodes last for minimum 2 weeks (Culpepper, 2014; McCormick, Murray, & McNew, 2015). Individuals suffering from BPD experience several intense emotions, changes in patterns of sleep, and unusual behaviors (Anderson & Bradley, 2013). Several studies reveal that factors including genetic, neuroendocrinological, and some of the environmental causes as the reasons behind the BPD (Lesch, 2004; Nabeshima & Kim, 2013).

Genetic factors play a significant role in mood disorders; hypothesized in a way that majority of the depressions are due to malfunctioning of the neurotrophic factors. Brain-derived neurotrophic factor (BDNF) is a major protein involved in the neuronal survival in addition to maintaining plasticity of dopaminergic, cholinergic, and serotonergic neurons in the CNS (Angelucci, Brenè, & Mathé, 2005). A malfunctioning of the protein has been associated with many types of mood disorders including MDD, BPD, schizophrenia, and depression (Angelucci et al., 2005; Autry & Monteggia, 2012; Hong et al., 2003; Lang et al., 2005; Soliman et al., 2010). It is encoded by *BDNF* gene and localized on the chromosome number 11 belonging to neurotrophin family. It is also one of the significant factors responsible for long-term memory and other functions of the brain such as neurogenesis (Binder & Scharfman, 2004; Calabrese et al., 2014; Cunha, Brambilla, & Thomas, 2010). Also, mice models with disrupted BDNF signaling show lesser response to the antidepressants. Very recently, infusion of single bilateral BDNF into ventricles has shown rapid and sustained antidepressant effect (Björkholm & Monteggia, 2016). Antidepressant drugs and electroconvulsive shock treatment methods are most common and immediate steps taken to help the patients with neurotic disorders. These treatment strategies have shown increase in the expression of BDNF molecules and its receptor TrkB, in association with neuronal plasticity (Castrén & Rantamäki, 2010). The most common genetic polymorphism V66M (Nassan et al., 2015; Tramontina et al., 2007; Zeni et al., 2016) results in blockage of the dentritic transport, thereby losing the antidepressant-like effect (Duman & Voleti, 2012). Also, this mutation has shown resistance toward the therapeutic action of escitalopram in mood disorder, thus making this polymorphism a genetic marker for therapeutic response in patients with MDD (Chang et al., 2012). Although several studies have reported the BDNF mutations in association with mood disorders, there is not even a single computational study reported at structural level till date.

In this context for the first time, we utilized a series of in silico tools, I Mutant 2.0 (Capriotti, Fariselli, & Casadio, 2005), Mutationassessor (Reva, Antipin, & Sander, 2011), SIFT (Sim et al., 2012), SNAP (Johnson et al., 2008), PolyPhen-2 (Adzhubei, Jordan, & Sunyaev, 2013), Align GVGD (Tavtigian, 2005), and PANTHER (Thomas et al., 2003), to validate the pathogenic effect of the V66M mutation on the protein.

Furthermore, to understand the structural and functional changes, molecular dynamics (MD) simulation studies and secondary structure element analysis were carried out using GROMACS 4.5.6 package (Pronk et al., 2013). We hope the findings of the study could provide valuable insights into mood disorder diagnosis and provide better treatment strategies toward personalized medicine.

## 2. MATERIALS AND METHODS

### 2.1 Analyzing the Deleterious Effect of the Mutation

The effect of the mutation V66M was analyzed based on the stability, pathogenicity, and change in the biochemical nature of the protein.

#### *2.1.1 I Mutant 2.0*

The stability changes were analyzed using I Mutant 2.0 (Capriotti et al., 2005). This tool is a support vector machine widely used to predict the stability changes in protein due to a single point mutation. The stability change predicted by classifier based on the $\Delta\Delta G$ values concludes the level of increase or decrease in the stability.

#### *2.1.2 Mutationassessor*

Mutationassessor predicts the level of the functional impact of the protein due to a substitution. The prediction is based on the evolutionary conservation of the affected amino acid. Given the sequence and mutation residue position, all the sequence homologs were searched to construct a multiple sequence alignment which is further clustered into subfamilies and scores a mutation by global conservation pattern. Thus, the level of functional impact is decided by the tool based on the degree of conservation of the protein (Reva et al., 2011). The results of this tool are represented as low impact, medium impact, or high impact over the protein's function.

#### *2.1.3 SIFT*

SIFT predicts the mutational effect based on the sequence homology to compute the likelihood that an amino acid substitution will have an adverse effect on protein's function (Sim et al., 2012). The prediction result of the tool measures the level of tolerance of a mutation over the protein. An intolerant mutation impairs the functioning of the protein, whereas a tolerant mutation necessarily does not interfere with the functioning of the protein.

#### 2.1.4 SNAP

Neural network (NN) based SNAP uses structural features such as secondary structure, solvent accessibility, and conservation to characterize a position in a protein. The NN prediction is made and reliability index is constructed which correlates with the severity of the mutation. The input is the FASTA format of the protein sequence along with the mutation position and substituting amino acid (Johnson et al., 2008).

#### 2.1.5 PolyPhen-2

In PolyPhen-2 prediction, the effect of the mutation relies on a number of criteria that includes sequence, phylogenetic, and structural features to characterize the substitution (Adzhubei et al., 2013). The mutations are predicted as possibly damaging, probably damaging, or benign based on the severity of the mutation's effect.

#### 2.1.6 Align GVGD

Along with the stability and pathogenic analyses, the biochemical analysis was predicted using Align GVGD. Freely available web-based program predicts the effect of missense mutations by combining the physiological and biological characters of amino acids and the multiple protein sequence alignments. The level of impact of the mutation is classified between Class 0 and Class 65, where mutation Class 0 is the least interfering and Class 65 mutation is most interfering with the function of the protein (Tavtigian, 2005).

#### 2.1.7 PANTHER

PANTHER predicts the functional impact of the mutation over the protein based on the conservational analysis (Thomas et al., 2003). The input is in the form of sequence along with the mutation position and substituting amino acid. The final results predict if the mutation is disease associated or neutral effect.

### 2.2 Conservational Analysis

Conservation analysis was performed using ConSurf tool (Glaser et al., 2003). This tool detects the conserved position of the amino acids in proteins and additionally identifies the structural and functional importance of each amino acid in particular regions of the protein. The tool primarily constructs a phylogenetic relationship between homologous sequences.

The conservational scores are projected with a score between 1 and 9, score 1 (blue) represents least conserved and 9 (dark pink) represents highest conserved amino acid.

## 2.3 Molecular Interaction Analysis

The protein interaction analysis was performed using Cytoscape 3.0 (Shannon et al., 2003) version that helps to understand the interacting proteins. The protein interaction analysis was made to understand the various proteins interacting with BDNF. The protein interaction was constructed by importing the (.tsv) file from the STRING database (von Mering et al., 2003).

## 2.4 Molecule Preparation

In the absence of the protein structure, we modeled the 3D structure of the protein using Phyre2 (Protein Homology/analogy Recognition Engine) server available at http://www.sbg.bio.ic.ac.uk/~phyre2/html/page.cgi?id=index (Kelley, Mezulis, Yates, Wass, & Sternberg, 2015). This server predicts the structure of the protein using four major steps that initially searches for homologous sequences followed by fold library scanning, loop modeling, and side chain placement. The input for the server is in the form of a FASTA sequence of the protein obtained from UniProt database with ID: P23560. PDB ID: 3IJ2 was considered as the template. The modeled protein was further validated using PROCHECK server (Morris, MacArthur, Hutchinson, & Thornton, 1992). The obtained structure was considered as the native protein molecule, and the mutant protein (V66M) was prepared using Swiss Pdb-Viewer tool (Guex & Peitsch, 1997). Both native and mutant structures were energy minimized with Gromos96 force field using the same tool.

## 2.5 Molecular Dynamics Simulation

The MD simulations were performed with the help of GROMACS 4.5.6 package (Pronk et al., 2013). The initial protein structure for the simulation process was subjected to GROMOS 43a1 force field and solvated in a box size of 10 Å with an explicit water box under periodic boundary conditions. Further, the system was neutralized using genion tool by adding sodium ($Na^+$) ions. MD simulation procedure comprises three crucial stages: energy minimization, equilibration, and production MD. The energy minimization process was made using the steepest descent method and continued till the system reached a maximum force <1000.0 kJ/mol/nm. Minimized system was subjected to two-phase equilibration step: canonical ensemble and isobaric ensemble performed for 50,000 steps each. This well-equilibrated system was taken for the production MD and carried until the protein reached convergence at 300 K temperature and recorded every 2 ps. A similar procedure was carried out for the native and mutant protein (V66M) for a time period of 30 ns. The results of the MD simulation are in the form of trajectories and further analyzed to understand the various structural differences that could have occurred throughout the simulation process.

### 2.6 Trajectory Analysis

The trajectories obtained as a result of MD simulations were subjected to the utilities available in the GROMACS. Of the various analyses available, relevant analyses were chosen that can specifically highlight the significant difference upon mutation in this study. g_rms, g_rmsf, g_gyrate, and g_hbond tools were used to plot the root mean square deviation (RMSD), root mean square fluctuation (RMSF), radius of gyration, and intramolecular hydrogen bonds to understand the fundamental differences. All the graphs were plotted using Xmgrace program.

### 2.7 Structural Analysis

The differences in the secondary structure elements between the native and mutant were estimated with the aid of iRDP-ICAPS tool (Panigrahi, Sule, Ghanate, Ramasamy, & Suresh, 2015) available at http://irdp.ncl.res.in/CompStabAna.html. The input for the server requires the protein in the form of 3D coordinates. Hence, to understand the changes in the secondary structure elements, 30 ns coordinate files of the native and mutant were given as the input. Various secondary structure elements were plotted as a graph to understand the overall change. The exact location of the mutant residue in the secondary structure was made using Phyre2 server.

## 3. RESULTS

### 3.1 In Silico Tool Analysis

The effect of V66M on protein function was predicted using various in silico tools. The tools were categorized into pathogenicity and stability analyses. I Mutant 2.0 predicted the mutation to be decreasing the stability of the

Table 1 In Silico Tool Results of the Mutation V66M Over the BDNF Protein

| Mutation | I Mutant 2.0 | Mutationassessor | SIFT | SNAP | PolyPhen-2 | Align GVGD | PANTHER |
|---|---|---|---|---|---|---|---|
| V66M | Decrease | Medium impact | Intolerant | Disease | Possibly damaging | Class 65 | Disease |

protein. Likewise all the pathogenic tools predicted mutation V66M to be showing a pathogenic effect. The mutation showed medium impact on the protein's function and possibly damaging effect as predicted by Mutationassessor and PolyPhen-2 tools. SNAP and PANTHER tools predicted the mutation to be associated with the disease. Finally, SIFT projected there was intolerance in the presence of V66M mutation. When analyzing the biochemical property using Align GVGD, we identified that the V66M mutation was found to be highly interfering with the protein's function with a score classified as Class 65 (Table 1).

## 3.2 Conservation Analysis

The level of conservation of each amino acid was analyzed using ConSurf tool. A more conserved position is possibly involved in the functionality of the protein; a mutation in the conserved position may have a direct effect on the functionality of the protein. The results from ConSurf explain that valine at 66th position showed conservation score of 6, predicting a comparatively conserved region across the species (Fig. 1). The tool also predicted valine to be in the buried region as well as in the hydrophobic core of the protein.

## 3.3 Protein Interaction Analysis

The BDNF protein was found to be interacting with several other proteins of the CNS and mostly involved in cell signaling pathway. Primarily, STRING database was used to extract the protein interactions, and further a detailed map was made using Cytoscape 3.0 (Fig. 2). We found 10 interacting proteins, of which 5 proteins such as NTRK1, NTRK2, NTRK3, NTF3, and NTF4 participate in neuronal differentiation. Although SORT1 is not directly involved in the development of the nervous system, its potential molecular function is stated to be having interaction with the BDNF and it is required for both neuronal and tumor cell survival. Meanwhile, CREB1 is directly involved in the neuronal development and cell survival process. PGLC1 shows intracellular signaling cascade association, and CDC42 is involved in the cell division control. Finally, NGFR is a tumor necrosis factor protein. Overall, the analysis indicates that BDNF protein is associated with the

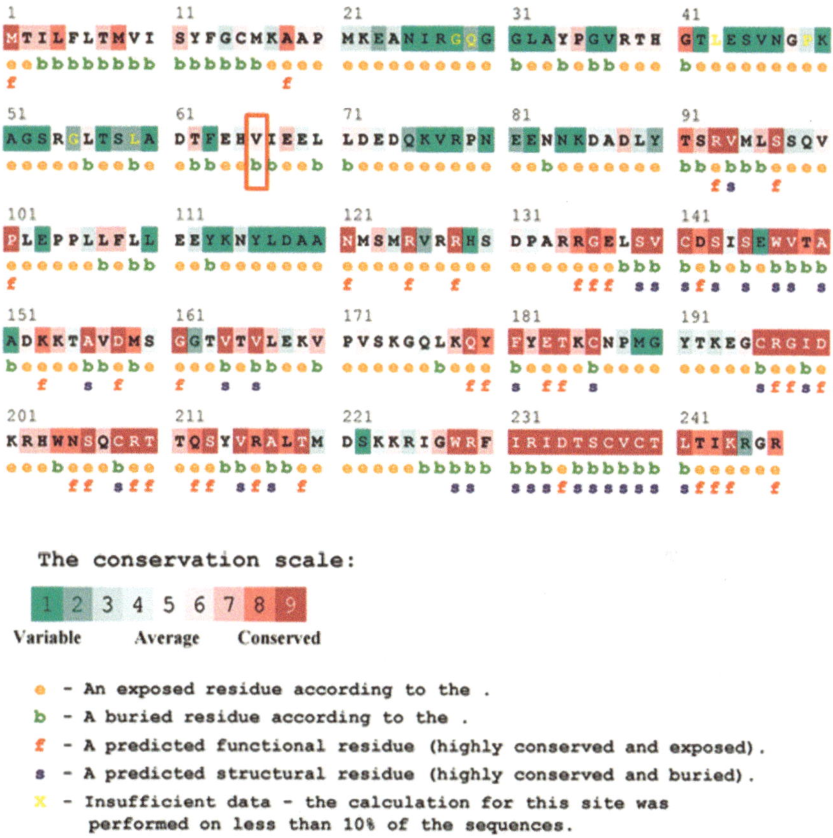


**Fig. 1** Conservation analysis of each amino acid across species using ConSurf tool. A *darker pink color* represents the most conserved amino acid and *darker blue* represents the most variable amino acid. The ConSurf result predicted V66 to be partially conserved amino acid with a conservation score of 6.

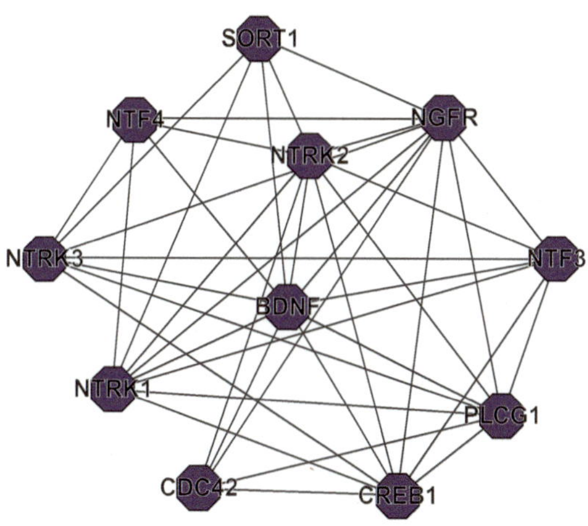


**Fig. 2** Protein interaction analysis of the BDNF protein using Cytoscape 3.0.

nervous system and neuronal development, although its association is also observed with various other proteins such as cellular signaling and cell death cascade. Of the different proteins, BDNF showed closer association with the nervous system and neuronal development.

## 3.4 Molecule Preparation

The structure of BDNF protein was modeled using Phyre2 server and the 3D structure was validated before analyses using PROCHECK server. The PROCHECK server implies the level of reliability of a modeled protein based on the Ramachandran plot. The plot explained that most of the residues reside in the favorable region (Fig. 3). Based on this, mutant modeling

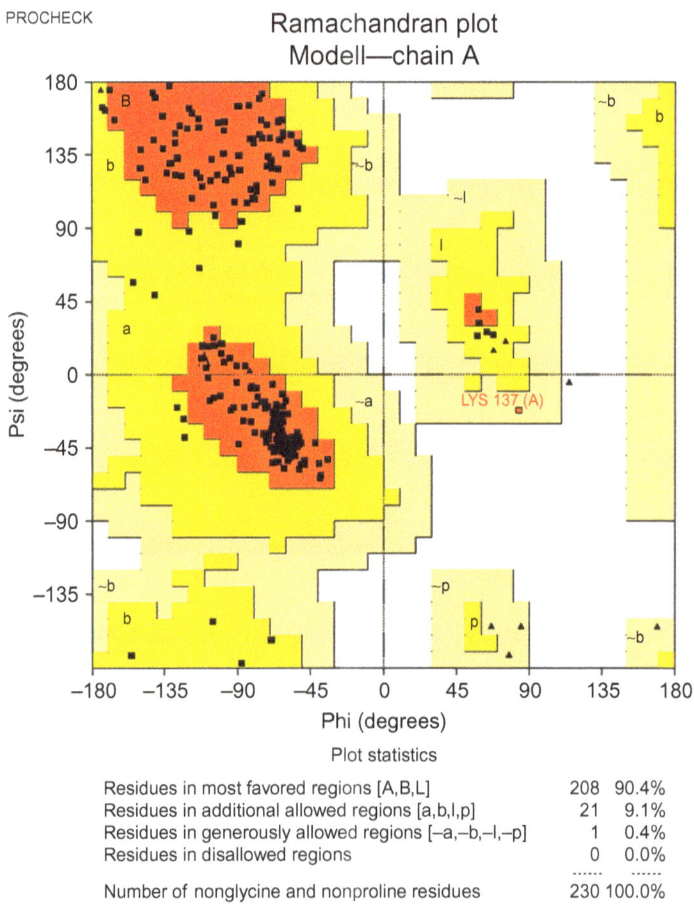


**Fig. 3** The Ramachandran plot analysis to validate the modeled protein. The Ramachandran plot obtained using PROCHECK tool.

was performed and the resultant structures were energy minimized using Swiss Pdb-Viewer.

## 3.5 Molecular Dynamics Simulation Analysis

The resultant trajectories were analyzed using various utilities available in the GROMACS package. The convergence of the native molecule was observed between 28 and 30 ns at 0.5 nm from the RMSD plot (Fig. 4). This indicates that the trajectory of the native molecule attained a stable conformation and can be a reliable model for further analysis. To have a better comparative analysis, we have simulated the mutant protein for 30 ns. A comparative native and mutant RMSF plot was made which indicated an increase in fluctuation patterns in the mutant protein throughout the simulation (Fig. 5). Thus, as a next step, hydrogen bond analysis and radius of gyration analysis were made to understand the plausible reason behind the stability loss. The number of intramolecular hydrogen bond analysis (Fig. 6) explains that the native protein possesses a higher number of hydrogen bonds at the end of 30 ns; meanwhile, the mutant showed a lesser number of hydrogen bonds within the molecule. Further, the compactness of the

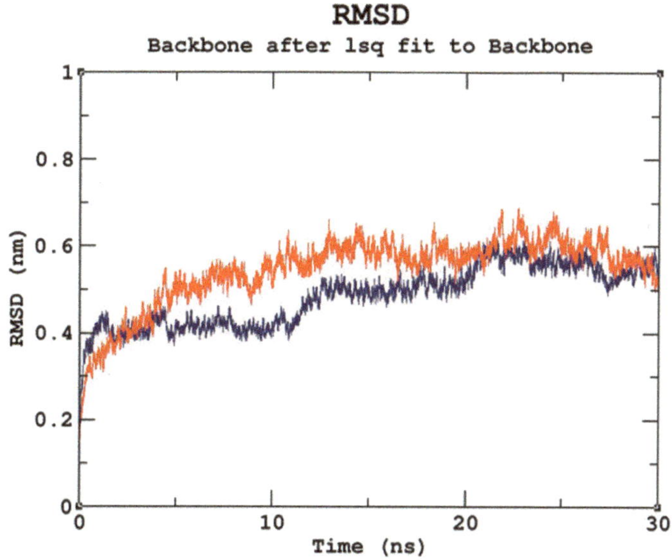


**Fig. 4** The root mean square deviation plot of the native and mutant V66M in *blue* and *red color*, respectively for 30 ns.

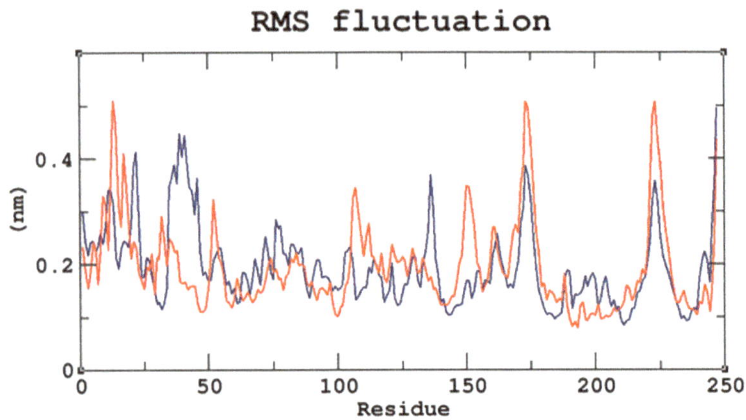


**Fig. 5** Root mean square fluctuation plot of the native and mutant V66M in *blue* and *red color*, respectively for 30 ns.

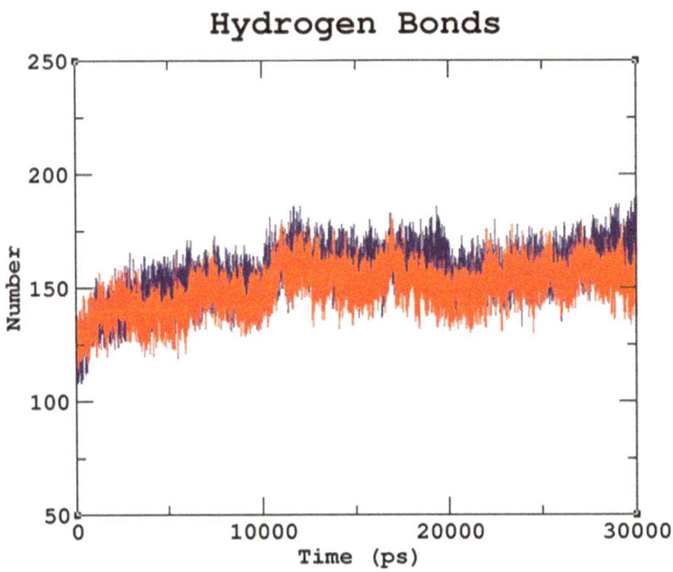


**Fig. 6** Hydrogen bond analysis of the native and mutant V66M in *blue* and *red color*, respectively for 30 ns.

protein was studied using radius of gyration (Rg), wherein the least values explain a higher compactness and vice versa. We found from our MD analysis that least Rg values were attained by the native molecule with higher degree of compactness and the mutant showed lesser compactness with higher Rg values (Fig. 7).

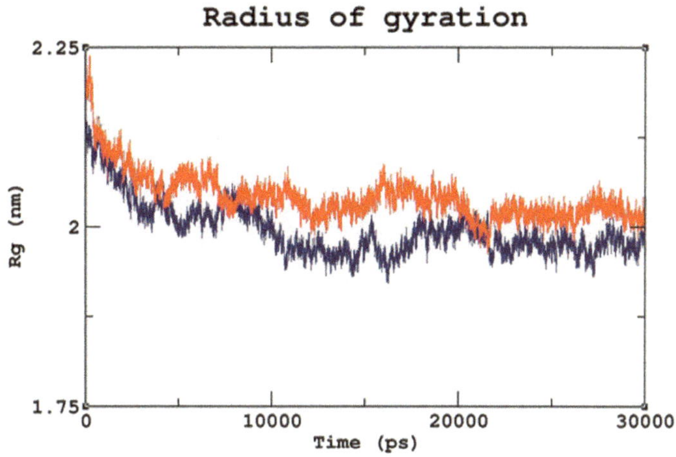


**Fig. 7** Radius of gyration (Rg) plot of the native and mutant V66M in *blue* and *red color*, respectively for 30 ns.

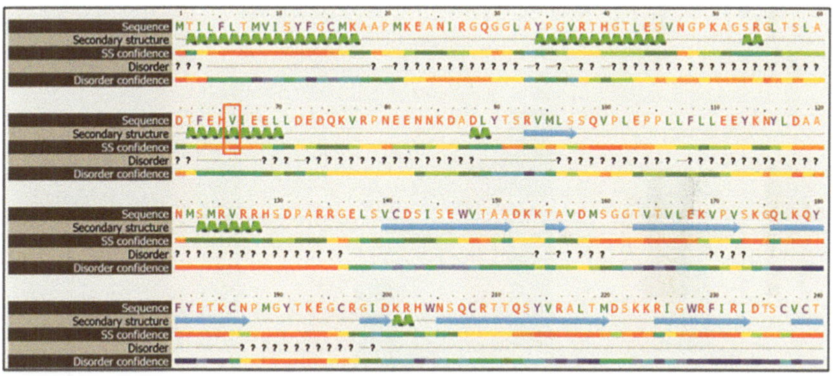


**Fig. 8** Contribution of each amino acid in BDNF protein toward secondary structure element. The plot was obtained using Phyre2 server.

## 3.6 Secondary Structure Analysis

The contribution of each amino acid in the formation of secondary structure was identified using Phyre2 server. We found that position 66 was found to be contributing toward the alpha-helix formation (Fig. 8). Hence to compare the shift in any of the secondary structure elements, the three-dimensional structure obtained at the end of 30 ns simulations was analyzed using the iRDP-ICAPS tool. The tool predicts various secondary structure elements such as alpha helix, random coil, bend, and hydrogen bond (Table 2). We found that there was a decrease in the number of alpha helix and a drastic increase in the number of random coils in the case of mutant

**Table 2** The Various Secondary Structure Elements Present in the Native and Mutant After 30 ns Simulation Generated Using iRDP-ICAPS Server

| Secondary Structure Element | Native | Mutant |
|---|---|---|
| Alpha helix (H) | 15.04 | 14.68 |
| Residues in isolated beta-bridge (B) | 5.28 | 2.38 |
| 3-Helix (G) | 0 | 0 |
| 5-Helix (I) | 0 | 0 |
| Hydrogen bond turn (T) | 7.54 | 5.69 |
| Random coil (C) | 28.46 | 42.46 |

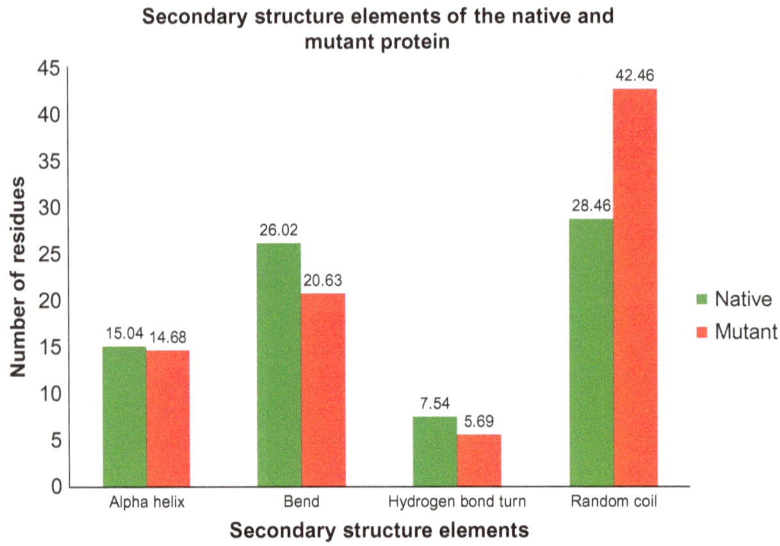


**Fig. 9** Explaining the various secondary structure elements present in the protein after 30 ns simulation. The value of the various secondary structure elements was obtained from iRDP-ICAPS server.

(Fig. 9). This further predicts that there may be an overall change in the secondary structure of the protein.

## 4. DISCUSSION

Preclinical studies have suggested that BDNF plays a major role in the pathophysiology of MDD (Hashimoto, 2010). Recent reports from family-based association studies have predicted that the *BDNF* gene may be a potential risk locus for the development of BPD (Hashimoto, Shimizu, & Iyo, 2004). Also from our protein interaction analysis, we found that of the 10 proteins, 5 proteins were associated with neuronal differentiation. The majority of human genetic variation is represented by single-nucleotide polymorphisms, and many of them are believed to cause genotypic and phenotypic differences among individuals. The most common form of polymorphism observed in BDNF is V66M mutation. The V66M mutation in BDNF protein strongly correlates to BPD (Lohoff et al., 2005; Neves-Pereira et al., 2002; Sklar et al., 2002), making this polymorphism a plausible biomarker. The identification of such biomarkers may aid both in the diagnosis and development of effective medications to treat mood disorders.

Although there is continuing research carried on the polymorphisms in BDNF, there are no structural studies recorded till date. Therefore, this study is first of its kind which attempts to reveal the structural and functional relationship of the mutant using various in silico tools and microscopic computational approach. The combination of various computation tools can improve the automated predictions accuracy and significantly outperform than just relying on the results from single tool (Bendl et al., 2014; Lopes et al., 2012; Masso & Vaisman, 2008; P et al., 2017). Furthermore, structural changes in the protein were studied at an atomistic level with the help of powerful MD simulation studies (Ali, Sneha, Priyadharshini Christy, Zayed, & George Priya Doss, 2016; Doss et al., 2016; Papaleo, 2015; Sneha & George Priya Doss, 2016; Zaki et al., 2017).

The in silico stability and pathogenic analyses along with biochemical analysis predicted V66M mutation with functional and stability impact over the protein. To further identify the structural changes in the protein, MD simulation was carried out for 30 ns. While analyzing the trajectories, the RMSD graph elucidated that the mutant protein showed higher deviating patterns in comparison to the native. A higher RMSD of the mutant from native predicts a decrease in the stability of the protein (Yun & Guy, 2011). This decrease in stability was found to be correlating with the earlier predictions made by the stability analysis tools (Fig. 4 and Table 1). To add to the plausible reason behind the reduction in the stability, hydrogen bond analysis and compactness of protein were performed. From the analysis, a decrease in the number of hydrogen bonds (Fig. 6) and reduction in the compactness were observed (Fig. 7) which correlates with the decrease in the stability as observed in the RMSD graph (Fig. 4). The stability difference caused in the protein was further studied by analyzing the shift in the secondary structure elements between the native and mutant proteins.

Subsequently, an in-depth structural analysis was made by examining the difference in the secondary structure elements using iRDP-ICAPS tool. Initially, Phyre2 tool was used to locate the position of the V66 in the BDNF protein and found that the V66 position contributes to the formation of the alpha helix (Fig. 8). Steif and coworkers reported the involvement of the mutation in the alpha helix to cause stability differences in the protein (Steif, Hinz, & Cesareni, 1995). To analyze this change, the overall secondary structure elements present in the native and mutant proteins at the end of 30 ns simulations were calculated using iRDP-ICAPS tool. Through this analysis we found a negligible difference in the alpha-helix region and increase in the number of random coils in the mutant (Fig. 9 and Table 2). Jha and Udgaonkar (2009) reported the loss in the secondary structure in the case of steady unfolding conditions where protein assumes random coil structure (Jha & Udgaonkar, 2009). This increase in random coil was observed with the mutant (Fig. 9), further confirming that the mutation may have lost its stability. This loss of stability due to the random coil could be the reason behind the increased RMSD values, decrease in hydrogen bond formation, and reduced compactness. Overall we found that the mutation V66M induces increased random coil conformation of the protein and further decreased the stability, thereby hindering the protein's function.

## 5. CONCLUSION

In conclusion, through a combinational in silico tools and MD analysis we analyzed the possible structural changes occurred due to the substitution of valine with methionine at position 66 of the BDNF protein. Since the BDNF signaling pathway consists of many proteins involved in the neuronal development and CNS, a mutation that inactivates this protein could lead to many brain-associated diseases and conditions such as mood disorders. A computational study to understand the changes occurred due to mutations could not only reduce the time-consuming experimental analysis but also aim at predicting the effects of a mutation on the protein at an atomistic level. This study can aid the determination of the effects caused by the mutation and can pave the way for better treatment strategies.

## ACKNOWLEDGMENTS

The authors acknowledge the management of VIT University, Vellore, India, and (BRAF) @ CDAC for providing the facilities required to perform this work.

*Disclosure statement*: No potential conflict of interest exists.

## REFERENCES


Adzhubei, I., Jordan, D. M., & Sunyaev, S. R. (2013). Predicting functional effect of human missense mutations using PolyPhen-2. *Current Protocol in Human Genetics*chapter 7, unit 7.20.

Ali, S. K., Sneha, P., Priyadharshini Christy, J., Zayed, H., & George Priya Doss, C. (2016). Molecular dynamics-based analyses of the structural instability and secondary structure of the fibrinogen gamma chain protein with the D356V mutation. *Journal of Biomolecular Structure and Dynamics, 27*, 1–11.

Anderson, K. N., & Bradley, A. J. (2013). Sleep disturbance in mental health problems and neurodegenerative disease. *Nature and Science of Sleep, 5*, 61–75.

Angelucci, F., Brenè, S., & Mathé, A. A. (2005). BDNF in schizophrenia, depression and corresponding animal models. *Molecular Psychiatry, 10*, 345–352.

Arpawong, T. E., Lee, J., Phillips, D. F., Crimmins, E. M., Levine, M. E., & Prescott, C. A. (2016). Effects of recent stress and variation in the serotonin transporter polymorphism (5-HTTLPR) on depressive symptoms: A repeated-measures study of adults age 50 and older. *Behavior Genetics, 46*, 72–88.

Autry, A. E., & Monteggia, L. M. (2012). Brain-derived neurotrophic factor and neuropsychiatric disorders. *Pharmacological Reviews, 64*, 238–258.

Bendl, J., Stourac, J., Salanda, O., Pavelka, A., Wieben, E. D., Zendulka, J., et al. (2014). PredictSNP: Robust and accurate consensus classifier for prediction of disease-related mutations. *PLoS Computational Biology, 10*, e1003440.

Binder, D. K., & Scharfman, H. E. (2004). Brain-derived neurotrophic factor. *Growth Factors, 22*, 123–131.

Björkholm, C., & Monteggia, L. M. (2016). BDNF—A key transducer of antidepressant effects. *Neuropharmacology, 102*, 72–79.

Calabrese, F., Rossetti, A. C., Racagni, G., Gass, P., Riva, M. A., & Molteni, R. (2014). Brain-derived neurotrophic factor: A bridge between inflammation and neuroplasticity. *Frontiers in Cellular Neuroscience, 8*, 430.

Capriotti, E., Fariselli, P., & Casadio, R. (2005). I-Mutant2.0: Predicting stability changes upon mutation from the protein sequence or structure. *Nucleic Acids Research, 33*, W306–W310.

Castrén, E., & Rantamäki, T. (2010). The role of BDNF and its receptors in depression and antidepressant drug action: Reactivation of developmental plasticity. *Developmental Neurobiology, 70*, 289–297.

Chang, H. S., Lee, H.-Y., Ham, B.-J., Park, Y. C., Hahn, S.-W., Jeong, Y.-J., et al. (2012). Association between brain-derived neurotrophic factor V66M and treatment responses to escitalopram in patients with major depressive disorder. *Asia-Pacific Psychiatry, 4*, 241–249.

Culpepper, L. (2014). The diagnosis and treatment of bipolar disorder: Decision-making in primary care. *The Primary Care Companion for CNS Disorder, 16*, PCC.13r01609.

Cunha, C., Brambilla, R., & Thomas, K. L. (2010). A simple role for BDNF in learning and memory? *Frontiers in Molecular Neuroscience, 3*, 1.

Doss, C. G., Alasmar, D. R., Bux, R. I., Sneha, P., Bakhsh, F. D., Al-Azwani, I., et al. (2016). Genetic epidemiology of glucose-6-dehydrogenase deficiency in the Arab world. *Scientific Reports, 6*, 37284.

Duman, R. S., & Voleti, B. (2012). Signaling pathways underlying the pathophysiology and treatment of depression: Novel mechanisms for rapid-acting agents. *Trends in Neurosciences, 35*, 47–56.

Glaser, F., Pupko, T., Paz, I., Bell, R. E., Bechor-Shental, D., Martz, E., et al. (2003). ConSurf: Identification of functional regions in proteins by surface-mapping of phylogenetic information. *Bioinformatics, 19*, 163–164.

Goldman, N., Glei, D. A., Lin, Y.-H., & Weinstein, M. (2010). The serotonin transporter polymorphism (5-HTTLPR): Allelic variation and links with depressive symptoms. *Depression and Anxiety, 27*, 260–269.

Guex, N., & Peitsch, M. C. (1997). SWISS-MODEL and the Swiss-PdbViewer: An environment for comparative protein modeling. *Electrophoresis*, *18*, 2714–2723.

Hashimoto, K. (2010). Brain-derived neurotrophic factor as a biomarker for mood disorders: An historical overview and future directions. *Psychiatry and Clinical Neurosciences*, *64*, 341–357.

Hashimoto, K., Shimizu, E., & Iyo, M. (2004). Critical role of brain-derived neurotrophic factor in mood disorders. *Brain Research Reviews*, *45*, 104–114.

Hong, C.-J., Huo, S.-J., Yen, F.-C., Tung, C.-L., Pan, G.-M., & Tsai, S.-J. (2003). Association study of a brain-derived neurotrophic-factor genetic polymorphism and mood disorders, age of onset and suicidal behavior. *Neuropsychobiology*, *48*, 186–189.

Jha, S. K., & Udgaonkar, J. B. (2009). Direct evidence for a dry molten globule intermediate during the unfolding of a small protein. *Proceedings of the National Academy of Sciences of the United States of America*, *106*, 12289–12294.

Johnson, A. D., Handsaker, R. E., Pulit, S. L., Nizzari, M. M., O'Donnell, C. J., & de Bakker, P. I. W. (2008). SNAP: A web-based tool for identification and annotation of proxy SNPs using HapMap. *Bioinformatics*, *24*, 2938–2939.

Kelley, L. A., Mezulis, S., Yates, C. M., Wass, M. N., & Sternberg, M. J. E. (2015). The Phyre2 web portal for protein modeling, prediction and analysis. *Nature Protocols*, *10*, 845–858.

Lang, U. E., Hellweg, R., Kalus, P., Bajbouj, M., Lenzen, K. P., Sander, T., et al. (2005). Association of a functional BDNF polymorphism and anxiety-related personality traits. *Psychopharmacology*, *180*, 95–99.

Lee, B.-H., & Kim, Y.-K. (2010). The roles of BDNF in the pathophysiology of major depression and in antidepressant treatment. *Psychiatry Investigation*, 7, 231–235.

Lesch, K. P. (2004). Gene-environment interaction and the genetics of depression. *Journal of Psychiatry & Neuroscience: JPN*, *29*, 174–184.

Lohoff, F. W., Sander, T., Ferraro, T. N., Dahl, J. P., Gallinat, J., & Berrettini, W. H. (2005). Confirmation of association between the Val66Met polymorphism in the brain-derived neurotrophic factor (BDNF) gene and bipolar I disorder. *American Journal of Medical Genetics. Part B, Neuropsychiatric Genetics*, *139B*, 51–53.

Lopes, M. C., Joyce, C., Ritchie, G. R. S., John, S. L., Cunningham, F., Asimit, J., et al. (2012). A combined functional annotation score for non-synonymous variants. *Human Heredity*, *73*, 47–51.

Marks, D. M., Shah, M. J., Patkar, A. A., Masand, P. S., Park, G.-Y., & Pae, C.-U. (2009). Serotonin-norepinephrine reuptake inhibitors for pain control: Premise and promise. *Current Neuropharmacology*, 7, 331–336.

Masso, M., & Vaisman, I. I. (2008). Accurate prediction of stability changes in protein mutants by combining machine learning with structure based computational mutagenesis. *Bioinformatics*, *24*, 2002–2009.

McCormick, U., Murray, B., & McNew, B. (2015). Diagnosis and treatment of patients with bipolar disorder: A review for advanced practice nurses. *Journal of American Association of Nurse Practitioners*, *27*, 530–542.

Morris, A. L., MacArthur, M. W., Hutchinson, E. G., & Thornton, J. M. (1992). Stereochemical quality of protein structure coordinates. *Proteins*, *12*, 345–364.

Nabeshima, T., & Kim, H.-C. (2013). Involvement of genetic and environmental factors in the onset of depression. *Experimental Neurobiology*, *22*, 235–243.

Nassan, M., Croarkin, P. E., Luby, J. L., Veldic, M., Joshi, P. T., McElroy, S. L., et al. (2015). Association of brain-derived neurotrophic factor (BDNF) Val66Met polymorphism with early-onset bipolar disorder. *Bipolar Disorders*, *17*, 645–652.

Neves-Pereira, M., Mundo, E., Muglia, P., King, N., Macciardi, F., & Kennedy, J. L. (2002). The brain-derived neurotrophic factor gene confers susceptibility to bipolar disorder: Evidence from a family-based association study. *The American Journal of Human Genetics*, *71*, 651–655.

P, S., D, T. K., Tanwar, H., R, S., C, G. P., & Zayed, H. (2017). Structural analysis of G1691S variant in the human filamin B gene responsible for Larsen syndrome: A comparative computational approach. *Journal of Cellular Biochemistry*.

Panigrahi, P., Sule, M., Ghanate, A., Ramasamy, S., & Suresh, C. G. (2015). Engineering proteins for thermostability with iRDP web server. *PLoS One*, *10*, e0139486.

Papaleo, E. (2015). Integrating atomistic molecular dynamics simulations, experiments, and network analysis to study protein dynamics: Strength in unity. *Frontiers in Molecular Biosciences*, *2*, 1–6.

Pronk, S., Páll, S., Schulz, R., Larsson, P., Bjelkmar, P., Apostolov, R., et al. (2013). GROMACS 4.5: A high-throughput and highly parallel open source molecular simulation toolkit. *Bioinformatics*, *29*, 845–854.

Reva, B., Antipin, Y., & Sander, C. (2011). Predicting the functional impact of protein mutations: Application to cancer genomics. *Nucleic Acids Research*, *39*, e118.

Shannon, P., Markiel, A., Ozier, O., Baliga, N. S., Wang, J. T., Ramage, D., et al. (2003). Cytoscape: A software environment for integrated models of biomolecular interaction networks. *Genome Research*, *13*, 2498–2504.

Sim, N.-L., Kumar, P., Hu, J., Henikoff, S., Schneider, G., & Ng, P. C. (2012). SIFT web server: Predicting effects of amino acid substitutions on proteins. *Nucleic Acids Research*, *40*, W452–W457.

Sklar, P., Gabriel, S. B., McInnis, M. G., Bennett, P., Lim, Y.-M., Tsan, G., et al. (2002). Family-based association study of 76 candidate genes in bipolar disorder: BDNF is a potential risk locus. *Molecular Psychiatry*, *7*, 579–593.

Sneha, P., & George Priya Doss, C. (2016). Molecular dynamics: New frontier in personalized medicine. *Advances in Protein Chemistry and Structural Biology*, *102*, 181–224.

Soliman, F., Glatt, C. E., Bath, K. G., Levita, L., Jones, R. M., Pattwell, S. S., et al. (2010). A genetic variant BDNF polymorphism alters extinction learning in both mouse and human. *Science*, *80*, 327.

Steif, C., Hinz, H.-J., & Cesareni, G. (1995). Effects of cavity-creating mutations on conformational stability and structure of the dimeric 4-α-helical protein ROP: Thermal unfolding studies. *Proteins: Structure, Function, and Genetics*, *23*, 83–96.

Tavtigian, S. V. (2005). Comprehensive statistical study of 452 BRCA1 missense substitutions with classification of eight recurrent substitutions as neutral. *Journal of Medical Genetics*, *43*, 295–305.

Thomas, P. D., Kejariwal, A., Campbell, M. J., Mi, H., Diemer, K., Guo, N., et al. (2003). PANTHER: A browsable database of gene products organized by biological function, using curated protein family and subfamily classification. *Nucleic Acids Research*, *31*, 334–341.

Tramontina, J., Frey, B. N., Andreazza, A. C., Zandona, M., Santin, A., & Kapczinski, F. (2007). Val66met polymorphism and serum brain-derived neurotrophic factor levels in bipolar disorder. *Molecular Psychiatry*, *12*, 230–231.

von Mering, C., Huynen, M., Jaeggi, D., Schmidt, S., Bork, P., & Snel, B. (2003). STRING: A database of predicted functional associations between proteins. *Nucleic Acids Research*, *31*, 258–261.

Yun, S., & Guy, H. R. (2011). Stability tests on known and misfolded structures with discrete and all atom molecular dynamics simulations. *Journal of Molecular Graphics & Modelling*, *29*, 663–675.

Zaki, O. K., Krishnamoorthy, N., El Abd, H. S., Harche, S. A., Mattar, R. A., Al Disi, R. S., et al. (2017). Two patients with Canavan disease and structural modeling of a novel mutation. *Metabolic Brain Disease*, *32*, 171–177.

Zeni, C. P., Tramontina, S., Aguiar, B. W., Salatino-Oliveira, A., Pheula, G. F., Sharma, A., et al. (2016). BDNF Val66Met polymorphism and peripheral protein levels in pediatric bipolar disorder and attention-deficit/hyperactivity disorder. *Acta Psychiatrica Scandinavica*, *134*, 268–274.

CHAPTER FIVE

# A Computational Approach to Identify the Biophysical and Structural Aspects of Methylenetetrahydrofolate Reductase (MTHFR) Mutations (A222V, E429A, and R594Q) Leading to Schizophrenia

**Himani Tanwar*, P. Sneha*, D. Thirumal Kumar*, R. Siva*[,1], Charles Emmanuel Jebaraj Walter†, C. George Priya Doss*[,1]**

*School of Biosciences and Technology, VIT University, Vellore, Tamil Nadu, India
†Faculty of Biomedical Sciences, Technology & Research, Sri Ramachandra University, Chennai, India
[1]Corresponding authors: e-mail address: siva.ramamoorthy@gmail.com; georgepriyadoss@vit.ac.in

## Contents



*Advances in Protein Chemistry and Structural Biology*, Volume 108
ISSN 1876-1623
http://dx.doi.org/10.1016/bs.apcsb.2017.01.007

## Abstract

The association between depression and methylenetetrahydrofolate reductase (MTHFR) has been continually demonstrated in clinical studies, yet there are sparse resources available to build a relationship between the mutations associated with MTHFR and depression. The common mutations found to be associated with schizophrenia and MTHFR are A222V, E429A, and R594Q. Although abundant research on structural and functional effects caused by A222V mutation is available, very less amount of studies have been done on the other two mutants (E429A and R594Q). Hence in this study, a comparative analysis was carried out between the most common A222V mutation, a prevalent E429A mutation, and a less prevalent and less deleterious R594Q mutation. To predict structural rearrangements upon mutation, we proposed a computational pipeline using in silico prediction tools, molecular docking, and molecular dynamics simulation analysis. Since the association of flavin adenine dinucleotide (FAD) is important for the functioning of the protein, binding analysis between protein and the coenzyme was performed. This would enable us to understand the interference level of each mutation over FAD-binding activity. Consequently, we found that two mutations (A222V and E429A) showed lesser binding activity and structural deviations when compared to the native molecule and mutant R594Q. Comparatively, higher structural changes were observed with A222V mutant complex in comparison to other mutant complexes. Computational studies like this could render better insights into the structural changes in the protein and their relationship with the disease condition.

## 1. INTRODUCTION

Bipolar disorders (BD) and schizophrenia (SCZ) are the two common psychiatric disorders associated with low folate levels and defects in folate metabolism. SCZ is a chronic and serious psychotic illness, with a lifetime incidence of 1%–2% in the global population characterized by abnormal social behavior, false believes (delusions), and perceptions (hallucinations) (American Psychiatric Association, 1994; Muntjewerff & Blom, 2005). BD is a condition in which a person has alternating periods of mania or elevated mood and depression with or without a history of major depression and lifetime prevalence of 0.3%–1.6%, equal among men and women (Smith & Weissman, 1992; Weissman et al., 1996). Various genome-wide association (GWA) studies suggested that both environmental and genetic factors contribute toward the etiology of these complex disorders (Leonardo & Hen, 2006; Prathikanti & Weinberger, 2005). The *MTHFR* (methylenetetrahydrofolate reductase) gene mutation could be one of the possible genetic risk factors resulting in SCZ and BDs (Peerbooms et al., 2011). In recent studies, the defect in *MTHFR* gene has been associated with some psychiatric symptoms, suggesting its desirable role in prognosis of these psychiatric disorders (Picker & Coyle, 2005). The cytogenetic location of the *MTHFR* gene is 1p36.3, short (p) arm of chromosome 1 at position 36.3 and 2.2 kb in length with a total of 11 exons (Goyette et al., 1994).

The MTHFR enzyme plays a crucial role in the folate metabolism pathway by irreversibly converting 5,10-methylenetetrahydrofolate to 5-methyltetrahydrofolate, which is the predominant circulatory form of folate. 5-MTHF plays a central role in one-carbon metabolism and DNA methylation. It donates a methyl group to homocysteine, a major source of methyl groups in the brain, when S-adenosylmethionine is generated and catalyzes the conversion of homocysteine to methionine (Arzaghi et al., 2011; Cantoni, 1952; Das & Singal, 2004; Ueland, Hustad, Schneede, Refsum, & Vollset, 2001). This conversion is important because homocysteine and its metabolites may have a direct effect on the *N*-methyl-D-aspartate glutamate receptors in the brain, which inhibits the methylation processes in the central nervous system (Bottiglieri, Hyland, & Reynolds, 1994; Cohen-Woods et al., 2010). Therefore, the normal activity of the enzyme MTHFR is to maintain the pool of circulating folate and methionine and to reduce the levels of homocysteine (Rozen, 1996). Folate deficiency has also been associated with a disturbed metabolism of dopamine, serotonin, and noradrenaline, that is, neurotransmitter abnormalities possibly contributing to the development of SCZ (Bottiglieri et al., 2000).

The mutation in MTHFR causes a deformation, leading to decreased enzyme activity. Although huge numbers of mutations have been reported in the protein, the most prevalent mutations were identified to be A222V, E429A, and R594Q (Leclerc, Sibani, & Rozen, 2013). However, the result of various analyses still proves that *MTHFR* gene mutation, especially A222V, are associated with a higher risk of BPD and SCZ (Allen et al., 2008; Gilbody, Lewis, & Lightfoot, 2006; Jönsson et al., 2008; Lewis, Zammit, Gunnell, & Smith, 2005; Muntjewerff, Kahn, Blom, & den Heijer, 2006; Shi, Gershon, & Liu, 2008; Zintzaras, 2006). The MTHFR A222V mutation results in reduced levels of bioavailability of folate and folate metabolites. Therefore, MTHFR polymorphisms can increase the serum homocysteine level and as a consequence result in an impairment of DNA methylation. These polymorphisms have been the target for various studies in analyzing neurological conditions, cardiovascular anomalies, and psychiatric disorders during recent years. Although various studies have tested the association of the three mutations with the psychiatric disorders (BDs and SCZ), the results have been uncertain (Nicol & Gottesman, 1983). Consequently, in this study, a comparison between the effect of the mutations (A222V, E429A, and R594Q) over protein's structure and function was elucidated.

For this, a computational pipeline was drawn into three steps to evaluate the impact of mutations in the MTHFR protein. Primarily, in silico tools such as PhD-SNP (Capriotti, Calabrese, & Casadio, 2006), PolyPhen-2 (Adzhubei et al., 2010), SIFT (sorts intolerant from tolerant) (Ng & Henikoff, 2003), SNAP (Johnson et al., 2008), MutationAssessor (Reva, Antipin, & Sander, 2011), and HANSA (Acharya & Nagarajaram, 2012) were used to analyze the pathogenic effect of the mutants. Further, the stability analysis was performed using MUPro (Cheng, Randall, & Baldi, 2005), mCSM (Pires, Ascher, & Blundell, 2014), and DUET (Pires et al., 2014). The activity of the enzyme MTHFR is greatly dependent on the FAD (Ames, Elson-Schwab, & Silver, 2002; Hustad, Schneede, & Ueland, 2013). Therefore, docking of the FAD with the native and mutant molecules was performed using AutoDock tools. Finally, the docked complexes were subjected to molecular dynamics (MD) simulation analysis to understand the impact of mutation at an atomistic level. MD simulations using GROMACS 4.5.6 (Pronk et al., 2013) helped us to understand and compare the structural changes occurred upon mutation.

# 2. MATERIALS AND METHODS

## 2.1 SNP Dataset

A literature study of the mutations in the MTHFR protein associated with SCZ was studied using the OMIM (Online Mendelian Inheritance in Man) (Amberger, Bocchini, Scott, & Hamosh, 2009) and NCBI PubMed databases. The protein sequence of MTHFR in the FASTA format was retrieved from the UniProt database (P42898).

## 2.2 In Silico Analysis of Deleterious SNPs

Three stepwise computational pipelines were used to evaluate the impact of mutations on the MTHFR protein. Primarily, tools such as PhD-SNP (Capriotti et al., 2006), PolyPhen-2 (Adzhubei et al., 2010), SIFT (Ng & Henikoff, 2003), SNAP (Johnson et al., 2008), MutationAssessor (Reva et al., 2011), and HANSA (Acharya & Nagarajaram, 2012) were used to analyze the pathogenic effect of the mutants. Also, the stability analysis was performed using MUPro (Cheng et al., 2005), mCSM (Pires et al., 2014), and DUET (Pires et al., 2014).

PhD-SNP (predictor of human deleterious single nucleotide polymorphisms) works on the algorithm of the SVM classifier with each SVM tested on the protein sequence (Capriotti et al., 2006). PolyPhen-2 is an advanced method of PolyPhen that uses the machine learning to produce high-quality multiple sequence alignments and also optimized for high-throughput analysis of next-generation sequencing (Adzhubei et al., 2010). SIFT is based on the evolution correlation. The position of amino acids is very crucial for the functioning of the protein. The evolution-based amino acid position helps in understanding the pathogenicity and separates intolerant mutations from tolerated ones (Ng & Henikoff, 2003). The SNAP (screening for non-acceptable polymorphisms) method based on neural networks applies an advanced machine-learning approach to predict any functional effects caused by nonsynonymous SNPs. The information about the sequence and structural components, such as secondary structure, solvent accessibility, and residue conservation within sequence families, is used to find whether a substitution in a protein confers a gain or a loss of protein's function (Johnson et al., 2008). HANSA, also a SVM (support vector machine)-based method classifies missense mutations as neutral or deleterious, using a novel set of features (referred to as nsSNP neutral disease (nsSNPND) discriminatory features) associated with nsSNPs. This nsSNPND discriminatory feature set includes position-specific probabilities, local protein structural status, and the intrinsic properties of the wild-type and mutated residues (Acharya & Nagarajaram, 2012). MUPro is a set of machine-learning programs, which predicts how a single-site amino acid mutation affects protein stability (Cheng et al., 2005). mCSM (mutation cutoff scanning matrix) is also a machine-learning method, which predicts the effect of missense mutations based on structural signatures. These techniques determine the distance patterns between the atoms used to represent the amino acid environment (Pires et al., 2014). DUET consolidates two complementary approaches such as mCSM and SDM in a consensus prediction, obtained by combining the results of the separate methods in an optimized predictor using SVM (Pires et al., 2014).

## 2.3 Conservation Analysis

The level of conservation of the amino acid sequence at each position was calculated with the help of ConSurf. The ConSurf tool predicts the level of conservation based on multiple sequence alignment. The input of the tool requires the protein sequence in FASTA format for which a multiple sequence was built using MAFFT. The evolutionary conservation of each amino acid position in the alignment was calculated using the Rate4Site algorithm. The phylogenetic relations between the aligned proteins and the stochastic nature of the evolutionary process are considered by the algorithm for which the conservation level for each residue is assigned. Finally, visual inspection of the conservation patterns on the three-dimensional structure (3D) structure enables the identification of functionally important regions (Glaser et al., 2003).

## 2.4 Protein Model Generation

The 3D structure of the protein was modeled using the threading approach with the aid of I-TASSER server (Yang et al., 2014) and the quality of the obtained structure was validated using the RAMPAGE server (Lovell et al., 2002). Among the obtained protein models, the protein with the best Ramachandran score was taken for further analysis. Three individual mutations (A222V, E429A, and R594Q) were introduced into the native model structure using a Swiss-PdbViewer package (Kaplan & Littlejohn, 2001). After inducing mutation, the structures were subjected to optimization and energy minimization using the same software.

## 2.5 Molecular Docking Analysis

The molecular docking analysis was performed using AutoDock tools 4.2 (Morris et al., 2009). Prior to docking, the flavin adenine dinucleotide (PubChem ID: 643975) was prepared using the Dock prep of the chimera tools (Pettersen et al., 2004). Additionally, the native and mutant proteins were energy minimized using Swiss-PdbViewer (Guex & Peitsch, 1997). To the protein structure, hydrogen bonds and charges were added. Additionally, the protein molecule was assigned the AD4 type. The prepared ligands were then torsion adjusted prior to applying autogrid function with the size of $60 \times 60 \times 60$ along the $x$-, $y$-, and $z$-axes. The grid box was set around the active site of the protein. After the autogrid, autodock was performed using a Lamarckian algorithm. The docked complex with the best binding energy was then visualized using a Maestro software of the Schrodinger.

## 2.6 MD Simulation

MD analysis was performed to study the possible effects that the amino acid may have on the MTHFR protein structure. All the MD simulations for docked complexes (one native and three mutant complexes) were carried out using the GROMACS 4.5.6 package (Pronk et al., 2013) along with the GROMOS9643a1 force field. Initially, all models were solvated utilizing simple point charge water model. The system was neutralized by adding the chlorine ions. Subsequently, the entire system was subjected to descent energy minimization until reaching a tolerance of 100 kJ/mol. After the equilibration of the system for 50,000 steps with the fixed protein at 300K for a while; the system was gradually relaxed and heated up to 300K. Finally, 30 ns MD simulations were performed under the normal temperature and pressure with the coupling time constant of 1.0 ps. The graphs were plotted using XMGRACE and the protein structure was visualized using PyMOL.

# 3. RESULTS

## 3.1 Pathogenic Prediction

The pathogenic predictions using various tools suggested mutation A222V to be most deleterious in comparison to the other two mutations (Table 1). Nevertheless, all the three mutations showed a destabilizing effect as predicted by DUET and mCSM tools (Table 2). MUPro predicted A222V with a decrease in stability and E429A and R594Q with an increase in the stability.

## 3.2 Conservation Analysis

The results from the ConSurf tool predicted the level of conservation of each amino acid at a particular position. When comparing the three mutation positions in understanding the level of conservation throughout an evolutionary process, A222 was found to be highly conserved with a

**Table 1** In Silico Tool Results of the Pathogenic Analysis

| Mutation | PhD-SNP | PolyPhen-2 | SIFT | SNAP | Mutation Assessor | HANSA |
|---|---|---|---|---|---|---|
| A222V | Deleterious | Deleterious | Deleterious | Deleterious | Deleterious | Deleterious |
| E429A | Neutral | Neutral | Neutral | Neutral | Neutral | Neutral |
| R594Q | Neutral | Neutral | Neutral | Neutral | Neutral | Neutral |

**Table 2** In Silico Tool Results of the Stability Analysis

| Mutation | DUET | mCSM | MUPro |
|---|---|---|---|
| A222V | Decrease | Decrease | Decrease |
| E429A | Decrease | Decrease | Decrease |
| R594Q | Decrease | Decrease | Increase |

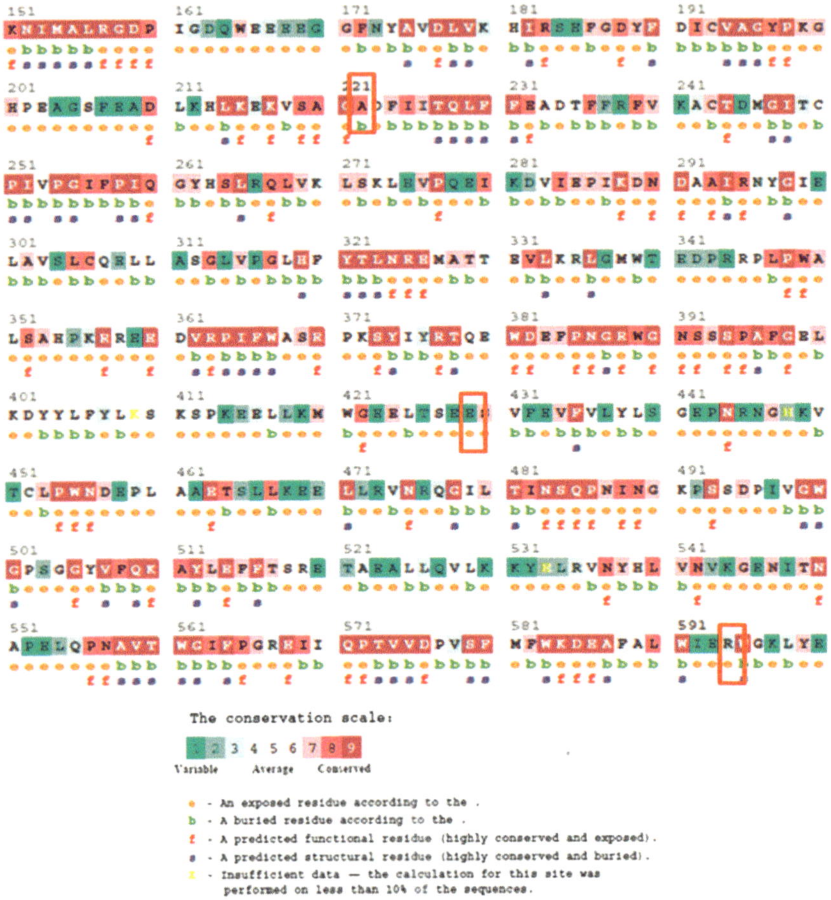


**Fig. 1** ConSurf tool predicts the level of conservation of each amino acid at a particular position.

conservation score of 8 and present in the buried region of the protein (Fig. 1), whereas E429 and R594 positions were comparatively less conserved with scores 1 and 3, respectively, and present in the exposed region. A mutation in the buried region has more impact than in the exposed region, and the intensity of the mutation increases with an increase in the level of conservation. Overall, a mutation at position 222 may have a greater deleterious effect than the other mutant positions.

## 3.3 Molecular Modeling

I-TASSER server uses a threading approach to model the protein. The tool searched and identified 10 templates to have better similarity, of which the template with the highest *Z*-score was taken for modeling the protein.

**Table 3** The Ramachandran Plot Analysis of the Models Obtained

| Protein Models | Favored Region (%) | Allowed Region (%) | Outlier Region (%) |
|---|---|---|---|
| Model 1 | 70.6 | 20.5 | 8.9 |
| Model 2 | 78.3 | 15.4 | 6.3 |
| Model 3 | 81.5 | 11.9 | 6.6 |
| Model 4 | 64.7 | 21.9 | 13.5 |
| Model 5 | 79.4 | 15.0 | 5.7 |

The modeling of the MTHFR protein was made considering the methylenetetrahydrofolate reductase from the *Thermus thermophilus* HB8 structure with a PDB ID: 3APT having a resolution of 1.85 Å as the template. The template showed a higher identity of 38% with the query and was further modeled using an I-TASSER server. Five models were obtained as a result of molecular modeling, and all the models were subjected to further validation using RAMPAGE. The Ramachandran scores of all the models are listed in Table 3. Among all the models, Model 3 was used for further analysis as it showed highest number of residues (81.5%) in the favored region. This model was taken as the native structure, and the mutations were induced using Swiss-PdbViewer and further energy minimized prior to structural analysis. The modeled structure with 656 amino acids was used for further analysis (Fig. 2).

## 3.4 Molecular Docking Analysis

Molecular docking analysis was performed to understand the binding efficiency of the FAD to the native and mutant proteins. The grid box was set around the active site residues THR94, TRP95, PRO97, HIS127, THR129, MET154, LEU156, ARG157, GLY158, ASP159, TYR174, ALA175, ALA195, TYR197, HIS201, ALA204, ASP210, HIS213, LYS217, ILE226, GLN228, and TYR321. When comparing the binding energy of FAD with the four protein molecules, the lowest binding energy was observed with native and the highest with A222V mutant complexes (Table 4). When analyzing the native docked complex, FAD formed three hydrogen bonds with TYR197, ARG157, and GLY158 (Fig. 3A). The A222V mutant protein did not show any hydrogen bond formation with FAD (Fig. 3B). E429A showed two-hydrogen bond formation with ASP159 and GLN228 (Fig. 3C). An increase in the number of hydrogen bonds formation was observed in R594Q. Four hydrogen bonds were

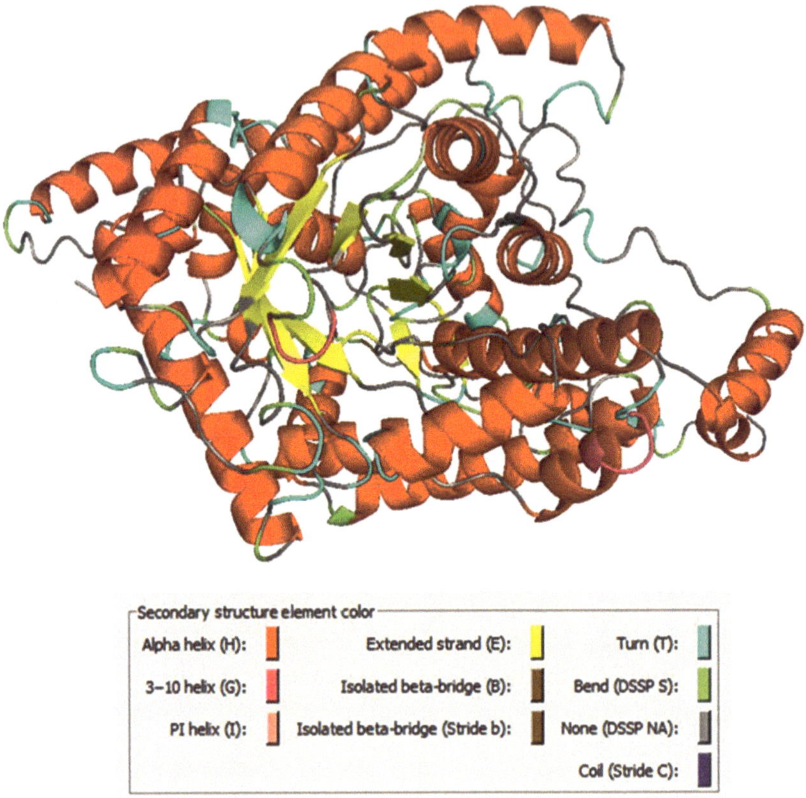


**Fig. 2** Modeled structure of MTHFR protein with 656 amino acids and marked with their secondary structure elements using DSSP plug-in of the PyMOL.

**Table 4** The Molecular Docking Results Using AutoDock Tools 4.2

| Protein Molecule | Binding Energy (kcal/mol) | Number of Hydrogen Bonds | Hydrogen Bond Formation |
|---|---|---|---|
| Native | −4.54 | 3 | Each with TYR197, ARG157, and GLY158 |
| A222V mutant | −1.39 | — | — |
| E429A | −2.09 | 2 | Each with ARG157, GLN228 |
| R594Q | −1.63 | 4 | Two hydrogen bonds with ASP159. Each with THR94 and TYR197 |

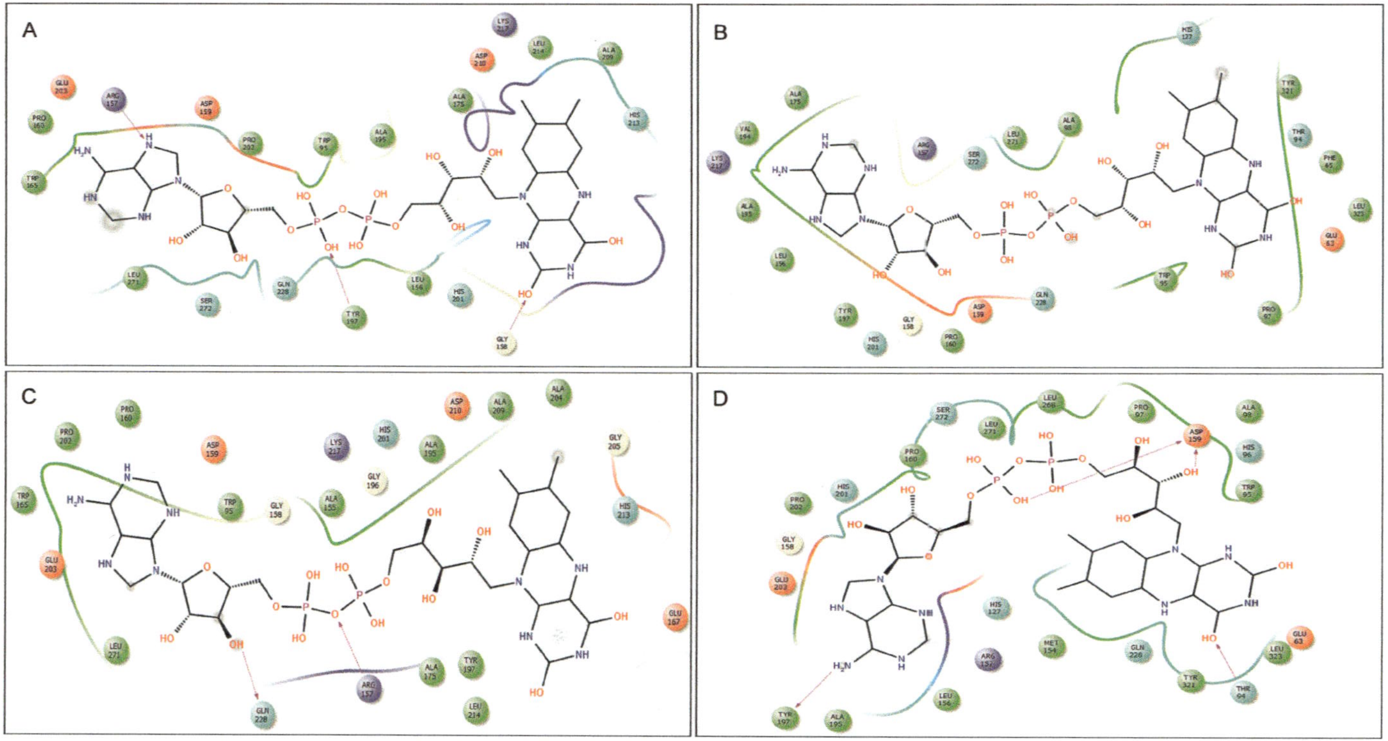


**Fig. 3** The docked images of the native and mutant protein with FAD. (A) Native, (B) A222V, (C) E429A, and (D) R594Q.

formed with ASP159, THR94, and TYR197 (Fig. 3D). This infers that there is a loss of interaction in case of E429A and A222V mutations.

## 3.5 MD Simulation Analysis

The trajectory files of the MD simulation were analyzed using various utilities of the GROMACS package. Initially, variable deviations were observed in the native complex at the beginning of the simulation and complete convergence was observed nearing 30 ns with an RMSD value of ~0.525, whereas the A222V mutant complex showed higher RMSD value of ~0.6 nm and E429A mutant complex showed a convergence at ~0.55 nm at the end of 30 ns (Fig. 4). Also, the R594Q mutant complex showed an RMSD value of ~0.545 nm, which is much in between the native and the E429A mutant complexes. From the RMSF plot, we observed a higher fluctuation pattern with respect to A222V mutant complex (Fig. 5). Additionally, compactness of the protein complexes was assessed using radius of gyration (Rg) utility. From the Rg plot, we observed the higher level of compactness with respect to native complex followed by R594Q mutant complex. Meanwhile, A222V and E429A mutant complexes showed a lesser compactness (Fig. 6). Finally, intermolecular

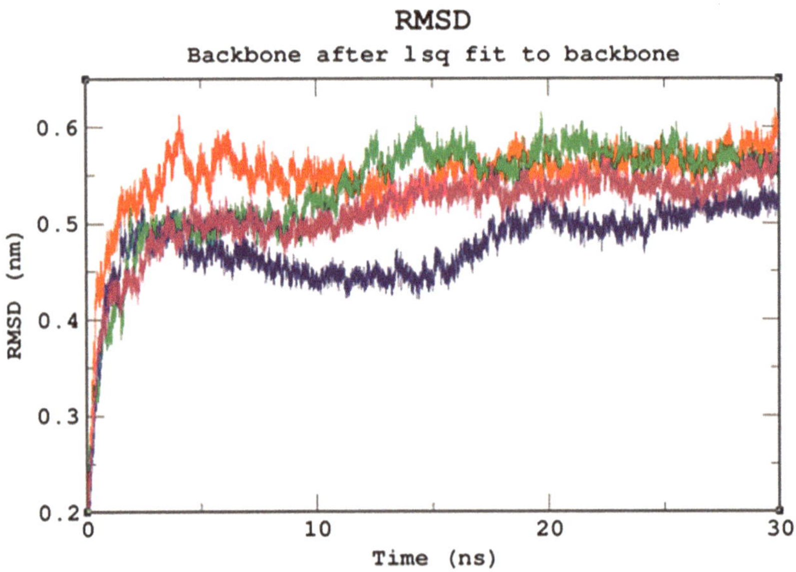


**Fig. 4** Root mean square deviation of the native and mutant complexes. *Blue*, native complex; *red*, A222V mutant complex; *green*, E429A mutant complex; and *magenta*, R594Q mutant complex.

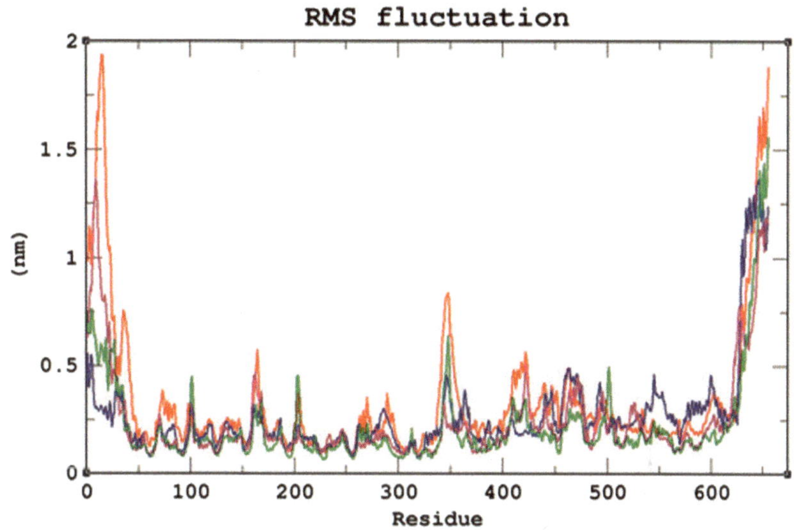


**Fig. 5** Root mean square fluctuation of the native and mutant complexes. *Blue*, native complex; *red*, A222V mutant complex; *green*, E429A mutant complex; and *magenta*, R594Q mutant complex.

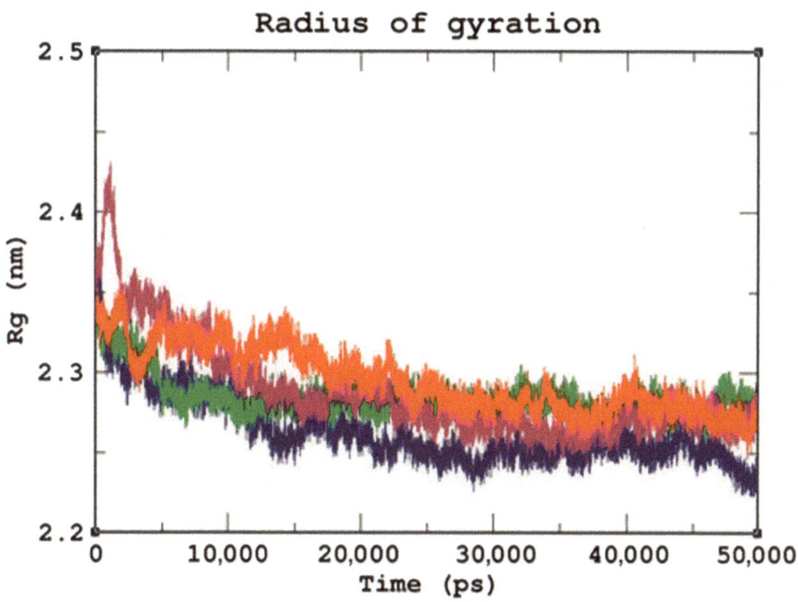


**Fig. 6** Radius of gyration of the native and mutant complexes. *Blue*, native complex; *red*, A222V mutant complex; *green*, E429A mutant complex; and *magenta*, R594Q mutant complex.

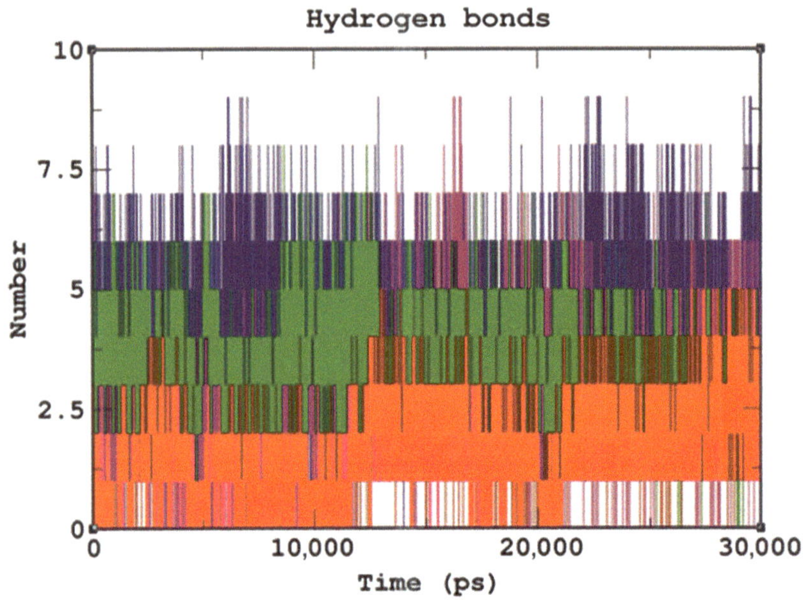


**Fig. 7** Hydrogen bond analysis of the native and mutant complexes. *Blue*, native complex; *red*, A222V mutant complex; *green*, E429A mutant complex; and *magenta*, R594Q mutant complex.

hydrogen bond analysis was performed to understand the difference in the number of hydrogen bonds formed between the protein and the FAD molecule. Lesser number of hydrogen bonds was observed in case of A222V mutant complex (Fig. 7) when compared to native, E429A and R594Q mutant complexes. This was further confirmed by analyzing the average number of hydrogen bonds formed throughout the simulation. It was observed that native and R594Q mutant complexes showed the highest average number of intermolecular hydrogen bonds formed with 8.63 and 7.94, respectively, whereas the E429A and A222V mutant complexes showed comparatively a lesser average number of hydrogen bonds with 6.83 and 4.59, respectively (Fig. 8).

## 4. DISCUSSION

Mood disorders are although associated with psychological significance; these disorders have much more intense reactions at the genetic level. The environmental exposure plays a major role that may result in changing the gene expression is epigenetic based rather than relying on the DNA

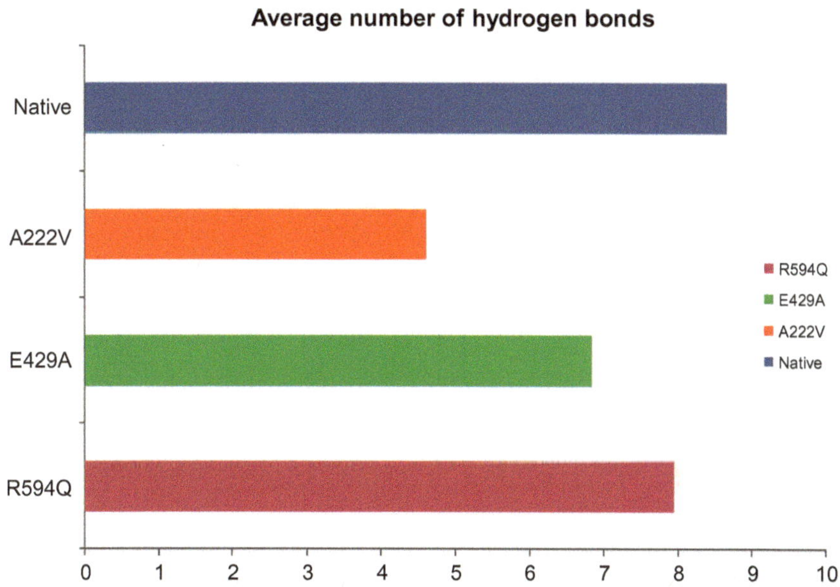


**Fig. 8** Average number of hydrogen bonds formed between the protein and FAD at the end of simulation. *Blue*, native complex; *red*, A222V mutant complex; *green*, E429A mutant complex; *magenta*, R594Q mutant complex.

sequence. This could be a plausible biological elucidation for correlations between the genetic variation in folate metabolism and SCZ and other functional psychiatric disorders (Dror & Allen, 2008; El-Hadidy et al., 2014; Gilbody et al., 2006). A recent study identified a mutation in a gene (*MTHFR*), leading to mood disorder, eventually affecting the methylation of the other gene (*SLC6A4*) (Devlin et al., 2010). Hence understanding these mutations on their structural and functional level could help in better treatment strategies. The most common form of mutation in the *MTHFR* gene is c.C677T that processes to produce A222V mutation at the protein level. Many reports have investigated the effect of A222V on enzyme function and structure both experimentally and computationally (Abhinand et al., 2016; Lewis et al., 2005). Although the earlier computational study discusses about the deleterious effects of A222V, the current study includes other mutations that are responsible in the development of SCZ. Their prevalence has been estimated, yet very less research has been done on this mutation toward mood disorder. The global prevalence of the mutations in MTHFR has been identified to be 29.3% and 23.6% for A222V and E429A, respectively (Marini et al., 2008). Consequently, in the current study, we have made an attempt to compare the mutations using computation tools to predict the differences in the structural behaviorness. The pathogenic and stability in silico tools predicted A222V mutation to be having higher deleterious impact in comparison to the other two mutations. As a next step, MD simulation analysis was carried out to understand the structural differences. This kind of combined computational approach may aid in understanding the structural differences (Ali, Sneha, Priyadharshini Christy, Zayed, & George Priya Doss, 2016; Sneha & George Priya Doss, 2016; Sudhakar et al., 2016; Thirumal Kumar et al., 2016) as well as in designing better therapeutic strategies (Sneha & Doss, 2016). The major barrier for the research in these mutations is that there are no templates available for developing a model for the MTHFR protein. According to the MODBASE and SWISS MODEL (3apta, 3aptb), the protein showed 39% identity covering up to 340 residues. However, there is only deposition between 48–338 and 48–340. Thus, this has become a limiting point in analyzing the E429A and R594Q mutations. Consequently, the threading approach for modeling the protein was performed which resulted in a protein structure with 656 amino acids (Fig. 2). Since the FAD activity is a crucial concept for the functioning of the enzyme, we have analyzed the level of interference of each mutation over the FAD-binding activity and found that there was least involvement of A222V mutation with no formation of hydrogen bonds, followed by E429A mutation with formation of two hydrogen bonds, whereas native showed two-hydrogen bonds formation. This analysis further confers that the A222V and E429 mutations may cause a higher effect on the protein than R594Q mutation. Finally, MD simulation was performed to elucidate the structural differences caused due to substitutions at an atomistic level. The RMSD value of A222V mutant complex was observed to be higher in comparison to that of native predicting a lesser stable conformation of the protein (Yun & Guy, 2011). The probable reason behind the loss of stability might be due to the reduction in the number of hydrogen bonds since these bonds are crucial for maintaining the stability of the protein (Hubbard & Kamran Haider, 2010; Pace et al., 2014) and also the loss of hydrogen bonds reduces the efficiency of ligand binding (Wade & Goodford, 1989). Consequently, there was a decrease in binding efficiency of FAD in the presence of mutation A222V for which the loss of hydrogen bonds might be a probable reason (Figs. 3B and 7). In the presence of E429A mutation, molecular docking analysis predicted a lesser number of hydrogen bonds when compared to native complex (Fig. 3C). Similarly a decrease in the number of intermolecular hydrogen bonds was observed for E429A mutant complex using MD simulation (Figs. 7 and 8). From the molecular docking and MD analyses, we infer that A222V and E429A mutations showed decreased binding activity with FAD, which might be due to the loss in hydrogen bonds and subsequent loss in the contact between the cofactor and the enzyme, thereby decreasing the activity of the MTHFR protein. Although the severity of E429A mutation is not as much as A222V, their involvement in the protein may affect the binding of FAD and further reduce the activity of the enzyme. However, R594Q mutation did not show a major effect on the protein. Through molecular docking analysis, we observed an increase in the number of hydrogen bonds formation (Fig. 3D) and stable deviation patterns in RMSD analysis when compared to the other two mutant complexes (Fig. 4). Finally, we calculated the average number of hydrogen bonds formed within complexes, in which similar number was observed in R594Q and native protein complexes (Figs. 7 and 8). The results from our MD simulation studies were in consistent with the previously analyzed experimental work wherein the mutant A222V highly interferes with the enzyme activity followed by E429A and least interference of R594Q mutation (Leclerc et al., 2013). Such structural insights over the differences caused due to the substitution in the protein structure may help in elucidating the plausible reason behind the loss of functioning of the enzyme and produce a genotype–phenotype correlation.

## 5. CONCLUSION

In conclusion, a combined study such as the inclusion of molecular docking with MD may further help in analyzing the effect of mutations on a broader perspective. From the tools and molecular dynamics analysis results, we find that both A222V and E429A showed a decreased affinity toward FAD, with more interference in the function observed with A222V mutation. As hydrogen bonds and stability of the protein directly influence the cofactor binding, a reduction in the hydrogen bonds with A222V and E429A mutations leads to decreased binding efficiency with the FAD, thereby affecting the overall enzyme activity. Overall, we observed a structural difference in case of A222V mutation than the E429A mutation. This study may further provide insights into the mutational effect caused over the enzyme's structure and function.

## ACKNOWLEDGMENTS

The authors acknowledge the management of VIT University, Vellore, India and (BRAF) @ CDAC for providing the facilities required to perform this work.

*Disclosure statement*: No potential conflict of interest exists.

## REFERENCES


Abhinand, P. A., Shaikh, F., Bhakat, S., Radadiya, A., Bhaskar, L. V. K. S., Shah, A., et al. (2016). Insights on the structural perturbations in human MTHFR Ala222Val mutant by protein modeling and molecular dynamics. *Journal of Biomolecular Structure & Dynamics*, *34*, 892–905.

Acharya, V., & Nagarajaram, H. A. (2012). Hansa: An automated method for discriminating disease and neutral human nsSNPs. *Human Mutation*, *33*, 332–337.

Adzhubei, I. A., Schmidt, S., Peshkin, L., Ramensky, V. E., Gerasimova, A., & Bork, P. (2010). A method and server for predicting damaging missense mutations. *Nature Methods*, 7, 248–249.

Ali, S. K., Sneha, P., Priyadharshini Christy, J., Zayed, H., & George Priya Doss, C. (2016). Molecular dynamics-based analyses of the structural instability and secondary structure of the fibrinogen gamma chain protein with the D356V mutation. *Journal of Biomolecular Structure and Dynamics*, 1–11. http://dx.doi.org/10.1080/07391102.2016.1229634.

Allen, N. C., Bagade, S., McQueen, M. B., Ioannidis, J. P. A., Kavvoura, F. K., Khoury, M. J., et al. (2008). Systematic meta-analyses and field synopsis of genetic association studies in schizophrenia: The SzGene database. *Nature Genetics*, *40*, 827–834.

Amberger, J., Bocchini, C. A., Scott, A. F., & Hamosh, A. (2009). McKusick's Online Mendelian Inheritance in Man (OMIM). *Nucleic Acids Research*, *37*, D793–D796.

American Psychiatric Association. (1994). *Diagnostic and statistical manual of mental disorders* (4th ed.). Washington, DC: American Psychiatric Press.

Ames, B. N., Elson-Schwab, I., & Silver, E. A. (2002). High-dose vitamin therapy stimulates variant enzymes with decreased coenzyme binding affinity (increased K(m)): Relevance to genetic disease and polymorphisms. *The American Journal of Clinical Nutrition*, *75*, 616–658.

Arzaghi, S. M., Hossein-Nezhad, A., Shariat, S. V., Ghodsipour, A., Shams, J., & Larijani, B. (2011). C677T methylenetetrahydrofolate reductase (MTHFR) gene polymorphism in schizophrenia and bipolar disorder: An association study in Iranian population. *Iranian Journal of Psychiatry*, *6*, 1–6.

Bottiglieri, T., Hyland, K., & Reynolds, E. H. (1994). The clinical potential of ademetionine (S-adenosylmethionine) in neurological disorders. *Drugs*, *48*, 137–152.

Bottiglieri, T., Laundy, M., Crellin, R., Toone, B. K., Carney, M. W., & Reynolds, E. H. (2000). Homocysteine, folate, methylation, and monoamine metabolism in depression. *Journal of Neurology, Neurosurgery, and Psychiatry*, *69*, 228–232.

Cantoni, G. L. (1952). The nature of the active methyl donor formed enzymatically from L-methionine and adenosinetriphosphate[1,2]. *Journal of the American Chemical Society*, *74*, 2942–2943.

Capriotti, E., Calabrese, R., & Casadio, R. (2006). Predicting the insurgence of human genetic diseases associated to single point protein mutations with support vector machines and evolutionary information. *Bioinformatics*, *22*, 2729–2734.

Cheng, J., Randall, A., & Baldi, P. (2005). Prediction of protein stability changes for single-site mutations using support vector machines. *Proteins*, *62*, 1125–1132.

Cohen-Woods, S., Craig, I., Gaysina, D., Gray, J., Gunasinghe, C., Craddock, N., et al. (2010). The Bipolar Association Case-Control Study (BACCS) and meta-analysis: No association with the 5,10-methylenetetrahydrofolate reductase gene and bipolar disorder. *The American Journal of Medical Genetics B: Neuropsychiatric Genetics*, *153*, 1298–1304.

Das, P. M., & Singal, R. (2004). DNA methylation and cancer. *Journal of Clinical Oncology*, *22*, 4632–4642.

Devlin, A. M., Brain, U., Austin, J., Oberlander, T. F., Oberlander, T., Warburton, W., et al. (2010). Prenatal exposure to maternal depressed mood and the MTHFR C677T variant affect SLC6A4 methylation in infants at birth. *PLoS One*, *5*, e12201.

Dror, D. K., & Allen, L. H. (2008). Effect of vitamin B12 deficiency on neurodevelopment in infants: Current knowledge and possible mechanisms. *Nutrition Reviews*, *66*, 250–255.

El-Hadidy, M. A., Abdeen, H. M., Abd El-Aziz, S. M., Al-Harrass, M., El-Hadidy, M. A., Abdeen, H. M., et al. (2014). MTHFR gene polymorphism and age of onset of schizophrenia and bipolar disorder. *Biomedical Research International*, *2014*, 318483.

Gilbody, S., Lewis, S., & Lightfoot, T. (2006). Methylenetetrahydrofolate reductase (MTHFR) genetic polymorphisms and psychiatric disorders: A huge review. *American Journal of Epidemiology*, *165*, 1–13.

Glaser, F., Pupko, T., Paz, I., Bell, R. E., Bechor-Shental, D., Martz, E., et al. (2003). ConSurf: Identification of functional regions in proteins by surface-mapping of phylogenetic information. *Bioinformatics*, *19*, 163–164.

Goyette, P., Sumner, J. S., Milos, R., Duncan, A. M., Rosenblatt, D. S., Matthew, R. G., et al. (1994). Human methylenetetrahydrofolate reductase: Isolation of cDNA, mapping and mutation identification. *Nature Genetics*, 7, 195–200.

Guex, N., & Peitsch, M. C. (1997). SWISS-MODEL and the Swiss-Pdb viewer: An environment for comparative protein modeling. *Electrophoresis*, *18*, 2714–2723.

Hubbard, R. E., & Kamran Haider, M. (2010). Hydrogen bonds in proteins: Role and strength. *Encyclopedia of Life Sciences*. Chichester, UK: John Wiley & Sons Ltd.

Hustad, S., Schneede, J., & Ueland, P. M. (2013). The methylenetetrahydrofolate reductase 677C→T polymorphisms as a modulator of a B vitamin network with major effects on homocysteine metabolism. *The American Journal of Human Genetics*, *80*, 846–855.

Johnson, A. D., Handsaker, R. E., Pulit, S. L., Nizzari, M. M., O'Donnell, C. J., & de Bakker, P. I. W. (2008). SNAP: A web-based tool for identification and annotation of proxy SNPs using HapMap. *Bioinformatics*, *24*, 2938–2939.

Jönsson, E. G., Larsson, K., Vares, M., Hansen, T., Wang, A. G., Djurovic, S., et al. (2008). Two methylenetetrahydrofolate reductase gene (MTHFR) polymorphisms, schizophrenia and bipolar disorder: An association study. *American Journal of Medical Genetics Part B*, *147B*, 976–982.

Kaplan, W., & Littlejohn, T. G. (2001). Swiss-PDB viewer (deep view). *Briefings in Bioinformatics*, *2*, 195–197.

Leclerc, D., Sibani, S., & Rozen, R. (2013). *Molecular biology of methylenetetrahydrofolate reductase (MTHFR) and overview of mutations/polymorphisms*. (pp. 2000–2013). Austin, TX: Landes Bioscience.

Leonardo, E. D., & Hen, R. (2006). Genetics of affective and anxiety disorders. *Annual Review of Psychology*, *57*, 117–137.

Lewis, S. J., Zammit, S., Gunnell, D., & Smith, G. D. (2005). A meta-analysis of the MTHFR C677T polymorphism and schizophrenia risk. *American Journal of Medical Genetics Part B*, *135B*, 2–4.

Lovell, S. C., Davis, I. W., Arendall, W. B., de Bakker, P. I., Word, J. M., Prisant, M. G., et al. (2002). Structure validation by Calpha geometry: phi, psi and Cbeta deviation. *Proteins*, *50*, 437–450.

Marini, N. J., Gin, J., Ziegle, J., Keho, K. H., Ginzinger, D., Gilbert, D. A., et al. (2008). The prevalence of folate-remedial MTHFR enzyme variants in humans. *Proceedings of the National Academy of Sciences of the United States of America*, *105*, 8055–8060.

Morris, G. M., Huey, R., Lindstrom, W., Sanner, M. F., Belew, R. K., Goodsell, D. S., et al. (2009). AutoDock4 and AutoDockTools4: Automated docking with selective receptor flexibility. *Journal of Computational Chemistry*, *30*, 2785–2791.

Muntjewerff, J.-W., & Blom, H. J. (2005). Aberrant folate status in schizophrenic patients: What is the evidence? *Progress in Neuro-Psychopharmacology & Biological Psychiatry*, *29*, 1133–1139.

Muntjewerff, J. W., Kahn, R. S., Blom, H. J., & den Heijer, M. (2006). Homocysteine, methylenetetrahydrofolate reductase and risk of schizophrenia: A meta-analysis. *Molecular Psychiatry*, *11*, 143–149.

Ng, P. C., & Henikoff, S. (2003). SIFT: Predicting amino acid changes that affect protein function. *Nucleic Acids Research*, *31*, 3812–3814.

Nicol, S. E., & Gottesman, I. I. (1983). Clues to the genetics and neurobiology of schizophrenia: Uncertainty about the etiology of schizophrenia persists, but compelling evidence implicates genes and links dopamine to the disorder. *American Scientist*, 7, 398–404.

Pace, C. N., Fu, H., Lee Fryar, K., Landua, J., Trevino, S. R., Schell, D., et al. (2014). Contribution of hydrogen bonds to protein stability. *Protein Science*, *23*, 652–661.

Peerbooms, O. L., van Os, J., Drukker, M., Kenis, G., Hoogveld, L., de Hert, M., et al. (2011). Meta-analysis of MTHFR gene variants in schizophrenia, bipolar disorder and unipolar depressive disorder: Evidence for a common genetic vulnerability? *Brain, Behavior, and Immunity*, *25*(8), 1530–1543.

Pettersen, E. F., Goddard, T. D., Huang, C. C., Couch, G. S., Greenblatt, D. M., Meng, E. C., et al. (2004). UCSF chimera—A visualization system for exploratory research and analysis. *Journal of Computational Chemistry*, *25*, 1605–1612.

Picker, J. D., & Coyle, J. T. (2005). Do maternal folate and homocysteine levels play a role in neurodevelopmental processes that increase risk for schizophrenia? *Harvard Review of Psychiatry*, *13*, 197–205.

Pires, D. E. V., Ascher, D. B., & Blundell, T. L. (2014). DUET: A server for predicting effects of mutations on protein stability using an integrated computational approach. *Nucleic Acids Research*, *42*, W314–W319.

Prathikanti, S., & Weinberger, D. R. (2005). Psychiatric genetics—The new era: Genetic research and some clinical implications. *British Medical Bulletin*, *73–74*, 107–122.

Pronk, S., Páll, S., Schulz, R., Larsson, P., Bjelkmar, P., Apostolov, R., et al. (2013). GROMACS 4.5: A high-throughput and highly parallel open source molecular simulation toolkit. *Bioinformatics*, *29*, 845–854.

Reva, B., Antipin, Y., & Sander, C. (2011). Predicting the functional impact of protein mutations: Application to cancer genomics. *Nucleic Acids Research*, *39*, e118.

Rozen, R. (1996). Molecular genetics of methylenetetrahydrofolate reductase deficiency. *Journal of Inherited Metabolic Disease*, *19*, 589–594.

Shi, J., Gershon, E. S., & Liu, C. (2008). Genetic associations with schizophrenia: Meta-analyses of 12 candidate genes. *Schizophrenia Research*, *104*, 96–107.

Smith, A. L., & Weissman, M. M. (1992). Epidemiology. In E. S. Paykel (Ed.), *Handbook of affective disorders* (pp. 111–129). Edinburgh, UK: Churchill Livingstone.

Sneha, P., & Doss, C. G. P. (2016). Gliptins in managing diabetes—Reviewing computational strategy. *Life Sciences*, *166*, 108–120.

Sneha, P., & George Priya Doss, C. (2016). Chapter seven—Molecular dynamics: New frontier in personalized medicine. *Advances in Protein Chemistry and Structural Biology*, *102*, 181–224.

Sudhakar, N., George Priya Doss, C., Thirumal Kumar, D., Chakraborty, C., Anand, K., & Suresh, M. (2016). Deciphering the impact of somatic mutations in exon 20 and exon 9 of PIK3CA gene in breast tumors among Indian women through molecular dynamics approach. *Journal of Biomolecular Structure and Dynamics*, *34*, 29–41.

Thirumal Kumar, D., George Priya Doss, C., Sneha, P., Tayubi, I. A., Siva, R., Chakraborty, C., et al. (2016). Influence of V54M mutation in giant muscle protein titin: A computational screening and molecular dynamics approach. *Journal of Biomolecular Structure and Dynamics*, 1–12.

Ueland, P. M., Hustad, S., Schneede, J., Refsum, H., & Vollset, S. E. (2001). Biological and clinical implications of the MTHFR C677T polymorphism. *Trends in Pharmacological Sciences*, *22*, 195–201.

Wade, R. C., & Goodford, P. J. (1989). The role of hydrogen-bonds in drug binding. *Progress in Clinical and Biological Research*, *289*, 433–444.

Weissman, M. M., Bland, R. C., Canino, G. J., Faravelli, C., Greenwald, S., Hwu, H. G., et al. (1996). Cross-national epidemiology of major depression and bipolar disorder. *Journal of American Medical Association*, *276*, 293–299.

Yang, J., Yan, R., Roy, A., Xu, D., Poisson, J., & Zhang, Y. (2014). The I-TASSER suite: Protein structure and function prediction. *Nature Methods*, *12*, 7–8.

Yun, S., & Guy, H. R. (2011). Stability tests on known and misfolded structures with discrete and all atom molecular dynamics simulations. *Journal of Molecular Graphics and Modelling*, *29*, 663–675.

Zintzaras, E. (2006). C677T and A1298C methylenetetrahydrofolate reductase gene polymorphisms in schizophrenia, bipolar disorder and depression: A meta-analysis of genetic association studies. *Psychiatric Genetics*, *16*, 105–115.

CHAPTER SIX

# Stress-Induced NLRP3 Inflammasome in Human Diseases

**Elísabet Alcocer-Gómez[1], Beatriz Castejón-Vega, Mario D. Cordero[1]**
Research Laboratory, University of Sevilla, Sevilla, Spain
[1]Corresponding authors: e-mail address: agelisabet@gmail.com; mdcormor@us.es

## Contents



## Abstract

Stress is a complex event that induces disturbances to physiological and psychological homeostasis, and it may have a detrimental impact on certain brain and physiological functions. In the last years, a dual role of the stress effect has been studied in order to elucidate the molecular mechanism by which can induce physiological symptoms after psychological stress exposition and vice versa. In this sense, inflammation has been proposed as an important starring. And in the same line, the inflammasome complex has emerged to give responses because of its role of stress sensor. The implication of the same complex, NLRP3 inflammasome, in different diseases such as cardiovascular, neurodegenerative, psychiatric, and metabolic diseases opens a door to develop new therapeutic perspectives.

*Advances in Protein Chemistry and Structural Biology*, Volume 108
ISSN 1876-1623
http://dx.doi.org/10.1016/bs.apcsb.2017.02.002

## 1. INTRODUCTION

Few experiences are more common among people than exposure to stressful life events. Stress is a common aspect of life, being a stimulant for some, but a burden for many others. Several definitions have been advanced for the concept of stress, each focusing on aspects of external or internal challenge or stimulus; on a physiological response to the stimulus; or on stimulus perception by an organism (Goldstein & McEwen, 2002; McEwen, 2002; Sapolsky, 2005).

The term "stress" was coined by Selye, who defined it as "the nonspecific response of the body to any demand for change" (Selye, 1936). For this author, a stressor is an agent that produces stress at any time. In 1936, Selye advanced the concept of the general adaptation syndrome. This syndrome represents the chronologic development of the physiological response to stressors when their action is prolonged. It followed a similar consistent pattern of three stages: the alarm reaction, the stage of resistance, and the stage of exhaustion (Selye, 1936). Selye developed these phases based on neural and hormonal processes that are taking place in the body. In this way, the alarm reaction stage involves a neural response of the autonomic sympathetic nervous system (SNS) which induces a rapid secretion of adrenaline. This phase is followed by a stage of resistance which brings to increased levels of cortisol changing the body metabolism. Finally, long-term exposure to the stage of exhaustion will lead to damage to body systems such as the immune, digestive, or kidney systems.

In numerous experiments, Selye observed that laboratory animals subject to different noxious physical and emotional stimuli (e.g., extremes of cold or heat, blaring light, perceptual frustration) all exhibited the same pathological changes or alterations, including stomach ulcerations, enlargement of the adrenals, and shrinkage of the lymphoid tissue. Later, Selye showed that persistent stress could cause these animals to suffer several diseases similar to those observed in humans, such as heart attacks, kidney disease, or stroke (The American Institute of Stress, 2015; https://www.stress.org).

Selye's ideas were not accepted positively by psychologists working in the stress field. Thus, Mason wrote that in the psychological stress field it had been noted that responses to any mental stimulus might vary strongly among individuals or from one time to another in the same person (Mason, 1975a,1975b). Selye tried to rebuttal this criticism and wrote that the fact that stressors, or even the same stressor, can induce different lesions in different individuals. This is what he has called "conditioning factors" (e.g., age, sex, genetic predisposition, exposure to environmental factors, treatment with drugs or hormones, etc.) that can selectively inhibit or enhance one or the other stress effects (Selye, 1975). Another psychologist wrote that the research on failure to adapt to persistent stress suggests that maladaptation comes from psychological exhaustion and not from physiological exhaustion as was proposed by Selye (Burchfield, 1979). Later, Reznick introduced the concept of the stress cycle to try to integrate many of the ideas that exist in the stress field into a unified model that would combine both psychological and physiological elements of stress and provide a comprehensive definition of the concept of stress (Reznick, 1989).

An integrated definition asserts that stress is a phenomenon that includes a stimulus (stressor), that induces a reaction in the brain (stress perception), that precipitates physiological fight-or-flight systems in the body (stress response) (Dhabhar & McEwen, 1997). It must be noted that the only way that a stressor can act on the body or the brain is through the biological stress response. Although several factors are implicated, the principal mediators of stress effects are epinephrine and norepinephrine that are released by the SNS, and corticotropin releasing hormone, adrenocorticotropin, and cortisol, which emanate following activation of the hypothalamic–pituitary–adrenal axis (HPA). Seeing that practically each cell in the body expresses receptors for one or more of these factors, stress hormones can provoke changes in virtually every cell and tissue and alert them to the existence of a stressor (Dhabhar, 2014).

An important distinctive characteristic of stress is the duration of the biological effects of stress. Short-term stress has been described as stress that persists for minutes to hours, and chronic or long-term stress has been defined as stress that continues for many hours per day for weeks or months (Dhabhar & McEwen, 1997). Whereas chronic stress usually has deleterious effects, short-term stress can be protective as it prepares the body to deal with demands.

From the first ideas of Selye, an ever-burgeoning body of biomedical research has established positive associations between persistent stress and the development of a wide range of pathological disease states, including psychiatric disorders such as major depressive disorder (MDD) (Maes et al., 2012; Weidenfeld & Yirmiya, 1996) and posttraumatic stress disorder (Pace & Heim, 2011; Yehuda, 2009); neurodegenerative disorders such as Parkinson's disease (PD) (Macht, Schwarz, & Ellgring, 2005) and Alzheimer's disease (AD) (Piirainen et al., 2017); systemic health disorders such as metabolic disorders (Janczura et al., 2015), cardiovascular diseases (CVDs) (Black & Garbutt, 2002), and inflammatory diseases (Elenkov & Chrousos, 2002).

## 2. INFLAMMATORY DISEASES AND STRESS

Psychological, physical, or metabolical stress can induce an inflammatory response. The relationship between stressful events and the development of inflammatory diseases (as neurological, digestive, cardiovascular, or immunologic diseases) is well reported (Baum & Posluszny, 1999). Thus, "stress hormone" production in the stress response is analogous to the inflammatory process produced in an organism after trauma or tissue damage or while is invaded by certain microorganisms. Next, we briefly describe the human inflammatory response.

Inflammation is a protective immune response mounted by the innate immune system to danger stimuli (dead cells, pathogens, irritants, etc.) and is strictly regulated by the host. In response to an infection, the immune system arranges an organized response to eliminate the pathogen and to repair tissue integrity. Innate immune cells comprise the first line of defense and are implicated in recognizing the initial danger, inhibiting the infection and facilitating the compilation of more immune cells through the release of cytokines and chemokines. Adaptive immune cells (lymphocytes B and T) then appear on the scene and promote the removal of infected cells. An incapacity to provide protection responses against pathogens can have harmful consequences for the host, and, for this reason, the immune system has developed numerous mechanisms to combat infection effectively. However, the same inflammatory tracks that are necessary for host protection against pathogens can also be triggered in response to tissue injury in the nonappearance of infection. Inflammation and tissue damage that occurs from metabolic and cellular dysfunction, trauma, ischemia–reperfusion damage, or environmental irritants usually results in the absence of disease and is frequently termed to as sterile inflammation (Lukens, Gross, & Kanneganti, 2012). Also, deficient inflammation can drive to chronic infection of pathogens, while high levels of inflammation can lead to persistent or systemic inflammatory diseases (Guo, Callaway, & Ting, 2015).

Innate immune function depends upon recognition of pathogen-associated molecular patterns (PAMPs), which are patterns associated with microbes that are pathogenic to the host, and danger-associated molecular patterns (DAMPs), which are ligands related to tissue damage and cellular

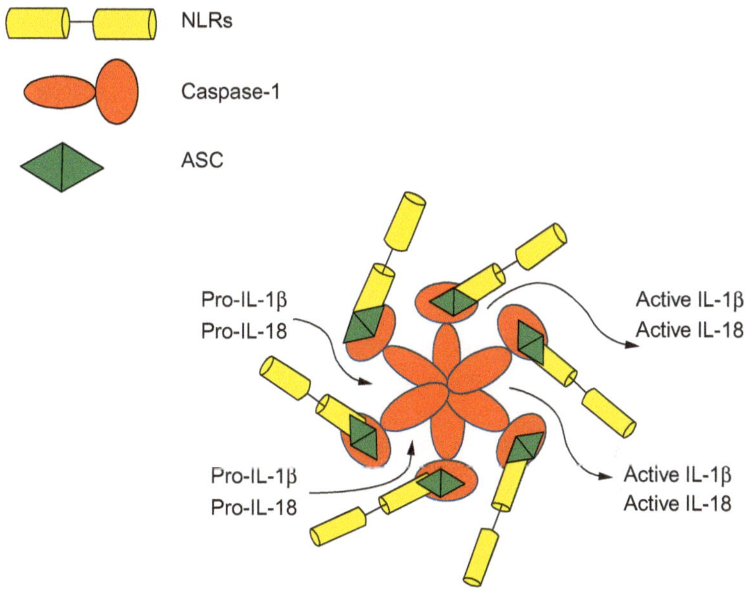


**Fig. 1** Molecular structure of the inflammasome complex with a NLR, caspase-1, and ASC. The molecular assembly of the inflammasome induces the activation and IL-1β and IL-18 release.

stress, by germline-encoded pattern recognition receptors (PRRs). PRRs are expressed by cells of the innate immune system such as monocytes, macrophages, neutrophils, and dendritic cells, as well as epithelial and adaptive immune system cells. An extensive range of PRRs has been characterized, such as Toll-like receptors (TLRs), retinoic acid-inducible gene (RIG)-I-like receptors (RLRs), the absent in melanoma 2-like receptors, and nucleotide-binding oligomerization domain (NOD)-like receptors (NLRs) (Fig. 1). The activation of PRRs by PAMPs or DAMPs results in the initiation of a cascade of signaling events that lead to the maturation and differentiation of immune cells and the production of proinflammatory cytokines and chemokines, which constitute a base for the adaptive immune response.

## 3. THE INFLAMMASOMES

Inflammasome activation is an important function mediated by the immune system. Various PPRs are important constituents in the inflammasome complex as well as intracellular PRRs denominated nucleotide-binding NLRs (Schroder & Tschopp, 2010) and the interferon-inducible protein AIM2 (absent in melanoma-2). When activated by DAMPs and PAMS, NLR or AIM2 forms a protein complex with pro-caspase-1, the effector molecule, with or without the help of an adapter molecule such as apoptosis-associated speck–caspase recruit domain (ASC). The complex so composed is termed the inflammasome (Fig. 1). The term inflammasome was introduced by Martinon and Tschopp to describe the multiprotein complex needed for caspase-1 processing and activation of the inflammatory cytokines such interleukin-1β (IL-1β) and IL-18 (Martinon, Burns, & Tschopp, 2002). Inflammasomes complex are implicated in regulating the secretion of nonconventional alarm molecules, the alarmins (Bianchi, 2007), lipid biogenesis, glycolysis, and in the inflammation-induced cell death with characteristics of apoptosis and necrosis called "pyroptosis" (McIntire, Yeretssian, & Saleh, 2009). Literature sustains that the aberrant activation of the inflammasomes is a risk factor for the emergence of autoimmune, autoinflammatory, chronic inflammatory, and metabolic diseases. Accordingly, activation of the inflammasomes should be controlled to avoid unwanted host damage and too much inflammation (Franchi, Munoz-Planillo, & Nunez, 2012; Menu & Vince, 2011).

As mentioned, the NLR family of proteins implicated in the innate immune responses produce the assembly of large caspase-1-activating multiprotein complexes termed inflammasomes (Schroder & Tschopp, 2010). Activation of caspase-1 through autoproteolytic maturation induces to a production of the proinflammatory cytokines IL-1β and IL-18. Four NLR-subset inflammasomes have been characterized: the NLRP1/NALP1b inflammasome (Boyden & Dietrich, 2006), the NLRC4/IPAF inflammasome (Zhao et al., 2011). the NLRP3/NALP3 inflammasome (Martinon, Petrilli, Mayor, Tardivel, & Tschopp, 2006), and the AIM2-containing inflammasome (Fernandes-Alnemri, Yu, Datta, Wu, & Alnemri, 2009). These types of inflammasomes contain a particular NLR protein or AIM2, the ASC and/or cardinal adaptor proteins, and pro-caspase-1. The NLRP3 inflammasome is the most investigated and the best characterized; this complex is formed by the NLR, the adapter molecule ASC, and the effector molecule pro-caspase-1. Activation of caspase-1 leads to the maturation and secretion of the proinflammatory cytokines IL-1β and IL-18, which induce several biological effects related to inflammation, infection, and autoimmune processes. Maturation of IL-1β and IL-18 by cleavage with caspase-1 is necessary to provoke the immune response.

To date, it has been demonstrated that inflammasome formation is thought to be regulated at both the transcriptional and posttranslational steps. A priming step in inflammasome formation results in the activation of the transcription factor NF-κB, which is crucial for upregulating the transcription of both pro-IL-1β and NLRP3 itself. This stage also serves to trigger posttranslational modifications of inflammasome components (Niebur, 2008; Wright, 2004). The second step triggers assembly of the NLRP3 inflammasome complex (Guo et al., 2015). Transcriptional modulation is not required to ASC and pro-caspase-1 (Kigerl, de Rivero Vaccari, Dietrich, Popovich, & Keane, 2014). NLRP3 inflammasome can be activated by both exogenous (including tissue damage, infection, and metabolic dysregulation) (Pressman, 1976) and endogenous molecules such as crystalline molecules, extracellular adenosine triphosphate (ATP; through its cell surface receptor P2X7R), A fibrils, lipopolysaccharide (LPS), hyaluronan, and uric acid crystals (Guo et al., 2015). For a large number of NLRP3 inflammasome activators, it seems improbable that they all bind to the NLRP3 structure to form the NLRP3 inflammasome complex. So far, there have been three primarily mechanisms concerning the activation of the NLRP3 inflammasome, such as the efflux of potassium, the generation of reactive oxygen species (ROS), and the rupture of lysosomal (Fig. 2) (Liston & Masters, 2017).

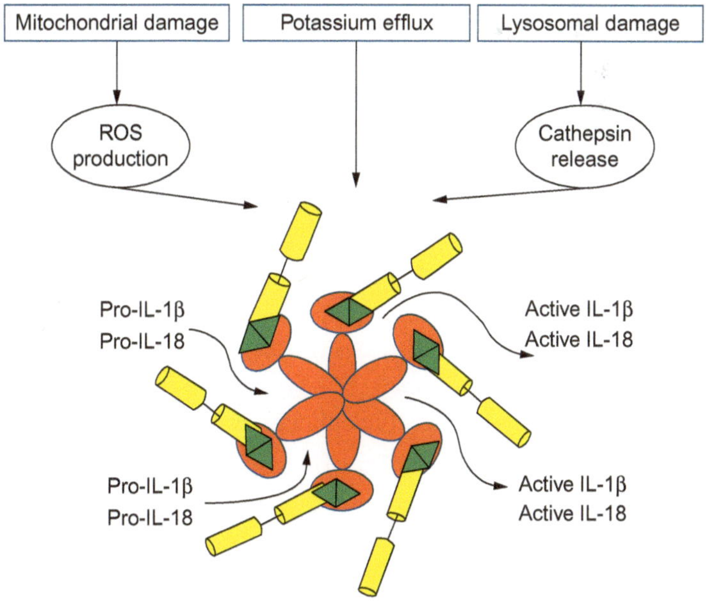


**Fig. 2** Inflammasome activation. Different pathways to induce the prime and activation of the inflammasome complex.

## 4. THE INFLAMMASOME IN THE PSYCHIATRIC DISEASES

The discovery of inflammasomes by Martinon et al. (2002) has incited substantial interest in the function of inflammasomes in the mechanism of inflammation and associated disease patterns. Recently, a large body of literature denotes to the presence of inflammasome-mediated inflammatory pathways in central nervous system (CNS) disorders.

Neuroinflammation, described by overproduction of proinflammatory cytokines in brain, is a recognized factor in the pathogenesis of psychiatric diseases and neurodegenerative illness (Frank Cannon, Alto, McAlpine, & Tansey, 2009) such as depression (Walker, Kavelaars, Heijnen, & Dantzer, 2014), AD (Pimplikar, 2014), Huntington's disease (Möller, 2010), PD (Hirsch, Vyas, & Hunot, 2012), and multiple sclerosis (MS) (Frohman, Racke, & Raine, 2006). It has also been involved in the decline of cognition (Ownby, 2010), sickness behavior (Biesmans et al., 2013), and diminished memory (Hein & O'Banion, 2009), as well as in age-related elevated sensitization of the immune system to intrinsic and extrinsic stimuli (Godbout et al., 2005; Sparkman & Johnson, 2008).

Activation of the inflammasome in neurons can induce directly to neuronal cell death (Liu et al., 2004), at the same time activation within microglial cells can lead to inflammatory responses or repair mechanisms. While four major types of inflammasomes are reported (NLRP1, NLRP3, NLRC4, and AIM2), NLRP3 inflammasome is involved in linkage to various CNS disorders because it plays a crucial role in the development of CNS inflammation (Liu, Mi, & Wang, 2013). Extensive evidence has indicated that NLRP3 inflammasome is activated in several psychiatric and neurodegenerative disorders such depression (Alcocer-Gómez & Cordero, 2014), AD (Masters & Selkoe, 2012), PD (Codolo et al., 2013; Zhou et al., 2016), frontotemporal dementia, amyotrophic lateral sclerosis (ALS), Huntington's disease, MS (Glass, Saijo, Winner, Marchetto, & Gage, 2010; Heneka, Kummer, & Latz, 2014), and prion diseases (Hafner-Bratkovic, Bencina, Fitzgerald, Golenbock, & Jerala, 2012; Shi et al., 2012) among others.

### 4.1 Major Depressive Disorder

Exposure to stress induces disturbances to physiological homeostasis, and it may cause an adverse impact on several brain functions. Uncontrollable stress is a crucial contributing factor for MDD, a severe neuropsychiatric disease that involves important changes in mood, self-esteem, eating, circadian rhythms, and cognition (Iwata, Ota, & Duman, 2013). While it affects up to 10% of the general population worldwide, its pathogenic pathway remains unresolved (Kessler, 2012). Accordingly, research is focused on characterizing its pathophysiology at the molecular and cellular level. Increasing evidence suggests that inflammation is a crucial factor in the pathophysiology of stress and MDD (Dowlati et al., 2010; Koo, Russo, Ferguson, Nestler, & Duman, 2010; Müller, 2014). Thus, literature shows that depressive symptoms are underlined by elevated levels of proinflammatory cytokines (that is, interleukin IL-6 and IL-1β and TNF-α), reduced levels of antiinflammatory cytokines (that is, IL-10 and IL-4), and are linked with polymorphisms in inflammation-related genes (Kokai, Kashiwamura, Okamura, Ohara, & Morita, 2002; Mikova, Yakimova, Bosmans, Kenis, & Maes, 2001; Wong, Dong, Maestre-Mesa, & Licinio, 2008). Increased proinflammatory interleukins associate with HPA axis dysregulation, central serotonin depletion, and microglial activation, as well as with brain structure alteration in murine models of depression (Rosenblat, Cha, Mansur, & McIntyre, 2014).

The role of persistent mild stress in neuroinflammation and consequent development of MDD has also been established (Farooq et al., 2012). Several kinds of research suggest that IL-1β is a critical factor in the proinflammatory response to psychological stress as it is a crucial mediator of the antineurogenic and depressive-like behaviors induced by stress (Koo et al., 2010). It is interesting that IL-1β is one of the two known interleukins, together with IL-18, stimulated by the NLRP3 inflammasome complex. The elevated IL-1β levels and neuroinflammation in the brain of depressive patients potentially suggest a role for NLRP3 inflammasome complex in depression. In recent times, inflammasome has emerged as an unexpected sensor for stress and metabolic danger. Accordingly, a new inflammasome hypothesis of depression and associated comorbid systemic diseases has been proposed by Iwata. This theory suggests that the inflammasome is a central mediator by which psychological and physical stressors could contribute to the development of depression (Alcocer-Gómez & Cordero, 2017). Thus, this protein complex is responsive both to cell damage and to pathogens and stress, including the sensing of extracellular ATP (Iwata et al., 2013). Later, Iwata and coworkers proposed new data in a murine model in further support of the inflammasome hypothesis. The authors demonstrated that psychological stress causes the release of extracellular ATP and that the interaction of ATP with purinergic type 2X7 receptor induces the activation of the NLRP3 inflammasome cascade resulting in the release of IL-1β and further activation of TNF-α (Iwata et al., 2016). So TNF-α may prolong and amplify the initial stress response. In recent studies in animal models have been shown the role of TNF-α in depression induced by stress exposition which was mediated by indoleamine 2,3-dioxygenase, the initial step in the tryptophan catabolism (Liu, Peng, et al., 2015; Liu, Xu, et al., 2015).

Several lines of evidence have demonstrated that NLRP3 inflammasome plays a critical role in the pathophysiology of depression. Recently, we have shown that the depressive patients exhibit increased levels of cytokines IL-1β and IL-18, as well as the NLRP3 inflammasome in peripheral blood mononuclear cells (Alcocer-Gómez et al., 2013). Similar findings were seen in a mouse model; this research has shown the involvement of the NLRP3 inflammasome in LPS-induced mouse depressive-like behavior (Zhang et al., 2013). Other researchers have demonstrated that antidepressant treatment could suppress NLRP3 inflammasome activation (Li, Wang, Qin, Qu, & Ma, 2016; Liu, Peng, et al., 2015; Liu, Xu, et al., 2015; Lu, Yang, Geng, Ding, & Hu, 2014; Xue et al., 2015). Also, stress-induced depressive behaviors require a functional NLRP3 inflammasome (Alcocer-Gómez et al., 2016), and NLRP3 deletion suppresses these stress-like effects (Zhang et al., 2015). Other study shows data that microglial NLRP3 inflammasome activation is a mediator of IL-1β-associated CNS inflammation during persistent stress (Pan, Chen, Zhang, & Kong, 2014).

Together, all findings point to that NLRP3 inflammasome could be a target for novel therapeutic interventions to MDD. In this line, various studies indicate that antidepressant drugs regulate inflammasome activation. For example, the tricyclic antidepressant amitriptyline downregulates NLRP3 and caspase-1 gene expression in patients suffering from depression (Alcocer-Gómez et al., 2013). Also, an antidiabetic drug such as glyburide, which suppresses caspase-1 activation and IL-1β secretion, is an efficient inhibitor of NLRP3 activation following stress-induced depression (Lamkanfi et al., 2009). Furthermore, an antagonistic relationship has been found an antagonistic relationship between antiinflammatory agents and antidepressant drugs in subjects with MDD (Warner-Schmidt, Vanover, Chen, Marshall, & Greengard, 2011). Thus, selective serotonin reuptake inhibitors elevated IFN-γ and TNF-α levels in the frontal cortex, which was suppressed by antiinflammatory agents. However, tricyclic antidepressants did not produce that effect (Warner-Schmidt et al., 2011).

Recent new data suggest that caspase-1 inhibition with minocycline (an antibiotic that inhibits caspase-1 and caspase-3 transcription and has antiinflammatory, antiapoptotic, and neuroprotective effects; Garrido-Mesa, Zarzuelo, & Galvez, 2013; Molina-Hernandez, Tellez-Alcantara, Perez-Garcia, Olivera-Lopez, & Jaramillo-Jaimes, 2008; Yong et al., 2004) has a preventive properties on the stress response by modulating the relationship between the gut microbiota and stress. This proposes the concept of a microbiota–gut–inflammasome–brain axis in which the gut microbiota through inflammasome signaling regulate several pathways which will change brain function and affect depressive- and anxiety-like behaviors. It may offer new therapeutic targets for the management of psychiatric disorders (Wong et al., 2016).

Various anti-IL-1β-targeted treatments have been recently developed, such as the IL-1β receptor antagonist (anakinra) (Burls & Jobanputra, 2004) and antibodies (gevokizumab or canakinumab) (Issafras, Corbin, Goldfine, & Roell, 2014; Rondeau, Ramage, Zurini, & Gram, 2015). Specific drugs targeting P2X7R are currently in development. Thus, in an LPS-induced mouse model of depression, the P2X7R antagonist Brilliant Blue G and the P2X7R inhibitor carbonate cage proved to be effective to reduce depressive-like behaviors (Ma, Ren, Zhang, & Hashimoto, 2014; Wilkinson et al., 2014). Again, no ongoing clinical trial has been registered to date but is being proposed in the last year (Ellul, Boyer, Groc, Leboyer, & Fond, 2016).

## 4.2 Bipolar Disorder and Schizophrenia

Bipolar disorder (BD) is the sixth leading cause of disability in worldwide (WHO) with chronic course in which between 25% and 50% of patients attempt suicide (Fountoulakis, Gonda, Siamouli, & Rihmer, 2009). Mitochondrial dysfunction and activation of the inflammatory pathways are two of the most reported findings in BD. Specifically, higher levels of inflammatory cytokines and decreased levels of mitochondrial chain complex I subunits have been found in the brain and peripheral cells from patients with BD, leading to increasing production of mitochondrial ROS. Interestingly, mitochondrial ROS and inflammation are closely linked through NLRP3. Several studies are demonstrating mitochondrial dysfunction, with the particular implication of complex I and inflammation in patients with BD (Kim, Chen, & Andreazza, 2015; Scola, Kim, Young, & Andreazza, 2013). According to this line, decreased levels of the complex I subunits are responsible for transportation of electrons and are observed in the patients with BD, which could result in increased leakage of electrons and ROS. Altered levels of inflammatory cytokines were also shown in the brain and periphery of patients with BD, including IL-6, TNF, IFN-γ, and IL-1 (Konradi, Sillivan, & Clay, 2012), suggesting that activation of the inflammatory system may also play a role in the pathophysiology of BD. It has been studied that mitochondrial ROS production can induce inflammatory activation because of, for example, the inhibition of complex I and subsequent ROS production which increased the release of inflammatory factors such as IL-1, caspase-1, and by NF-κB activation (López-Armada, Riveiro-Naveira, Vaamonde-García, & Valcárcel-Ares, 2013). In this sense, it is possible to think about the implication of inflammasome in BD because NLRP3 is a redox sensor. Different inhibition of the mitochondrial chain complex has been shown to activate the NLRP3 inflammasome, with a reduction of the inflammasome activation by inhibition of mitochondrial ROS production (Zhou, Yazdi, Menu, & Tschopp, 2011). Recently, Kim and coworkers examined NLRP3 inflammasome in postmortem frontal cortex from patients with BD and schizophrenia and observed lower levels of complex I and NDUFS7, a subunit of complex I, accompanied by increased NLRP3 and ASC expression only in patients with BD. But increased levels of caspase-1, IL-1β, IL-6, TNF-α, and IL-10 in both BD and schizophrenia patients (Kim, Andreazza, Elmi, Chen, & Young, 2016).

## 4.3 Autism Spectrum Disorder

Autistic spectrum disorders (ASD) is a neurodevelopmental disorder characterized by persistent deficits in social interaction and social communication in multiple contexts, accompanied by repetitive patterns of behavior, interests, or activities (Diagnostic and statistical manual of mental disorders DSM 5). ASD is associated with psychiatric and nonpsychiatric comorbidities including immune alterations. In this context, increased production of cytokines by peripheral blood immune cells has been described in patients with ASD as responsible for many features of autism, such as mood and sleep disturbances (Ashwood & Wakefield, 2006). Gastrointestinal (GI) alterations have been described in ASD children. These alterations also result in immune activation, directly with the LPS, the endotoxin of bacteria TLR 4 ligand, or indirectly via the release of DAMPs. Many of these exogenous and endogenous danger signals activate the inflammasomes complex such as damaged cells, release ATP or PAMPs such as LPS both primes, and activate the NLRP3 inflammasome, causing secretion of proinflammatory cytokines and pyroptosis (Nadatani et al., 2016). In ASD patients, proinflammatory cytokines have been studied, and TNF-α, IFN-c, IL-6, IL-8, and IL-12 have been found increased in serum and cerebrospinal fluid (Alabdali, Al-Ayadhi, & El-Ansary, 2014; Suzuki et al., 2011). According to this evidence inflammatory profile in ASD patients, Saresella and coworkers investigated whether the inflammasome system is activated in cells of ASD children, by stimulating their cells with ATP. They observed upregulation of the AIM2 and the NLRP3 inflammasomes, activated caspase-1, and increased production of IL-1β and IL-18 (Saresella et al., 2016).

AD is a progressive neurodegenerative disease that leads to cognitive decline, disability, neuropsychiatric symptoms, and premature death. It is the most prevalent cause of dementia, and this prevalence changes according to age (Ziegler-Graham, Brookmeyer, Johnson, & Arrighi, 2008). The disorder is estimated to affect 1 in 85 people globally worldwide by 2050 (Brookmeyer, Johnson, Ziegler-Graham, & Arrighi, 2007). The major pathological traits of AD are characterized by damage to the hippocampus, entorhinal cortex, and basal forebrain, leading to spatial and temporal disorientation, altered cognitive function and memory loss. The neuropathological hallmarks of AD are diffuse amyloid plaques composed of three Ab peptides, and intracellular neurofibrillary tangles (NFTs), consisting mainly of aberrantly folded tau protein (Lalla & Donmez, 2013). Furthermore, dominant mutations were located in the amyloid precursor protein (APP) gene and in the presenilin 1 and 2 genes (PSEN1 and PSEN2), which encode for constituents of gamma-secretase, leading to early-onset AD (Giri, Zhang, & Leu, 2016). APP is cleaved by beta-secretase and gamma-secretase consecutively generating Ab1–40 and Ab1–42 amyloid peptides, respectively, which form and accumulate the amyloid plaques (Mendiola-Precoma, Berumen, Padilla, & Garcia-Alcocer, 2016). Studies since the discoveries of tau and Ab have offered relevant information about molecular pathogenetic events, yet little is known about the cause and cure of this disorder (Scheltens et al., 2016). Recent data suggest that neuroinflammation has a significant role in the onset of some chronic degenerative illness of the CNS characterized by a persistent neuronal death in specific areas of the CNS causing the cognitive and motor deficits that characterize neurodegenerative disorders.

## 5. NLRP3 INFLAMMASOME AND NEUROINFLAMMATION IN NEURODEGENERATIVE DISEASES

The inflammasome complex in the CNS is implicated in the activation of an innate immune inflammatory response through IL-1 cytokine release by glial cells. Glial cells include oligodendrocytes, astrocytes, and microglia. Reliable results show that an increase of proinflammatory cytokine expression by microglia and astrocytes induces neuroinflammation followed by neurodegeneration (Pennisi et al., 2016).

Microglia, which are the immune effector cells of the CNS, express various receptors of the innate immune system, such as NLRs, TLRs, CTL receptors, and RLRs (Furr, Chauhan, Sterka, Grdzelishvili, & Marriott, 2008; Hu et al., 2012; Shah et al., 2008). New findings have implicated NLR-mediated activation of microglia in various neurodegenerative (Heneka et al., 2013; Salminen, Ojala, Suuronen, Kaarniranta, & Kauppinen, 2008) and infectious CNS illness (Hafner-Bratkovic et al., 2012; Jamilloux & Henry, 2013). Inflammasome-mediated activation of microglia is involved in both infectious (Chang et al., 2012; Lee et al., 2012) and noninfectious (Abulafia et al., 2009) neurodegenerative disorders. Components of the inflammasome complex, including NALP1, NALP3, ASC, and caspase-1, are expressed by microglia. Thus, in response to pathogens or cellular stressors, microglia release great amount of proinflammatory cytokines (Dilger & Johnson, 2008; Walsh, Muruve, & Power, 2014; Walsh, Reinke, et al., 2014).

## 5.1 Inflammasomes and AD

Some research has shown that Ab accumulation initiates an immune response by activating brain-resident microglia (Pennisi et al., 2016). The amyloid deposition in the parenchyma of the brain is related to a strong inflammatory response (Akiyama et al., 2000). Recently, in vitro and in vivo studies have shown that Ab activates the NLRP3 inflammasome in microglial cells, describing a novel pathway that could lead to progression of AD (Halle et al., 2008; Heneka et al., 2013). Ab peptide and NFTs release DAMPs, which are identified by the NLR domain on glial cells. NLRP3 activation causes the formation of the inflammasome, which starts a cascade of events that lead to processing and release of pro-IL-1β, IL-18. Also, Ab interaction with neuronal membrane produces K+ ions efflux and activates inflammasome complex which, in turn, induces interleukins secretion. The second pathway to induce inflammasome activation in AD brains implicates extracellular ATP and the purinergic receptor P2X7. Thus, P2X7 receptor activation reduces intracellular K+ levels. Impaired activity of Na+/K+ ATPase decreases ions' gradients by the cell membrane, producing cytotoxicity and causing neuronal cell death, that in turn, activates the NLRP3 inflammasome complex (Rubartelli, 2014). Accordingly, the P2X7 expression is increased in human AD brains (McLarnon, Ryu, Walker, & Choi, 2006). Several studies have shown a direct relationship between inflammasome activation and pathogenesis of AD, suggesting inflammasome as a novel therapeutic target for this disease management. This notion was proved in APP/PS1 mice (transgenic mice with continuous deposition of Ab), where reduced levels of the NLRP3 inflammasome and caspase-1 activity were detected (Heneka et al., 2013). This NLRP3 inflammasome deficiency caused the production of microglia with phenotype M2 (antiinflammatory phenotype), resulting in decreased deposition of Ab (Heneka et al., 2013). Moreover, elevated expression of caspase-1 was observed in the brain of patients with AD, principally in the hippocampal region (Heneka et al., 2013). Also, in an AD murine model, administration of cathepsin B (which are released by microglia after Ab phagocytosis) inhibitors significantly decreased amyloid plaque in the brain and improved memory deficit. Conversely, inhibition of cathepsin B has been demonstrated to produce therapeutic benefit in a mouse model of AD (Hook, Kindy, & Hook, 2008). Considering the role played by cathepsin B in caspase-1 activation, this data show substantial evidence for the involvement of the NLRP3 inflammasome complex in AD.

In summary, the NLRP3 inflammasome complex emerges as an important factor in the pathogenesis and progression of AD making an appealing target for therapeutic intervention.

## 5.2 Multiple Sclerosis

MS is an unpredictable chronic progressive inflammatory demyelinating disorder of the CNS that disrupts the flow of information within the brain, and between the brain and body. Inflammation has been confirmed in patients with MS and the experimental animal model, experimental autoimmune encephalomyelitis (EAE) (Dowson, Simpson, Duffy, & O'Reilly, 2017). Inflammation induced by innate immunity influences the development of T cell-mediated autoimmunity in MS and EAE. Because NLRP3 constitutes a crucial component of the innate immunity system, the inflammasome complex has been studied in this disease. High activation of NLRP3 and high sensitivity to IFN-β treatment was observed in EAE (Inoue et al., 2016). NLRP3 inflammasome modulates control of disease by promoting migration of inflammatory cells to the CNS (Inoue, Williams, Gunn, & Shinohara, 2012). Inflammasome gene expression, NLRP3, caspase-1, and IL-1β were increased in peripheral blood cells from MS patients compared to healthy controls (Peelen et al., 2015), and NLRP3 inflammasome and its related cytokine IL-1β have been associated with the response to interferon beta in patients with relapsing-remitting MS (Malhotra et al., 2015).

## 5.3 Amyotrophic Lateral Sclerosis

ALS is a devastating and rapidly progressing neurodegenerative disease that involves upper- and lower motor neurons, which cause death usually from neurogenic muscle weakness. Recent studies reported upregulation of IL-18 cytokine in the sera of ALS patients, suggesting an implication of the inflammasome in the progress of the ALS (Italiani et al., 2014). Kadhim and coworkers have detected high expression of NLRP3 and caspase-1 in brain tissues from patients with ALS (Kadhim, Deltenre, Martin, & Sébire, 2016). NLRP3, ASC, IL-18, and active caspase-1 high levels were also found in postmortem tissue from ALS patients by Johann and coworkers who also explored the inflammasome complex in the spinal cord of humanized animal model of ALS carrying SOD1 (G93A) human mutation. In this experiment, spinal cord astrocytes were identified as the major cell type expressing NLRP3 components (Johann et al., 2015). This implication of the inflammasome in ALS has offered an interesting therapeutical target because recently, 17β-estradiol, a steroid hormone, has been shown to improve motoneuron survival in male SOD1 (G93A) ALS mice by down-regulating inflammasome activation (Heitzer et al., 2016).

## 5.4 Parkinson's Disease

PD is one of the most common neurodegenerative diseases. It is characterized by the loss of dopaminergic neurons in the substantia nigra and the formation in these cells of inclusions called Lewy bodies (LB) constituted by fibrillar α-synuclein (αSyn). Inflammation pathways have been related to the pathogenesis and/or progression of PD (Codolo et al., 2013); moreover, it is known that aggregated αSyn, released by neurons, activates microglia cells to produce proinflammatory mediators, such as IL-1β. IL-1β is one of the strongest proinflammatory cytokines; it is produced as an inactive mediator, and its maturation and activation require inflammasome activation. Aggregated αSyn activates microglia cells to produce IL-1β (Litteljohn et al., 2010). Recently, fibrillar αSyn has been shown to induce synthesis of IL-1β, through TLR2 interaction and involves NLRP3 inflammasome activation (Codolo et al., 2013). This is a fascinating finding since neuronal degeneration likely releases fibrillar αSyn.

In other study, NLRP3 inflammasomes were found activated in the serum of PD patients by elevation of serum IL-1β levels accompanied by an enhancement of caspase-1 activity and the midbrain of PD model mice. The activation of the NLRP3 inflammasome was through microglial endocytosis and lysosomal cathepsin B release. However, the microRNA-7 (miR-7) transfection demonstrated to inhibit microglial NLRP3 inflammasome activation (Zhou et al., 2016).

On the other hand, activation of the inflammasome with known stimuli such as LPS, the aluminum salt, or nigericin induces the aggregation of αSyn in a neuronal cell model of PD (Wong et al., 2016). Furthermore, truncated αSyn, found in LB of the PD brain, generated a highly aggregation-prone species very toxic to neuron, and the inhibition of caspase-1 improved the survival of a neuronal PD cell model (Wong et al., 2016). This is an impressive result with therapeutic perspective.

## 5.5 Other Neurological Conditions

Rasmussen's encephalitis (RE) is an inflammatory encephalopathy of unknown cause defined by seizures with progressive neurological disabilities. Inflammasome-associated genes (IL-1β, IL-18, NLRP1, NLRP3, and CASP1) have been shown increased in both cortex and white matter samples from RE patients (Ramaswamy et al., 2013).

HIV-1 infection of both the CNS and the peripheral nervous system leads to neurological alterations in more than 30% of the infected individuals (Nightingale et al., 2014). Associated CNS neuropathological features in patients with HIV-1 include activated perivascular macrophages and resident microglia in the brain, with the release of proinflammatory cytokines (e.g., IL-1β and TNF-α) (Rao, Ruiz, & Prasad, 2014), leading to chronic neuroinflammation. Recently, this high levels of IL-1β confirmed an implication of the inflammasome because IL-1β, IL-18, and caspase-1 gene expression were induced in brains of HIV-infected persons, and HIV-1 infection induced pro-IL-1β in human microglia accompanied by intracellular ASC translocation and caspase-1 activation which was inhibited by NLRP3 deficiency (Walsh, Muruve, et al., 2014; Walsh, Reinke, et al., 2014). Furthermore, the viral protein R (Vpr), an accessory protein that is released from HIV-infected cells, has been shown to induce in macrophages a caspase-1 cleavage and IL-1β release in a manner dependent on NLRP3 expression (Mamik et al., 2016).

Another important neurological condition is the prion disease. Prion diseases are neuroinflammatory and neurodegenerative disorders characterized by the accumulation of the abnormal disease-associated prion protein. The accumulation of prion protein leads to the activation of microglia with high production of proinflammatory cytokines. The NLRP3 inflammasome was found to be essential for IL-1β release upon exposure to aggregates of different presentations of prion protein (Nuvolone, Sorce, Schwarz, & Aguzzi, 2015), by an activation of the microglia (Shi et al., 2012). According to this, the role of inflammasome in the pathogenesis of misfolded protein diseases, such as prion diseases, PD, or AD, shows a very interesting topic to evaluate the potential of inflammasome-targeting therapeutic interventions in the management of these diseases.

## 6. INFLAMMASOME AND CANCER

The role of the immune system is to protect against diseases or other potentially damaging foreign bodies. Most of the times, it performs an excellent job of preventing diseases and infections, keeping us healthy. However, in a few cases, problems can occur, which can lead to the occurrence of numerous illnesses and diseases.

Inflammation is a first innate immune response to impaired tissue homeostasis. It is believed that most of the cancers arise from an acute inflammatory process. This is well documented for gastric cancer, hepatic cancer, and colorectal cancer (Zitvogel, Kepp, Galluzzi, & Kroemer, 2012). Usually, tumor development is linked to chronic infection, dietary factors, inhaled pollutants, obesity, tobacco use, or autoimmunity (Elinav et al., 2013). Chronic inflammation is the linked principle that underlines these processes, which is an aberrantly prolonged form of a protective respond to the loss of tissue homeostasis. Therefore, cancer development can be regard as a dysregulated form of protective tissue repair and growth respond (Elinav et al., 2013).

As a result, inflammation is recognized as a major hallmark of cancer, and it was added to the list within the 10 last years. Tumor microenvironment due to chronic inflammation has been linked to tumor initiation, progression, angiogenesis, metastasis, tumor cell proliferation and survival, apoptosis evasion, malignant transformation, and alterations in the antitumor adaptive immune response (Janowski, Kolb, Zhang, & Sutterwala, 2013; Wong et al., 2016). It is highly accepted that up to 25% of malignancies are related to chronic inflammation and chronic infection (Wong et al., 2016).

On top of that, carcinogenesis is the result of the interplay of multiple cell-intrinsic and cell-extrinsic processes, which include genomic instability, proliferative abnormality, reprogramming of the stromal environment, and aberrant differentiation between epithelial and mesenchymal states (Elinav et al., 2013). The immune system has been described to have different roles depending on which stage of cancer progression is under study (Zhiyu et al., 2016). As their names imply, inflammasomes have a critical role in cancer by influencing the formation, progression, and therapeutics response through their contribution to tissue homeostasis, inflammation, and immune responses (Zitvogel et al., 2012). It is believed that inflammasomes have the dual role in cancer: (i) developing cancer growth or (ii) eliminating malignant cells via the pyroptosis pathway (Zhiyu et al., 2016). That phenomenon might be attributed to NLRs difference, cell and tissue type, cancer stage, and experimental condition (Janowski et al., 2013).

The immune interaction with cancer was set by Schreiber and colleagues and was established in three stages of immunoediting. At first, due to an oncogenic stimulus such as chronic inflammation, the first transformed cells emerge, so the NLRs try to repair the damaged tissue. However, over- or underexpression of NLRs and their downstream signaling molecules can lead to increase the incidence of cancer. Then, NLRs are thought to contribute to immunosurveillance and destruction of these newly transformed cells. Finally, once a cancer cell escapes suppressive mechanisms, NLRs support the tumor by facilitating neovascularization, metastasis, and promoting immunosuppressive functions. This model is called "The three E's model": Elimination–Equilibrium–Escape (Kent & Blander, 2014).

A link between tumorigenesis and inflammation has been shown in many cancers such as gastric cancer which is associated with *Helicobacter pylori* and chronic gastritis, and rectal colon cancer, frequently related to inflammatory bowel disease (IBD) (Kolb, Liu, Janowski, Sutterwala, & Weizhou Zhang, 2014). Consequently, cancers of the gastrointestinal tract are often related to chronic inflammation. At the point of NLRP3 activation, the release of IL-1β and IL-18 occurs. These cytokines are proinflammatory and have protumorigenic effects on gastrointestinal cancers. In colitis-associated colon cancer (CAC) several studies have said that IL-1β may promote tumor growth and invasion by inducing an epithelial to mesenchymal transition and stem cell phenotype, but IL-18 seems to play a tumor suppressive role in CAC. It can be seen in several studies in which mice deficient in IL-18 increased the inflammation and tumorigenesis in an azoxymethane/dextran sulfate model, which mimics the pathogenesis of IBD driven colon tumorigenesis in human. Besides, it has been suggested that IL-18 take part in the maintenance of the epithelial barrier integrity after an injury in the colon epithelium (Kolb et al., 2014). IL-18 has supposed to have a role in IFN-mediated activation of STAT1, which is known to have a role in tumor suppression. So, the expression of NLRP3 is critical for the tumor suppression activity (Kolb et al., 2014).

It is important the fact that if there are single-nucleotide polymorphisms within NLRP3 region may play different roles in cancer. There are some examples of polymorphism which modulated the susceptibility of developing cancer. Genetic changes such as polymorphisms of IL-1β and IL-1βR which influence the possibility to develop gastric cancers or polymorphisms encoding in genes of IL-1β and IL-1R determine the susceptibility of suffering from Crohn's disease (Kolb et al., 2014; Menu & Vince, 2011).

It is interesting a discovery that suggest a different role for NLRP3 inflammasome activation in metastatic tumor cells and how its activation is altered depending on the progression of nonmetastatic cells to ones with metastatic capacity. Metastatic cells of melanoma cells show a high release of IL-1β. However, nonmetastatic ones only secrete a little amount of this cytokine (Janowski et al., 2013).

The third most diagnosed cancer globally is colorectal cancer, and its associated deaths occur due to its dissemination to secondary organs, primarily the liver. It has been studied that NLRP3 inflammasome components KO mice have developed colorectal cancer due to the lack of IL-18. It is required for an active inflammasome-mediated suppression of CRC growth in the liver. IL-1β, another component downstream the activation of the inflammasome, commonly exerts tumorigenic effects, conferring growth advantage and invasion. However, it is necessary after chemotherapy for anticancer immunosurveillance. Like IL-1β, IL-18 has dual functions in tumorigenesis. It is known that IL-18 is associated with a T helper 1 by immunostimulatory antitumorigenic response through its ability to induce IFN-γ and its effects on enhancing the cytolytic activity of T cells and NK cells. Control of metastatic liver growth occurred independently of T, B, and NKT cells but required NK cells. Its cells respond needs the continuous paracrine signaling by IL-18 for cell maturation and activation. However, IL-18 enhances hepatic melanoma metastasis. Also, a recent study reported an inhibitory effect of IL-18 on NK cell tumoricidal function through the induction of PD-1, an immunostimulatory costimulatory molecule, on their surface. As a result, the duties of these cytokines may be context dependent and tissue specific. In CRC it has been demonstrated that the signal is mediating the communication between innate immunity, such as Kupffer cells, and NK cell tumoricidal activity. It has been proposed a model in which the NLRP3 inflammasome in Kupffer cells cooperates with NK cells in sensing carcinogenesis and exerts their tumoricidal activity through FasL-induced apoptosis, which required FasL-sensitive tumor cells. FasL is controlled by IL-18, and it is expressed on the surface of NK cells playing a crucial role in cellular homeostasis and immune surveillance by inducing cell death through its receptor Fas. Thus, NK cells have been shown to correlate with patient survival in many cancers positively. However, in a model of lung metastasis, IL-1β production by cancer cells induces the development of NK cells which promoted metastasis (Dagenais & Saleh, 2016; Dupaul-Chicoine et al., 2015).

In contrast to the protective role of NLRP3 in CAC, it was shown to promote tumor formation in a chemical-induced fibrosarcoma model. It has seen that NLRP3 mice treated with methylcholanthrene exhibited prolonged tumor-free survival compared to wild-type mice. This decrease in tumor formation was due to an increased in natural killer; it demonstrates the role for NLRP3 in the suppression of NK cell activation and promotion of a suppressive tumor microenvironment (Janowski et al., 2013).

In gastrointestinal cancer, IL-1β and IL-18 are the primary products of the inflammasome activation and have tumorigenic effects. GC is considered a multifactorial process, in which bacterial, environmental, and host genetics factors are involved. It is associated with chronic inflammation of the gastric mucosa of consequence, commonly, of the existence of bacteria such as *H. pylori*, which actives the inflammasome. As it said before, several polymorphisms are involved in the development of this cancer. IL-1β may promote carcinogenesis, and IL-18 reduces tumorigenesis through the activation of NK cells. However, an inappropriate production of this cytokine may end in the development of cancer. Overexpression of IL-1β was found to induce gastric inflammation and cancer in mice (Castaño-Rodríguez, Kaakoush, Goh, Fock, & Mitchell, 2014; Xu et al., 2013).

Hepatocellular carcinoma is commonly developed in the context of chronic hepatitis which is characterized by hepatic inflammation. Wei and coworkers showed that inflammasome expression has low levels in normal hepatocytes, high levels in inflammatory hepatic setting and is downregulated in malignant liver cancer cells. Thus, this downregulation of the NLRP3 inflammasome indicated that NLRP3 has a protective role. However, these data are in disagreement with several studies because they have demonstrated that high level of IL-1β secretion in the tumor microenvironment is correlated with the cancer progression. That is because of the pleiotropic and different role of the NLRP3 inflammasome in multiple facets of oncogenesis (Wei et al., 2014).

In malignant mesothelioma (MM) it has shown a decrease in the expression of the NLRP3 inflammasome and component, such as caspase-1. Many studies have tried to enhance the NLRP3 inflammasome activity, but it has antitumor effects, through the activation of caspase-1 which lead to pyroptosis and cell death, and tumor promoting effects through enhancing cytokines production, the most important is IL-1β. Therefore, combining strategies with inhibitors of that cytokines may play a beneficial role in MM.

Malignant gliomas are the most common and deadly primary brain tumors in adults. It has demonstrated that NLRP3 inflammasome is an important molecular link between brain aging and glioma progression. NLRP3-KO mice have shown to reduce age-related inflammation in the hippocampus, protecting mice from age-related decline in cognition and memory and extending life health span. Age is strongly associated with increased glioma incidence, malignancy, and therapy resistance. It is thought that NLRP3 inflammasome may serve as a possible molecular link between aging and glioma tumorigenesis and progression. Therefore, inhibition of NLRP3 inflammasome can provide a therapeutics strategy for radiation-resistant glioma (Li & Liu, 2014).

Oral cavity squamous cell carcinoma is another cancer type, where NLRP3 inflammasome and its related protein are involved. It has been demonstrated that they are overexpressed in OSCC tumor cells. ACS is the leading promoter of tumor metastasis, so it has an oncogenic activity. ASC induced cell migration and invasion as well as induced metastasis in mice. As a result, ASC can be used as a biomarker of metastasis in OSCC (Wu et al., 2016).

One of the best clinically characterized examples of the association between inflammation and carcinogenesis is the development of colon cancer in patients suffering from ulcerative colitis, which is a common form of IBD. It has been shown a protective role for PYCARD, caspase-1, and NLRP3 against CAC development. They might function as inflammation tumor suppressors in the inflammation-driven colon tumorigenesis model. Also, IL-1β levels are significantly altered in patients who are suffering from acute or chronic gastrointestinal inflammation and have been implicated in angiogenesis, progression, and tumor metastasis. We can see that findings in several studies where it has been demonstrated that NLRP3 deficiency enhances gastrointestinal tumorigenesis and, said in other words, NLRP3 attenuates the development of gastrointestinal disease. These findings propose a protective role for the inflammasome during chronic colitis and CAC (Allen et al., 2010).

In nasopharyngeal carcinoma, it has shown that the NLRP3 inflammasomes are overexpressed, and expression levels correlate with patient survival. Overexpression of NLRP3 inflammasome and IL-1β in tumor cells likely contributes to local tumor control by neutrophils recruitment. It has been demonstrated a link between neutrophils and tumor suppression through inflammasome activation. The recruitments of the neutrophils significantly inhibit tumor growth in mice. As well as its presence, IL-1β production and the expression of inflammasome were associated with better survival in NOC patients. Furthermore, tumor growth and recurrence are inhibited by IL-1β, suggesting that the activation of the inflammasome is critical for tumor suppression because it produces the active form of IL-1β. Also, tumor-derived IL-1β potentially induces dendritic cell-dependent adaptive immunity against tumors. It is thought that TANs are recruited by tumor-produced cytokines and have cytotoxic effect against the tumor (Chen et al., 2012).

These all studies demonstrate opposing roles of the inflammasome in different types of cancer. The activation of the inflammasome can control tumor progression, or on the contrary, can contribute to chronic inflammation which may lead to carcinogenesis. NLRs can be regard as a fine balance between inflammation and repair to maintain the homeostasis (Kent & Blander, 2014). As a result, inflammasomes may provide a promising target for cancer therapy and prevention. However, due to their different roles depending on the type of tissue, cell, cancer stage, type of inducer, etc., further studies are required (Kolb et al., 2014).

## 7. NLRP3 AND CVD

CVDs are the number one cause of death globally. An estimated 17.3 million people died from CVDs in 2008, representing 30% of all global deaths. There is a prevalence of 80% of CVD deaths take place in low- and middle-income countries and occur almost equally in men and women. It is thought that by the year 2030, approximately 23.6 million people will die from CVDs, mainly from heart disease and stroke. In this sense, it is highly important to discover new biomarkers to improve the CVD risk prediction.

Bullon et al. performed a study to evaluate the NLRP3 inflammasome complex as a potential Cardiovascular Risk indicator in healthy males and post-AMI patients comparing both groups by conventional cardiovascular risk factors. The results showed that NLRP3 inflammasome and associated IL-1β release have potential as CVR biomarkers, most in post-AMI patients with otherwise low-risk scores. As well as target inflammasome is a potential therapeutic for the prevention of CVD. The pathophysiology of coronary disease has been linked to lipid peroxidation and other indicators of oxidative stress, autophagy impairment, and inflammatory markers such as IL-1β. It has shown higher levels of NLRP3 gene expression and lower level of the autophagy-related genes MAP-LC3 than the control group, thus suggesting an implication of both genes in the pathophysiology of coronary artery disease. Finally, inflammation has been proposed as a cardiovascular risk factor, with the IL-1β strongly implicated in CVD, which was seen in acute myocardial infarction group (Bullon et al., 2016).

Inflammation is a process where our body responds to an injury or infection, and it is triggered by components of the immune system. That tissue injury induces a local and systemic inflammatory response without microbial pathogen; it is referred to as "sterile inflammation." Inflammasomes are the key players in diseases such as gout, pseudogout, type 2 diabetes mellitus, metabolic syndrome, arteriosclerosis, asbestosis, silicosis, and AD. While these processes are highly necessary for the recruitments of leucocytes and healing, it may also induce additional injury, representing a mechanism of illness progression (Abbate, 2013; Takahashi, 2014).

This applies to the heart, where the inflammasomes occupy a central role in the development of cardiomyopathy, and it is involved in the pathogenesis of arteriosclerosis. Many stimuli can activate NLRP3, one of them are the intracellular cholesterol crystals which produce endothelial dysfunctions and may initiate or exacerbate vascular injury during hypercholesterolemia (Abbate, 2013; Ajdukovic, 2015). When NLRP3 is active, many cytokines are released, the two largest are IL-1β and IL-18. Both of them act as mediators that promote the cascade release of other cytokines. It has been discovered that NLRP3 and downstream cytokines are correlated with the severity of coronary artery disease (Ajdukovic, 2015).

Heart failure (HF) is another illness where NLRP3 inflammasome is implicated. It occurs when the heart muscle is much weakened and cannot pump enough blood and oxygen to satisfy our body needs. The importance of inflammation in HF can be understood by the increasing of inflammatory cytokines such as tumor necrosis factor alpha, IL-1β and IL-6, and c-reactive protein, whose levels are related to HF severity and prognosis. These cytokines may modulate myocardial remodeling, myocyte hypertrophy and apoptosis, decreased contractility, increased fibrosis, and other adverse structural changes (Butts, Gary, Dunbar, & Butler, 2015). IL-1β induces calcium leakage from sarcoplasmic reticulum in myocytes, impairing cardiac contractility. Furthermore, there is an increase in circulating inducible oxide nitric synthase, leading to cardiomyocyte apoptosis and tissue remodeling. On the other hand, IL-18 is increased during acute HF and remains elevated after discharge. IL-18 is found to be precursor in the nonfailing myocardium, however, in the failing heart is in their active form, inducing the production of TNF-α. It is a proinflammatory cytokine that is not expressed in healthy myocardium and induced cardiomyocyte hypertrophy and apoptosis (Ajdukovic, 2015).

Myocardial infarction (MI) is a very common disease condition which is associated with sterile inflammation. This inflammatory process is necessary for tissue healing, however, may also cause excessive damage and maladaptive ventricular remodeling leading to impaired myocardial function and HF (Mezzaroma, Toldo, & Abbate, 2013; Sandanger et al., 2013). MI happens when the flow of oxygen-rich blood to a section of heart muscle suddenly becomes blocked. If blood flow is not restored quickly, the section of heart muscle begins to die. The most common cause of MI is arteriosclerosis of the coronary arteries. It is a multifactorial, chronic inflammatory disease where IL-1β and IL-18 are overexpressed. When arteriosclerotic plaque become unstable, then they start to rupture and, then, facilitate thrombus formation, resulting in the occlusion of the coronary arteries, leading to MI. Many revascularization procedures can decrease the infarct size in the myocardium, improving clinical outcomes. However, paradoxically the reperfusion leads to exacerbated and accelerated the injury; it is called myocardial ischemia–reperfusion injury. MI is accompanied by inflammatory responses leading to leukocyte accumulation, and subsequent myocardial damage, healing, and scar formation (Takahashi, 2014).

Several studies have demonstrated that following an ischemic injury to the heart during acute MI, a significant inflammatory response occurs and leads to a cardiomyopathy. During MI cell debris is released and starts to activate the inflammasome, resulting in the production of the IL-1β and IL-18 (Abbate, 2013).

It is fully agreed that NLRP3 is central in MI because target deletion of NLRP3 using genetic KO models or with silencing RNA reduces apoptosis, infarct size, and cardiac dysfunctions during MI (Sandanger et al., 2013). In other words, targeting for inflammatory responses improves MI. The most important inflammatory mediator is IL-1, so using a neutralizing antibody against IL-1β and Anakinra, which is a recombinant IL-1β receptor antagonist exerts beneficial effects on acute MI. Furthermore, using ASC-KO, NLRP3-KO, or caspase-1 KO mice exhibited a significant decline of inflammatory responses, reducing the infarct size, myocardial fibrosis, and left ventricular dysfunctions after MI (Abbate, 2013; Takahashi, 2014).

The heart is mainly composed of two cell types: cardiac fibroblast and cardiomyocytes. Cardiac fibroblasts are the most abundant cell type in the human adult heart and serve as a sentinel cell that sense injury and activate inflammatory and matrix regulating responses following MI, being involved in myocardial ischemia–reperfusion injury (I/R) as well as in infarct repair and scar formation (Sandanger et al., 2013). So, this novel role of cardiac fibroblast means that they are implicated in cardiac disease development due to the production of autocrine/paracrine factors, such as cytokines, interacting with other cells such as cardiomyocytes, vascular cells, and inflammatory cells in a paracrine manner.

Myocardial I/R and MI activate the NLRP3 inflammasome through ROS generation and potassium efflux; this induces IL-1β release. In cardiac fibroblast this IL-1β recruits active inflammatory cells such as monocytes/macrophages and neutrophils to the ischemic myocardium. When these leucocytes are active, they provide the largest source of proinflammatory mediators. On the other hand, cardiomyocytes NLRP3 activation lead to a caspase-1-dependent cell death known as pyroptosis, but not IL-1β release. To sum up, cardiomyocytes and cardiac fibroblast have different roles, which ultimately result in cardiac inflammation and remodeling in MI (Fig. 3) (Sandanger et al., 2013; Takahashi, 2014).

The work with NLRP3-Knockout and ASC-Knockout mice has contributed to show different roles of both gene regarding myocardial damage after I/R. While NLRP3-KO mice show an improvement on MI and I/R, ASC-KO mice show it mildly. These findings indicate that these inflammasome components can function independently of the complex and that NLRP3 and ASC may have intrinsic cell roles in different cell types such as cardiac fibroblast, cardiomyocytes, and leucocytes. Furthermore, while the function of NLRP3 is restricted to one inflammasome, ASC acts as an adaptor protein in at least five inflammasomes. However, the effects of blocking all these activators are still unclear (Sandanger et al., 2013; Takahashi, 2014).

Lots of evidence indicates that the inflammatory response driven by the NLRP3 inflammasome contribute to the pathophysiology of MI. So, targeting the inflammasome or IL-1β may be a potential and effective therapeutic strategy

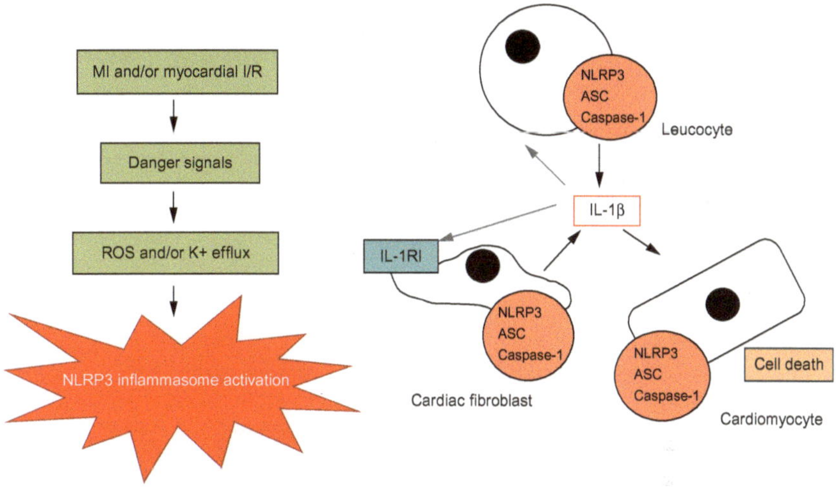


**Fig. 3** NLRP3 inflammasome activation and IL-1β release after myocardial damage.

to treat MI. Nowadays, lots of IL-1β blockades are available, such as Anakinra, Rilonacept, Canakinumab, and Gevokizumab. In addition, there are some NLRP3 inhibitor such as Apigenin which specifically targets the oligomerization of ASC; Parthenolide inhibits the activation of caspase-1 and the ATPase activity of NLRP3; Scropolioside B decreases the expression of NLRP3; Resveratrol, which is an activator of Sirt-1, an enzyme that contributes to cellular regulation, inhibits NF-κB, which suppressed NLRP3 transcription and subsequent IL-1β production (Ajdukovic, 2015). MCC950 is another compound which leads to the inhibition of NLRP3 by reducing infarct size and preserve cardiac functions (Bullon et al., 2016).

## 8. FUTURE PERSPECTIVES

As we have commented, we can understand the stress as a condition or process which can induce a detriment of the health by two form: (i) psychological stress which can cause pathological conditions in metabolism with systemic symptoms and signs and where immunological changes are observed with microglial activation and inflammatory pathways involved, and (ii) metabolic stress in which usually we can to find increased levels of inflammation and psychological symptoms. In this context, the implication of the inflammasome complex has been demonstrated by different experimental and clinical approximation; in vitro assays, animal models and patients, and NLRP3 could be a frequent target in both types to stress

(Alcocer-Gómez & Cordero, 2017). So, the therapeutic implication is a necessary research line which is being studied. Different compounds have been shown to inhibit NLRP3 with the protective effect in cardiovascular, Alzheimer's, pain, and inflammasome diseases. In the coming years, therefore, we could see similar pharmacological treatments for psychological and metabolic stress, as for example, depression and CVD. According to this, it could be interesting to evaluate the potential inhibitory effect of known drugs. Antidepressants and metformin are a very clear example of this.

## REFERENCES

Abbate, A. (2013). The heart on fire: Inflammasome and cardiomyopathy. *Experimental Physiology*, *98*(2), 385.

Abulafia, D. P., de Rivero Vaccari, J. P., Lozano, J. D., Lotocki, G., Keane, R. W., & Dietrich, W. D. (2009). Inhibition of the inflammasome complex reduces the inflammatory response after thromboembolic stroke in mice. *Journal of Cerebral Blood Flow and Metabolism*, *29*, 534–544.

Ajdukovic, J. (2015). The role of NLRP3 inflammasome in cardiovascular diseases. *Journal of Clinical & Experimental Cardiology*, *6*, 399.

Akiyama, H., Arai, T., Kondo, H., Tanno, E., Haga, C., & Ikeda, K. (2000). Cell mediators of inflammation in the Alzheimer disease brain. *Alzheimer Disease and Associated Disorders*, *14*, 47–53.

Alabdali, A., Al-Ayadhi, L., & El-Ansary, A. (2014). Association of social and cognitive impairment and biomarkers in autism spectrum disorders. *Journal of Neuroinflammation*, *8*(11), 4.

Alcocer-Gómez, E., & Cordero, M. D. (2014). NLRP3 inflammasome: A new target in major depressive disorder. *CNS Neuroscience & Therapeutics*, *20*(3), 294–295.

Alcocer-Gómez, E., & Cordero, M. D. (2017). NLRP3 inflammasome: Common nexus between depression and cardiovascular diseases. *Nature Reviews. Cardiology*, *14*(2), 124.

Alcocer-Gómez, E., de Miguel, M., Casas-Barquero, N., Núñez-Vasco, J., Sánchez-Alcazar, J. A., Fernández-Rodríguez, A., et al. (2013). NLRP3 inflammasome is activated in mononuclear blood cells from patients with major depressive disorder. *Brain, Behavior, and Immunity*, *36*, 111–117.

Alcocer-Gómez, E., Ulecia-Morón, C., Marín-Aguilar, F., Rybkina, T., Casas-Barquero, N., Ruiz-Cabello, J., et al. (2016). Stress-induced depressive behaviors require a functional NLRP3 inflammasome. *Molecular Neurobiology*, *53*(7), 4874–4882.

Allen, I. C., TeKippe, E. M., Woodford, R. M., Uronis, J. M., Holl, E. K., Rogers, A. B., et al. (2010). The NLRP3 inflammasome functions as a negative regulator of tumorigenesis during colitis-associated cancer. *The Journal of Experimental Medicine*, *207*(5), 1045–1056.

Ashwood, P., & Wakefield, A. J. (2006). Immune activation of peripheral blood and mucosal CD3+ lymphocyte cytokine profiles in children with autism and gastrointestinal symptoms. *Journal of Neuroimmunology*, *173*(1–2), 126–134.

Baum, A., & Posluszny, D. M. (1999). Health psychology: Mapping biobehavioral contributions to health and illness. *Annual Review of Psychology*, *50*, 137–163.

Bianchi, M. E. (2007). DAMPs, PAMPs and alarmins: All we need to know about danger. *Journal of Leukocyte Biology*, *81*, 1–5.

Biesmans, S., Meert, T. F., Bouwknecht, J. A., Acton, P. D., Davoodi, N., DeHaes, P., et al. (2013). Systemic immune activation leads to neuroinflammation and sickness behavior in mice. *Mediators of Inflammation*, *2013*, 271359.

Black, P. H., & Garbutt, L. D. (2002). Stress, inflammation and cardiovascular disease. *Journal of Psychosomatic Research*, *52*(1), 1–23.

Boyden, E. D., & Dietrich, W. F. (2006). Nalp1b controls mouse macrophage susceptibility to anthrax lethal toxin. *Nature Genetics*, *38*, 240–244.

Brookmeyer, R., Johnson, E., Ziegler-Graham, K., & Arrighi, H. M. (2007). Forecasting the global burden of Alzheimer's disease. *Alzheimer's & Dementia*, *3*, 186–191.

Bullon, P., Cano-García, F. J., Alcocer-Gómez, E., Varela-Lopez, A., Román-Malo, L., Ruiz-Salmerón, R. J., et al. (2016). Could NLRP3-inflammasome be a cardiovascular risk biomarkers in acute myocardial infarction patients? *Antioxidants & Redox Signaling*. http://dx.doi.org/10.1089/ars.2016.6970.

Burchfield, S. R. (1979). The stress response: A new perspective. *Psychosomatic Medicine*, *41*, 661–672.

Burls, A., & Jobanputra, P. (2004). The trials of anakinra. *Lancet*, *364*(9437), 827–828.

Butts, B., Gary, R. A., Dunbar, S. B., & Butler, J. (2015). The importance of NLRP3 inflammasome in heart failure. *Journal of Cardiac Failure*, *21*(7), 586–593.

Castaño-Rodríguez, N., Kaakoush, N. O., Goh, K. L., Fock, K. M., & Mitchell, H. M. (2014). The NOD-like receptor signalling pathway in Helicobacter pylori infection and related gastric cancer: A case-control study and gene expression analyses. *PloS One*, *9*(6), e98899.

Chang, J., Yang, L., Kouadir, M., Peng, Y., Zhang, S., Shi, F., et al. (2012). Antibody-mediated inhibition of integrin alpha5beta1 blocks neurotoxic prion peptide PrP106-126-induced activation of BV2 microglia. *Journal of Molecular Neuroscience*, *48*, 248–252.

Chen, L. C., Wang, L. J., Tsang, N. M., Ojcius, D. M., Chen, C. C., Ouyang, C. N., et al. (2012). Tumour inflammasome-derived IL-1β recruits neutrophils and improves local recurrence-free survival in EBV-induced nasopharyngeal carcinoma. *EMBO Molecular Medicine*, *4*(12), 1276–1293.

Codolo, G., Plotegher, N., Pozzobon, T., Brucale, M., Tessari, I., Bubacco, L., et al. (2013). Triggering of inflammasome by aggregated α-synuclein, an inflammatory response in synucleinopathies. *PloS One*, *8*(1), e55375.

Dagenais, M., & Saleh, M. (2016). Linking cancer-induced Nlrp3 inflammasome activation to efficient NK cell-mediated immunosurveillance. *Oncoimmunology*, *5*(5), e1129484.

Dhabhar, F. S. (2014). Effects of stress on immune function: The good, the bad, and the beautiful. *Immunologic Research*, *58*, 193–210.

Dhabhar, F. S., & McEwen, B. S. (1997). Acute stress enhances while chronic stress suppresses immune function in vivo: A potential role for leukocyte trafficking. *Brain, Behavior, and Immunity*, *11*, 286–306.

Dilger, R. N., & Johnson, R. W. (2008). Aging, microglial cell priming, and the discordant central inflammatory response to signals from the peripheral immune system. *Journal of Leukocyte Biology*, *84*, 932–939.

Dowlati, Y., Herrmann, N., Swardfager, W., Liu, H., Sham, L., Reim, E. K., et al. (2010). A meta-analysis of cytokines in major depression. *Biological Psychiatry*, *67*, 446–457.

Dowson, C., Simpson, N., Duffy, L., & O'Reilly, S. (2017). Innate immunity in systemic sclerosis. *Current Rheumatology Reports*, *19*(1), 2.

Dupaul-Chicoine, J., Arabzadeh, A., Dagenais, M., Douglas, T., Champagne, C., Morizot, A., et al. (2015). The Nlrp3 inflammasome suppresses colorectal cancer metastatic growth in the liver by promoting natural killer cell tumoricidal activity. *Immunity*, *43*(4), 751–763.

Elenkov, I. J., & Chrousos, G. P. (2002). Stress hormones, proinflammatory and antiinflammatory cytokines, and autoimmunity. *Annals of the New York Academy of Sciences*, *966*, 290–303.

Elinav, E., Nowarski, R., Thaiss, C. A., Hu, B., Jin, C., & Flavell, R. A. (2013). Inflammation-induced cancer: Crosstalk between tumours, immune cells and microorganisms. *Nature Reviews. Cancer*, *13*, 759–771.

Ellul, P., Boyer, L., Groc, L., Leboyer, M., & Fond, G. (2016). Interleukin-1 β-targeted treatment strategies in inflammatory depression: Toward personalized care. *Acta Psychiatrica Scandinavica*, *134*(6), 469–484.

Farooq, R. K., Isingrini, E., Tanti, A., LeGuisquet, A. M., Arlicot, N., Minier, F., et al. (2012). Is unpredictable chronic mild stress (UCMS) a reliable model to study depression-induced neuroinflammation? *Behavioural Brain Research*, *231*, 130–137.

Fernandes-Alnemri, T., Yu, J. W., Datta, P., Wu, J., & Alnemri, E. S. (2009). AIM2 activates the inflammasome and cell death in response to cytoplasmic DNA. *Nature*, *458*, 509–513.

Fountoulakis, K. N., Gonda, X., Siamouli, M., & Rihmer, Z. (2009). Psychotherapeutic intervention and suicide risk reduction in bipolar disorder: A review of the evidence. *Journal of Affective Disorders*, *113*(1–2), 21–29.

Franchi, L., Munoz-Planillo, R., & Nunez, G. (2012). Sensing and reacting to microbes through the inflammasomes. *Nature Immunology*, *13*, 325–332.

Frank Cannon, T. C., Alto, L. T., McAlpine, F. E., & Tansey, M. G. (2009). Does neuroinflammation fan the flame in neurodegenerative diseases. *Molecular Neurodegeneration*, *4*, 1–13.

Frohman, E. M., Racke, M. K., & Raine, C. S. (2006). Multiple sclerosis—The plaque and its pathogenesis. *The New England Journal of Medicine*, *354*, 942–955.

Furr, S. R., Chauhan, V. S., Sterka, D., Jr., Grdzelishvili, V., & Marriott, I. (2008). Characterization of retinoic acid-inducible gene-I expression in primary murine glia following exposure to vesicular stomatitis virus. *Journal of Neurovirology*, *14*, 503–513.

Garrido-Mesa, N., Zarzuelo, A., & Galvez, J. (2013). What is behind the non-antibiotic properties of minocycline? *Pharmacological Research*, *67*, 18–30.

Giri, M., Zhang, M., & Leu, Y. (2016). Genes associated with Alzheimer's disease: An overview and current status. *Clinical Interventions in Aging*, *11*, 665–681.

Glass, C. K., Saijo, K., Winner, B., Marchetto, M. C., & Gage, F. H. (2010). Mechanisms underlying inflammation in neurodegeneration. *Cell*, *140*(6), 918–934.

Godbout, J., Chen, J., Abraham, J., Richwine, A., Berg, B., Kelley, K., et al. (2005). Exaggerated neuroinflammation and sickness behavior in aged mice following activation of the peripheral innate immune system. *FASEB Journal*, *19*, 1329–1331.

Goldstein, D. S., & McEwen, B. (2002). Allostasis, homeostats, and the nature of stress. *Stress*, *5*, 55–58.

Guo, H., Callaway, J. B., & Ting, P. Y. (2015). Inflammasomes: Mechanism of action, role in disease, and therapeutics. *Nature Medicine*, *21*(7), 677–687.

Hafner-Bratkovic, I., Bencina, M., Fitzgerald, K. A., Golenbock, D., & Jerala, R. (2012). NLRP3 inflammasome activation in macrophage cell lines by prion protein fibrils as the source of IL-1beta and neuronal toxicity. *Cellular and Molecular Life Sciences*, *69*, 4215–4228.

Halle, A., Hornung, V., Petzold, G. C., Stewart, C. R., Monks, B. G., Reinheckel, T., et al. (2008). The NALP3 inflammasome is involved in the innate immune response to amyloid-beta. *Nature Immunology*, *9*, 857–865.

Hein, A. M., & O'Banion, M. K. (2009). Neuroinflammation and memory: The role of prostaglandins. *Molecular Neurobiology*, *40*, 15–32.

Heitzer, M., Kaiser, S., Kanagaratnam, M., Zendedel, A., Hartmann, P., Beyer, C., et al. (2016). Administration of 17β-estradiol improves motoneuron survival and down-regulates inflammasome activation in male SOD1(G93A) ALS mice. *Molecular Neurobiology*. http://dx.doi.org/10.1007/s12035-016-0322-4. 12.

Heneka, M. T., Kummer, M. P., & Latz, E. (2014). Innate immune activation in neurodegenerative disease. *Nature Reviews. Immunology*, *14*(7), 463–477.

Heneka, M. T., Kummer, M. P., Stutz, A., Delekate, A., Schwartz, S., Vieira-Saecker, A., et al. (2013). NLRP3 is activated in Alzheimer's disease and contributes to pathology in APP/PS1 mice. *Nature*, *493*, 674–678.

Hirsch, E. C., Vyas, S., & Hunot, S. (2012). Neuroinflammation in Parkinson's disease. *Parkinsonism & Related Disorders*, *18*, S210–S212.

Hook, V. Y., Kindy, M., & Hook, G. (2008). Inhibitors of cathepsin B improve memory and reduce beta-amyloid in transgenic Alzheimer disease mice expressing the wild-type, but not the Swedish mutant, beta-secretase site of the amyloid precursor protein. *The Journal of Biological Chemistry*, *283*(12), 7745–7753.

Hu, X., Liu, G., Hou, Y., Shi, J., Zhu, L., Jin, D., et al. (2012). Induction of M2-like macrophages in recipient NOD-scid mice by allogeneic donor CD4(1)CD25(1) regulatory T cells. *Cellular & Molecular Immunology*, *9*, 464–472.

Inoue, M., Chen, P. H., Siecinski, S., Li, Q. J., Liu, C., Steinman, L., et al. (2016). An interferon-β-resistant and NLRP3 inflammasome-independent subtype of EAE with neuronal damage. *Nature Neuroscience*, *19*(12), 1599–1609.

Inoue, M., Williams, K. L., Gunn, M. D., & Shinohara, M. L. (2012). NLRP3 inflammasome induces chemotactic immune cell migration to the CNS in experimental autoimmune encephalomyelitis. *PNAS*, *109*(26), 10480–10485.

Issafras, H., Corbin, J. A., Goldfine, I. D., & Roell, M. K. (2014). Detailed mechanistic analysis of gevokizumab, an allosteric anti-IL-1b antibody with differential receptor-modulating properties. *The Journal of Pharmacology and Experimental Therapeutics*, *348*, 202–215.

Italiani, P., Carlesi, C., Giungato, P., Puxeddu, I., Borroni, B., Bossù, P., et al. (2014). Evaluating the levels of interleukin-1 family cytokines in sporadic amyotrophic lateral sclerosis. *Journal of Neuroinflammation*, *23*(11), 94.

Iwata, M., Ota, K. T., & Duman, R. S. (2013). The inflammasome: Pathways linking psychological stress, depression, and systemic illnesses. *Brain, Behavior, and Immunity*, *31*, 105–114.

Iwata, M., Ota, K. T., Li, X. Y., Sakaue, F., Li, N., Dutheil, S., et al. (2016). Psychological stress activates the inflammasome via release of adenosine triphosphate and stimulation of the purinergic type 2X7 receptor. *Biological Psychiatry*, *80*, 12–22.

Jamilloux, Y., & Henry, T. (2013). The inflammasomes: Platforms of innate immunity. *Medical Science*, *29*, 975–984.

Janczura, M., Bochenek, G., Nowobilski, R., Dropinski, J., Kotula-Horowitz, K., Laskowicz, B., et al. (2015). The relationship of metabolic syndrome with stress, coronary heart disease and pulmonary function—An occupational cohort-based study. *PLoS One*, *10*(8), e0133750.

Janowski, A. M., Kolb, R., Zhang, W., & Sutterwala, F. S. (2013). Beneficial and detrimental roles of NLRs in carcinogenesis. *Frontiers in Immunology*, *4*, 370.

Johann, S., Heitzer, M., Kanagaratnam, M., Goswami, A., Rizo, T., Weis, J., et al. (2015). NLRP3 inflammasome is expressed by astrocytes in the SOD1 mouse model of ALS and in human sporadic ALS patients. *Glia*, *63*(12), 2260–2273.

Kadhim, H., Deltenre, P., Martin, J. J., & Sébire, G. (2016). In-situ expression of interleukin-18 and associated mediators in the human brain of sALS patients: Hypothesis for a role for immune-inflammatory mechanisms. *Medical Hypotheses*, *86*, 14–17.

Kent, A., & Blander, J. M. (2014). Nod-like receptors: Key molecular switches in the conundrum of cancer. *Frontiers in Immunology*, *5*, 185.

Kessler, R. C. (2012). The costs of depression. *The Psychiatric Clinics of North America*, *35*, 1–14.

Kigerl, K. A., de Rivero Vaccari, J. P., Dietrich, W. D., Popovich, P. G., & Keane, R. W. (2014). Pattern recognition receptors and central nervous system repair. *Experimental Neurology, 258*, 5–16.

Kim, H. K., Andreazza, A. C., Elmi, N., Chen, W., & Young, L. T. (2016). Nod-like receptor pyrin containing 3 (NLRP3) in the post-mortem frontal cortex from patients with bipolar disorder: A potential mediator between mitochondria and immune-activation. *Journal of Psychiatric Research, 72*, 43–50.

Kim, H. K., Chen, W., & Andreazza, A. C. (2015). The potential role of the NLRP3 inflammasome as a link between mitochondrial complex I dysfunction and inflammation in bipolar disorder. *Neural Plasticity, 2015*, 408136.

Kokai, M., Kashiwamura, S., Okamura, H., Ohara, K., & Morita, Y. (2002). Plasma interleukin-18 levels in patients with psychiatric disorders. *Journal of Immunotherapy, 25*(1), S68–S71.

Kolb, R., Liu, G. H., Janowski, A. M., Sutterwala, F. S., & Weizhou Zhang, W. (2014). Inflammasomes in cancer: A double-edged sword. *Protein & Cell, 5*(1), 12–20.

Konradi, C., Sillivan, S. E., & Clay, H. B. (2012). Mitochondria, oligodendrocytes and inflammation in bipolar disorder: Evidence from transcriptome studies points to intriguing parallels with multiple sclerosis. *Neurobiology of Disease, 45*(1), 37–47.

Koo, J. W., Russo, S. J., Ferguson, D., Nestler, E. J., & Duman, R. S. (2010). Nuclear factor-kappaB is a critical mediator of stress-impaired neurogenesis and depressive behavior. *Proceedings of the National Academy of Sciences of the United States of America, 107*, 2669–2674.

Lalla, R., & Donmez, G. (2013). The role of sirtuins in Alzheimer's disease. *Frontiers in Aging Neuroscience, 5*, 16.

Lamkanfi, M., Mueller, J. L., Vitari, A. C., Misaghi, S., Fedorova, A., Deshayes, K., et al. (2009). Glyburide inhibits the Cryopyrin/Nalp3 inflammasome. *The Journal of Cell Biology, 187*(1), 61–70.

Lee, H. J., Lee, J. K., Lee, H., Carter, J. E., Chang, J. W., Oh, W., et al. (2012). Human umbilical cord blood-derived mesenchymal stem cells improve neuropathology and cognitive impairment in an Alzheimer's disease mouse model through modulation of neuroinflammation. *Neurobiology of Aging, 33*, 588–602.

Li, L., & Liu, Y. (2014). Aging-related gene signature regulated by Nlrp3 predicts glioma progression. *American Journal of Cancer Research, 5*(1), 442–449.

Li, R., Wang, X., Qin, T., Qu, R., & Ma, S. (2016). Apigenin ameliorates chronic mild stress-induced depressive behavior by inhibiting interleukin-1beta production and NLRP3 inflammasome activation in the rat brain. *Behavioural Brain Research, 296*, 318–325.

Liston, A., & Masters, S. L. (2017). Homeostasis-altering molecular processes as mechanisms of inflammasome activation. *Nature Reviews Immunology, 17*(3), 208–214.

Litteljohn, D., Mangano, E., Clarke, M., Bobyn, J., Moloney, K., & Hayley, S. (2010). Inflammatory mechanisms of neurodegeneration in toxin-based models of Parkinson's disease. *Parkinson's Disease, 2011*, 713517.

Liu, F., Lo, C. F., Ning, X., Kajkowski, E. M., Jin, M., Chiriac, C., et al. (2004). Expression of NALP1 in cerebellar granule neurons stimulates apoptosis. *Cellular Signalling, 16*, 1013–1021.

Liu, S. B., Mi, W. L., & Wang, Y. Q. (2013). Research progress on the NLRP3 inflammasome and its role in the central nervous system. *Neuroscience Bulletin, 29*, 779–787.

Liu, Y. N., Peng, Y. L., Liu, L., Wu, T. Y., Zhang, Y., Lian, Y. J., et al. (2015). TNFalpha mediates stress-induced depression by upregulating indoleamine 2,3-dioxygenase in a mouse model of unpredictable chronic mild stress. *European Cytokine Network, 26*, 15–25.

Liu, B., Xu, C., Wu, X., Liu, F., Du, Y., Sun, J., et al. (2015). Icariin exerts an antidepressant effect in an unpredictable chronic mild stress model of depression in rats and is associated with the regulation of hippocampal neuroinflammation. *Neuroscience*, *294*, 193–205.

López-Armada, M. J., Riveiro-Naveira, R. R., Vaamonde-García, C., & Valcárcel-Ares, M. N. (2013). Mitochondrial dysfunction and the inflammatory response. *Mitochondrion*, *13*(2), 106–118.

Lu, M., Yang, J. Z., Geng, F., Ding, J. H., & Hu, G. (2014). Iptakalim confers an antidepressant effect in a chronic mild stress model of depression through regulating neuroinflammation and neurogenesis. *The International Journal of Neuropsychopharmacology*, *17*, 1501–1510.

Lukens, J. R., Gross, J. M., & Kanneganti, T. D. (2012). IL-1 family cytokines trigger sterile inflammatory disease. *Frontiers in Immunology*, *3*, 315.

Ma, M., Ren, Q., Zhang, J. C., & Hashimoto, K. (2014). Effects of brilliant blue G on serum tumor necrosis factor-α levels and depression-like behavior in mice after lipopolysaccharide administration. *Clinical Psychopharmacology and Neuroscience*, *12*, 31–36.

Macht, M., Schwarz, R., & Ellgring, H. (2005). Patterns of psychological problems in Parkinson's disease. *Acta Neurologica Scandinavica*, *111*(2), 95–101.

Maes, M., Berk, M., Goehler, L., Song, C., Anderson, G., Gałecki, P., et al. (2012). Depression and sickness behavior are Janus-faced responses to shared inflammatory pathways. *BMC Medicine*, *10*, 66. http://dx.doi.org/10.1186/1741-7015-10-66.

Malhotra, S., Río, J., Urcelay, E., Nurtdinov, R., Bustamante, M. F., Fernández, O., et al. (2015). NLRP3 inflammasome is associated with the response to IFN-β in patients with multiple sclerosis. *Brain*, *138*(3), 644–652.

Mamik, M. K., Hui, E., Branton, W. G., McKenzie, B. A., Chisholm, J., Cohen, E. A., et al. (2016). HIV-1 viral protein R activates NLRP3 inflammasome in microglia: Implications for HIV-1 associated neuroinflammation. *Journal of Neuroimmune Pharmacology*. http://dx.doi.org/10.1007/s11481-016-9708-3.

Martinon, F., Burns, H., & Tschopp, J. (2002). The inflammasome: A molecular platform triggering activation of inflammatory caspases and processing of proIL-beta. *Molecular Cell*, *10*, 417–426.

Martinon, F., Petrilli, V., Mayor, A., Tardivel, A., & Tschopp, J. (2006). Gout associated uric acid crystals activate the NALP3 inflammasome. *Nature*, *440*, 237–241.

Mason, J. W. (1975a). A historical view of the stress field. *Journal of Human Stress*, *1*(1), 6–12.

Mason, J. W. (1975b). A historical view of the stress field. *Journal of Human Stress*, *1*(2), 22–36.

Masters, C. L., & Selkoe, D. J. (2012). Biochemistry of amyloid *β*-protein and amyloid deposits in Alzheimer disease. *Cold Spring Harbor Perspectives in Medicine*, *2*(6), a006262.

McEwen, B. S. (2002). *The end of stress as we know it*. Washington, DC: Dana Press.

McIntire, C. R., Yeretssian, G., & Saleh, M. (2009). Inflammasomes in infection and inflammation. *Apoptosis*, *14*, 522–535.

McLarnon, J. G., Ryu, J. K., Walker, D. G., & Choi, H. B. (2006). Upregulated expression of purinergic P2X(7) receptor in Alzheimer disease and amyloid-beta peptide-treated microglia and in peptide-injected rat hippocampus. *Journal of Neuropathology and Experimental Neurology*, *65*(11), 1090–1097.

Mendiola-Precoma, J., Berumen, L. C., Padilla, K., & Garcia-Alcocer, G. (2016). Therapies for prevention and treatment of Alzheimer's disease. *BioMed Research International*, *2016*, 2589276. http://dx.doi.org/10.1155/2016/2589276.

Menu, P., & Vince, J. E. (2011). The NLRP3 inflammasome in health and disease: The good, the bad and the ugly. *Clinical and Experimental Immunology*, *166*, 1–15.

Mezzaroma, E., Toldo, E., & Abbate, A. (2013). NLRP3 (cryopyrin) in acute myocardial infarction. *Cardiovascular Research*, *99*(1), 225–226.

Mikova, O., Yakimova, R., Bosmans, E., Kenis, G., & Maes, M. (2001). Increased serum tumor necrosis factor alpha concentrations in major depression and multiple sclerosis. *European Neuropsychopharmacology*, *11*, 203–208.

Molina-Hernandez, M., Tellez-Alcantara, N. P., Perez-Garcia, J., Olivera-Lopez, J. I., & Jaramillo-Jaimes, M. T. (2008). Antidepressant-like actions of minocycline combined with several glutamate antagonists. *Progress in Neuro-Psychopharmacology & Biological Psychiatry, 32*, 380–386.

Möller, T. (2010). Neuroinflammation in Huntington's disease. *Journal of Neural Transmission (Vienna, Austria: 1996), 117*, 1001–1008.

Müller, N. (2014). Immunology of major depression. *Neuroimmunomodulation, 21*, 123–130.

Nadatani, Y., Huo, X., Zhang, X., Yu, C., Cheng, E., Zhang, Q., et al. (2016). NOD-like receptor protein 3 inflammasome priming and activation in Barrett's epithelial cells. *Cellular and Molecular Gastroenterology and Hepatology, 2*(4), 439–453.

Niebur, E. (2008). Electrical properties of cell membranes. *Scholarpedia, 3*(6), 7166.

Nightingale, S., Winston, A., Letendre, S., Michael, B. D., McArthur, J. C., Khoo, S., et al. (2014). Controversies in HIV-associated neurocognitive disorders. *Lancet Neurology, 13*, 1139–1151.

Nuvolone, M., Sorce, S., Schwarz, P., & Aguzzi, A. (2015). Prion pathogenesis in the absence of NLRP3/ASC inflammasomes. *PloS One, 10*(2), e0117208.

Ownby, R. L. (2010). Neuroinflammation and cognitive aging. *Current Psychiatry Reports, 12*, 39–45.

Pace, T. W., & Heim, C. M. (2011). A short review on the psychoneuroimmunology of posttraumatic stress disorder: From risk factors to medical comorbidities. *Brain, Behavior, and Immunity, 25*(1), 6–13.

Pan, Y., Chen, X. Y., Zhang, Q. Y., & Kong, L. D. (2014). Microglial NLRP3 inflammasome activation mediates IL-1b-related inflammation in prefrontal cortex of depressive rats. *Brain, Behavior, and Immunity, 41*, 90–100.

Peelen, E., Damoiseaux, J., Muris, A. H., Knippenberg, S., Smolders, J., Hupperts, R., et al. (2015). Increased inflammasome related gene expression profile in PBMC may facilitate T helper 17 cell induction in multiple sclerosis. *Molecular Immunology, 63*(2), 521–529.

Pennisi, M., Crupi, R., Di Paola, R., Ontario, M. L., Bella, R., Calabrese, E. J., et al. (2016). Inflammasomes, hormesis, and antioxidants in neuroinflammation: Role of NRLP3 in Alzheimer disease. *Journal of Neuroscience Research*. http://dx.doi.org/10.1002/jnr.23986.

Piirainen, S., Youssef, A., Song, C., Kalueff, A. V., Landreth, G. E., Malm, T., et al. (2017). Psychosocial stress on neuroinflammation and cognitive dysfunctions in Alzheimer's disease: The emerging role for microglia? *Neuroscience & Biobehavioral Reviews*. pii: S0149-7634(16)30485-7.

Pimplikar, S. W. (2014). Neuroinflammation in Alzheimer's disease: From pathogenesis to a therapeutic target. *Journal of Clinical Immunology, 34*, 64–69.

Pressman, B. C. (1976). Biological applications of ionophores. *Annual Review of Biochemistry, 45*, 501–530.

Ramaswamy, V., Walsh, J. G., Sinclair, D. B., Johnson, E., Tang-Wai, R., Wheatley, B. M., et al. (2013). Inflammasome induction in Rasmussen's encephalitis: Cortical and associated white matter pathogenesis. *Journal of Neuroinflammation, 13*(10), 152.

Rao, V. R., Ruiz, A. P., & Prasad, V. R. (2014). Viral and cellular factors underlying neuropathogenesis in HIV associated neurocognitive disorders (HAND). *AIDS Research and Therapy, 11*, 13.

Reznick, A. Z. (1989). The cycle of stress—A circular model for the psychobiological response to strain and stress. *Medical Hypotheses, 30*, 217–222.

Rondeau, J. M., Ramage, P., Zurini, M., & Gram, H. (2015). The molecular mode of action and species specificity of canakinumab, a human monoclonal antibody neutralizing IL-1b. *MAbs*, 7, 1151–1160.

Rosenblat, J. D., Cha, D. S., Mansur, R. B., & McIntyre, R. S. (2014). Inflamed moods: A review of the interactions between inflammation and mood disorders. *Progress in Neuro-Psychopharmacology & Biological Psychiatry, 53*, 23–34.

Rubartelli, A. (2014). DAMP-mediated activation of NLRP3-inflammasome in brain sterile inflammation: The fine line between healing and neurodegeneration. *Frontiers in Immunology*, *5*, 99.

Salminen, A., Ojala, J., Suuronen, T., Kaarniranta, K., & Kauppinen, A. (2008). Amyloid-beta oligomers set fire to inflammasomes and induce Alzheimer's pathology. *Journal of Cellular and Molecular Medicine*, *12*, 2255–2262.

Sandanger, Ø., Ranheim, T., Vinge, L. E., Bliksøen, M., Alfsnes, K., Finsen, A. V., et al. (2013). The NLRP3 inflammasome is up-regulated in cardiac fibroblasts and mediates myocardial ischaemia-reperfusion injury. *Cardiovascular Research*, *99*(1), 164–174.

Sapolsky, R. M. (2005). The influence of social hierarchy on primate health. *Science*, *308*, 648–652.

Saresella, M., Piancone, F., Marventano, I., Zoppis, M., Hernis, A., Zanette, M., et al. (2016). Multiple inflammasome complexes are activated in autistic spectrum disorders. *Brain, Behavior, and Immunity*, *57*, 125–133.

Scheltens, P., Blennow, K., Breteler, M. M., de Strooper, B., Frisoni, G. B., Salloway, S., et al. (2016). Alzheimer's disease. *Lancet*, *388*, 505–517.

Schroder, K., & Tschopp, J. (2010). The inflammasomes. *Cell*, *140*, 821–832.

Scola, G., Kim, H. K., Young, L. T., & Andreazza, A. C. (2013). A fresh look at complex I in microarray data: Clues to understanding disease-specific mitochondrial alterations in bipolar disorder. *Biological Psychiatry*, *73*(2), e4–e5.

Selye, H. (1936). A syndrome produced by diverse nocuous agents. *Nature*, *138*(3479), 32.

Selye, H. (1975). Confusion and controversy in the stress field. *Journal of Human Stress*, *1*(2), 37–44.

Shah, V. B., Huang, Y., Keshwara, R., Ozment-Skelton, T., Williams, D. L., & Keshvara, L. (2008). Beta-glucan activates microglia without inducing cytokine production in Dectin-1-dependent manner. *Journal of Immunology*, *180*, 2777–2785.

Shi, F., Yang, L., Kouadir, M., Yang, Y., Wang, J., Zhou, X., et al. (2012). The NALP3 inflammasome is involved in neurotoxic prion peptide-induced microglial activation. *Journal of Neuroinflammation*, *9*, 73.

Sparkman, N. L., & Johnson, R. W. (2008). Neuroinflammation associated with aging sensitizes the brain to the effects of infection or stress. *Neuroimmunomodulation*, *15*, 323–330.

Suzuki, K., Matsuzaki, H., Iwata, K., Kameno, Y., Shimmura, C., Kawai, S., et al. (2011). Plasma cytokine profiles in subjects with high-functioning autism spectrum disorders. *PloS One*, *6*(5), e20470.

Takahashi, M. (2014). NLRP3 inflammasome as a novel player in myocardial infarction. *International Heart Journal*, *55*(2), 101–105.

The American Institute of Stress. (2015). *What is stress?* Available online: http://www.stress.org/what-is-stress/ (accessed on 29 September 2015).

Walker, A., Kavelaars, A., Heijnen, C., & Dantzer, R. (2014). Neuroinflammation and comorbidity of pain and depression. *Pharmacological Reviews*, *66*, 80–101.

Walsh, J. G., Muruve, D. A., & Power, C. (2014). Inflammasomes in the CNS. *Nature Reviews. Neuroscience*, *15*, 84–97.

Walsh, J. G., Reinke, S. N., Mamik, M. K., McKenzie, B. A., Maingat, F., Branton, W. G., et al. (2014). Rapid inflammasome activation in microglia contributes to brain disease in HIV/AIDS. *Retrovirology*, *11*, 35.

Warner-Schmidt, J. L., Vanover, K. E., Chen, E. Y., Marshall, J. J., & Greengard, P. (2011). Antidepressant effects of selective serotonin reuptake inhibitors (SSRIs) are attenuated by antiinflammatory drugs in mice and humans. *Proceedings of the National Academy of Sciences of the United States of America*, *108*(22), 9262–9267.

Wei, Q., Mu, K., Li, T., Zhang, Y., Yang, Z., Jia, X., et al. (2014). Deregulation of the NLRP3 inflammasome in hepatic parenchymal cells during liver cancer progression. *Laboratory Investigation*, *94*(1), 52–62.

Weidenfeld, J., & Yirmiya, R. (1996). Effects of bacterial endotoxin on the glucocorticoid feedback regulation of adrenocortical response to stress. *Neuroimmunomodulation*, *3*(6), 352–357.
Wilkinson, S. M., Gunosewoyo, H., Barron, M. L., Boucher, A., McDonnell, M., Turner, P., et al. (2014). The first CNS-active carborane: A novel P2X7 receptor antagonist with antidepressant activity. *ACS Chemical Neuroscience*, *5*, 335–339.
Wong, M. L., Dong, C., Maestre-Mesa, J., & Licinio, J. (2008). Polymorphisms in inflammation related genes are associated with susceptibility to major depression and antidepressant response. *Molecular Psychiatry*, *13*, 800–812.
Wong, M. L., Inserra, A., Lewis, M. D., Mastronardi, C. A., Leong, L., Choo, J., et al. (2016). Inflammasome signaling affects anxiety- and depressive-like behavior and gut microbiome composition. *Molecular Psychiatry*, *21*, 797–805.
Wright, S. H. (2004). Generation of resting membrane potential. *Advances in Physiology Education*, *28*(1–4), 139–142.
Wu, C. S., Chang, K. P., OuYang, C. N., Kao, H. K., Hsueh, C., Chen, L. C., et al. (2016). ASC contributes to metastasis of oral cavity squamous cell carcinoma. *Oncotarget*, 7(31), 50074–50085.
Xu, Y., Li, H., Chen, W., Yao, X., Xing, Y., Wang, X., et al. (2013). Mycoplasma hyorhinis activates the NLRP3 inflammasome and promotes migration and invasion of gastric cancer cells. *PloS One*, *8*(11), e77955.
Xue, J., Li, H., Deng, X., Ma, Z., Fu, Q., & Ma, S. (2015). L-Menthone confers antidepressant-like effects in an unpredictable chronic mild stress mouse model via NLRP3 inflammasome-mediated inflammatory cytokines and central neurotransmitters. *Pharmacology, Biochemistry, and Behavior*, *134*, 42–48.
Yehuda, R. (2009). Status of glucocorticoid alterations in post-traumatic stress disorder. *Annals of the New York Academy of Sciences*, *1179*, 56–69.
Yong, V. W., Wells, J., Giuliani, F., Casha, S., Power, C., & Metz, L. M. (2004). The promise of minocycline in neurology. *Lancet Neurology*, *3*, 744–751.
Zhang, Y., Liu, L., Liu, Y. Z., Shen, X. L., Wu, T. Y., Zhang, T., et al. (2015). NLRP3 inflammasome mediates chronic mild stress-induced depression in mice via neuroinflammation. *The International Journal of Neuropsychopharmacology*, *20*, 18.
Zhang, Y., Liu, L., Peng, Y. L., Liu, Y. Z., Wu, T. Y., Shen, X. L., et al. (2013). Involvement of inflammasome activation in lipopolysaccharide-induced mice depressive-like behaviors. *CNS Neuroscience & Therapeutics*, *20*, 119–124.
Zhao, Y., Yang, J., Shi, J., Gong, Y. N., Lu, Q., Xu, H., et al. (2011). The NLRC4 inflammasome receptors for bacterial flagellin and type III secretion apparatus. *Nature*, *477*, 596–600.
Zhiyu, W., Wang, N., Wang, Q., Peng, C., Zhang, J., Liu, P., et al. (2016). The inflammasome: An emerging therapeutic oncotarget for cancer prevention. *Oncotarget*, 7(31), 50766–50780.
Zhou, Y., Lu, M., Du, R. H., Qiao, C., Jiang, C. Y., Zhang, K. Z., et al. (2016). MicroRNA-7 targets Nod-like receptor protein 3 inflammasome to modulate neuroinflammation in the pathogenesis of Parkinson's disease. *Molecular Neurodegeneration*, *11*, 28.
Zhou, R., Yazdi, A. S., Menu, P., & Tschopp, J. (2011). A role for mitochondria in NLRP3 inflammasome activation. *Nature*, *469*, 221–225.
Ziegler-Graham, K., Brookmeyer, R., Johnson, E., & Arrighi, H. M. (2008). Worldwide variation in the doubling time of Alzheimer's disease incidence rates. *Alzheimer's & Dementia*, *4*, 316–323.
Zitvogel, L., Kepp, O., Galluzzi, L., & Kroemer, G. (2012). Inflammasomes in carcinogenesis and anticancer immune responses. *Nature Immunology*, *13*(4), 343–351.

CHAPTER SEVEN

# Stress-Adaptive Response in Ovarian Cancer Drug Resistance: Role of TRAP1 in Oxidative Metabolism-Driven Inflammation

**Maria Rosaria Amoroso*, Danilo Swann Matassa*, Ilenia Agliarulo*, Rosario Avolio*, Francesca Maddalena†, Valentina Condelli†, Matteo Landriscina†,‡,1, Franca Esposito*,1**

*Università di Napoli Federico II, Napoli, Italy
†Laboratorio di ricerca preclinica e traslazionale, IRCCS-CROB, Centro di Riferimento Oncologico della Basilicata, Rionero in Vulture, Italy
‡Università degli Studi di Foggia, Foggia, Italy
[1]Corresponding authors: e-mail address: matteo.landriscina@unifg.it; franca.esposito@unina.it

## Contents



## Abstract

Metabolic reprogramming is one of the most frequent stress-adaptive response of cancer cells to survive environmental changes and meet increasing nutrient requirements during their growth. These modifications involve cellular bioenergetics and cross talk

*Advances in Protein Chemistry and Structural Biology*, Volume 108
ISSN 1876-1623
http://dx.doi.org/10.1016/bs.apcsb.2017.01.004

with surrounding microenvironment, in a dynamic network that connect different molecular processes, such as energy production, inflammatory response, and drug resistance. Even though the Warburg effect has long been considered the main metabolic feature of cancer cells, recent reports identify mitochondrial oxidative metabolism as a driving force for tumor growth in an increasing number of cellular contexts. In recent years, oxidative phosphorylation has been linked to a remodeling of inflammatory response due to autocrine or paracrine secretion of interleukines that, in turn, induces a regulation of gene expression involving, among others, molecules responsible for the onset of drug resistance. This process is especially relevant in ovarian cancer, characterized by low survival, high frequency of disease relapse and chemoresistance. Recently, the molecular chaperone TRAP1 (tumor necrosis factor-associated protein 1) has been identified as a key junction molecule in these processes in ovarian cancer: in fact, TRAP1 mediates a metabolic switch toward oxidative phosphorylation that, in turn, triggers cytokines secretion, with consequent gene expression remodeling, finally leading to cisplatin resistance and epithelial-to-mesenchymal transition in ovarian cancer models. This review summarizes how metabolism, chemoresistance, inflammation, and epithelial-to-mesenchymal transition are strictly interconnected, and how TRAP1 stays at the crossroads of these processes, thus shedding new lights on molecular networks at the basis of ovarian cancer.

## 1. INTRODUCTION

To survive adverse environmental conditions, tumor cells have to remodel their signaling pathways through a regulation of main biological processes such as transcription, translation, and posttranslational modifications. This adaptation is regulated in a spatial and temporal manner and gives rise to individual tumor cells with distinct gene expression and metabolic signatures. This phenotypic heterogeneity is the result of tumor cell plasticity that—together with the genetic background of the tumor—determines whether cells resist environmental stress, enter dormancy, or metastasize.

The characteristic traits of tumor cells, defined "hallmarks of cancer," were identified and described as integrated circuits that, through parallel axes or intertwined networks, contribute to tumor growth and development (Hanahan & Weinberg, 2000). This list has been recently expanded to include novel features that are involved in the pathogenesis of almost all cancers (Hanahan & Weinberg, 2011). Among those, deregulation of cellular energetics, avoidance of immune destruction, and tumor-promoting inflammation have attracted special attention in recent years. Indeed, tumors have the capability to modify, or reprogram, cellular metabolism to support neoplastic proliferation. Rapidly growing cancers have been demonstrated to be highly dependent on glucose for their survival (the so-called Warburg effect) even in the presence of oxygen (Warburg, 1956); however, during the past years, evidence is accumulating on a fundamental role for mitochondrial metabolism in several neoplasms such as cervical, renal and bladder cancer, and melanomas (Moreno-Sánchez, Rodríguez-Enríquez, Marín-Hernández, & Saavedra, 2007). Moreover, oxidative phosphorylation (OXPHOS) appears to be fundamental to sustain survival and proliferation of chemoresistant cells (Viale, Corti, & Draetta, 2015). Second, cancer cells are able to escape immunological destruction, in particular by T and B lymphocytes, macrophages, and natural killer cells, thus rendering inflammation by innate immune cells a new fuel to support cancerous capability to growth and disseminate (Balkwill, 2004). Increasing evidence suggests that chemokines are produced by tumor cells as well as by cells of the tumor microenvironment (TME), including cancer-associated fibroblasts, mesenchymal stem cells, endothelial cells, tumor-associated macrophages, and tumor-associated neutrophils (Grivennikov, Greten, & Karin, 2010). Besides affecting tumor cell proliferation, angiogenesis, and metastasis, chemokines also seem to modulate senescence and cell survival (Grivennikov et al., 2010). Even if the role of inflammatory processes in cancer progression is becoming more and more relevant, less is known about molecular mechanisms and players that link this feature to cancer progression and metabolic remodeling that accompanies all steps of tumor growth. The understanding of this intricate network would be of great interest in ovarian cancer (OC), a tumor that is characterized by a complex immune environment and a high rate of relapses after traditional chemotherapy.

In this chapter, we will focus on recent findings demonstrating that oxidative metabolism plays a predominant role in several cancer types not only in supporting tumor growth but also contributing to disease progression and aggressiveness through chemoresistance and induction of epithelial-to-mesenchymal transition (EMT). We will also describe how inflammatory responses are involved in these mechanisms, supporting a relevant role of the TME in metabolic remodeling of tumors, to escape multiple stresses (nutrient deprivation, low oxygen availability, oxidative damage) induced by surrounding milieu, thus leading to tumor growth and spreading. In the end, we will focus on the molecular chaperone TRAP1, a key molecule at the crossroad between inflammation, metabolic rewiring, and chemoresistance. Historically known for its antiapototic function and found upregulated in several cancers, this chaperone is recently emerging as a possible prognostic factor in OC for its role in modulating metabolic features that are responsible for inflammatory signaling, the onset of platinum resistance and induction of EMT.

# 2. LINKING OXPHOS, INFLAMMATION, EMT, AND CHEMORESISTANCE IN CANCER

## 2.1 Metabolic Remodeling as a Survival Strategy and Escape From Chemotherapy in Cancer: The Importance of OXPHOS

Metabolic rewiring has been recognized as a novel hallmark of cancer (Hanahan & Weinberg, 2011). Tumor cell growth and survival rely on altered metabolic pathways such as aerobic glycolysis, fatty acid synthesis, and glutamine metabolism. Cancer cells remodel their metabolic pathways to cope with environmental stresses and increased bioenergetic needs required by unchecked cell proliferation (Pavlova & Thompson, 2016). For decades, tumors have been considered to be mainly glycolytic (the "Warburg effect"), being defective in mitochondrial function (Warburg, 1956); however, during last years, many studies highlighted the importance of OXPHOS for ATP supply in a large number of cancers, shedding new lights on metabolic requirements of cancer cells. In fact, small subpopulations of slow-cycling cells characterized by high tumorigenic potential and multidrug resistance have been isolated from different tumors and their metabolic characterization revealed a dependency on mitochondrial respiration rather than glycolysis, suggesting the existence of a common metabolic program (Viale et al., 2015).

A paradigmatic case is represented by melanomas. These aggressive and drug-resistant tumors are also metabolically heterogeneous (Roesch, 2015). Activating mutations in BRAF are the most common genetic alterations in melanoma, and inhibition of BRAF by small molecules leads to cell-cycle arrest and apoptosis (Roesch, 2015). However, BRAF inhibition induces transcriptional activation of OXPHOS genes, mitochondrial biogenesis, and increased expression of the mitochondrial master regulator PGC1α; therefore, treatment of BRAF-mutated melanomas with BRAF inhibitors (e.g., Vemurafenib) renders them addicted to OXPHOS, thus identifying an adaptive metabolic program that limits the efficacy of BRAF inhibitors (Haq et al., 2013). Additionally, melanoma cells resistant to Vemurafenib reprogram their metabolism toward glutamine dependence. Indeed, Vemurafenib-resistant cells uptake higher amounts of glutamine and have increased sensitivity to glutamine deprivation and to glutamine synthase inhibitors (BPTES or L-DON) than their Vemurafenib-sensitive counterparts, both in vivo and in vitro, thus providing a proof of concept for the potential of targeting the glutamine metabolism as an alternative strategy to suppress acquired drug resistance in melanoma (Hernandez-Davies et al., 2015). Concomitantly, it has been demonstrated that a subset of human melanoma tumors expresses high levels of PGC1α and mitochondrial genes of oxidative metabolism, with higher rates of survival under oxidative stress compared to PGC1α-negative melanomas, proposing its use as a negative prognostic factor for patient stratification (Vazquez et al., 2013).

However, melanomas do not represent a special case. Several other solid tumor types have shown preference for OXPHOS in certain circumstances. Pancreatic ductal adenocarcinoma, one of the most deadly disease worldwide, is addicted to KRAS oncogene, whose signaling is considered one of the main targets to design novel pharmacological therapy in this cancer. In pancreatic ductal adenocarcinoma, cells surviving to drugs that abolish this oncogenic signaling and is responsible for tumor relapse have cancer stem cell features and rely on OXPHOS for survival, with decreased dependence on glycolysis for cellular energetics, thus suggesting a therapeutic strategy to combine targeting of KRAS pathway and mitochondrial respiration in chemoresistant cancers (Viale et al., 2014).

The dependency on mitochondrial pathways for survival is proving crucial even in colorectal cancer, a disease with poor outcome and frequent intrinsic chemoresistance. The comparison between transcriptomics data of liver metastases from chemotherapy-treated and chemo-naïve colon cancer patients reveals that, upon chemotherapy, cancer cells shift their metabolism from glycolysis toward OXPHOS via the SIRT1–PGC1α signaling pathway and, therefore, increases resistance of cancer cells to chemotherapy, thus highlighting OXPHOS as a metabolic target to prevent disease progression by overcoming drug resistance (Vellinga et al., 2015).

Moreover, mitochondrial metabolism and consequent reactive oxygen species (ROS) generation are essential to drive anchorage-independent growth through regulation of ERK/MAPK signaling pathway in KRAS-mutated models of lung cancer, and, consequently, disrupting mitochondrial function by loss of the mitochondrial transcription factor A gene reduced tumorigenesis in an oncogenic KRAS-driven mouse model of lung cancer (Weinberg et al., 2010).

Glioblastoma is another challenging neoplasia, with about 90% of patients becoming resistant after a first-line therapy with the alkylating agent temozolomide (TMZ). Oliva et al. (2010) showed that TMZ induces profound changes in the activities of the mitochondrial electron transport chain (ETC) and cellular bioenergetic function. Moreover, TMZ-resistant glioma cells display enhanced bioenergetic reserve capacity and, consequently, pharmacological inhibition of complexes II–III and IV activities abrogates TMZ chemoresistance. Most strikingly, these changes are clinically relevant, since they were recapitulated in patient biopsies obtained after adjuvant therapy with TMZ.

For their part, nonsolid tumors are often characterized by a nonglycolytic metabolism, thanks to high oxygen availability. One example is represented by diffuse large B-cell lymphomas (DLBCLs), in which a specific subtype is OXPHOS addicted and highly dependent on mitochondrial energy production and nutrient utilization pathways. Recent studies by Norberg et al. (in press) identified a crucial role for the mitochondrial translation apparatus in a survival mechanism that supports a central metabolic feature of OXPHOS-DLBCLs. Genetic or pharmacologic perturbations of the mitochondrial translation apparatus by the selective inhibitor tigecycline lead to subtype-selective toxicity in OXPHOS-DLBCL cell lines and primary OXPHOS-DLBCLs, with significant reduction in mitochondrial supercomplex abundance, complex I activity, and elevated mitochondrial ROS, thus opening new therapeutic strategies for OXPHOS-type DLBCLs. In agreement with this study, it has been recently demonstrated that quiescent acute myelogenous leukemia-initiating cells have a lower energetic level than proliferating cells and rely mainly on OXPHOS for energy production, lacking leukemia-initiating cells the compensatory glycolytic mechanism upon OXPHOS inhibition (Lagadinou et al., 2013). These data mechanistically explain why tigecycline was identified in a chemical screen as an agent that selectively kills leukemic stem and progenitor cells. Notably, tigecycline is FDA approved and is being actively developed for its potential therapeutic benefits in several diseases (Skrtić et al., 2011).

Targeting glutamine uptake also leads to antileukemic responses in acute myeloid leukemia (Jacque et al., 2015). Primary acute myeloid leukemia cells have an increased mitochondrial mass and oxygen consumption rate compared with normal hematopoietic cells, and targeting mitochondrial translation yields antileukemic effects. The inhibition of mitochondrial respiratory chain complex I by metformin can reduce oxygen consumption and induce cytotoxicity in this disease (Wheaton et al., 2014).

In keeping with these findings, a large-scale synthetic lethal RNAi screening allowed to identify pyruvate dehydrogenase, the limiting enzyme for pyruvate entry into the mitochondrial tricarboxylic acid cycle (TCA), as critical gene for the survival of chronic myelogenous leukemia cells upon BCR-ABL inhibition (Alvarez-Calderon et al., 2015). As a consequence, mitochondrial ATP synthase inhibitor oligomycin-A greatly sensitized leukemia cells to tyrosine kinase inhibitors (TKIs) in vitro, suggesting that targeting mitochondrial metabolism as an adjuvant therapy could improve therapeutic responses to TKI in patients with BCR-ABL(+) leukemias.

Metabolic remodeling is therefore a wide-spread survival strategy to adapt to adverse environments, but also to elude anticancer drugs; this generates chemoresistant phenotypes with specific bioenergetic addiction that could be highly druggable. Hence, targeting cancer metabolism is emerging as a promising new strategy for the development of anticancer agents, either used alone or in combination therapies, thus highlighting new vulnerabilities that can be exploited to eradicate cancer cells responsible for tumor relapse. Moreover, selected targeting of mitochondrial bioenergetics may have broader implications to bypass resistance mechanisms that allow tumors to escape inhibition of canonical growth factor signaling, including those mediated by RAS and BRAF pathways.

## 2.2 Inflammation: An Intricate Relationship With Tumor Progression and Drug Resistance

Among emerging hallmarks, the two distinct attributes of cancer cells that are particularly compelling are the reprogramming of cellular energy metabolism, that has been examined earlier, and tumor-promoting inflammation. Several studies have recently highlighted links between inflammation and cancer, showing that the identification of underlying molecular mechanisms may be highly relevant for cancer therapy. The presence of leukocytes within tumors was first observed by Rudolf Virchow in the 19th century, but only in the last decade, clear evidence has been obtained on inflammation playing a critical role in tumorigenesis, and some of the causal mechanisms have been elucidated (Karin, Lawrence, & Nizet, 2006). A robust link between inflammation and cancer emerges by the use of antiinflammatory therapies that have shown efficacy in cancer prevention and treatment (Gonda, Tu, & Wang, 2009). Moreover, inflammation owing to bacterial or viral infection is a well-reported precursor of human carcinogenesis. For instance, *Helicobacter pylori* is associated with gastric cancer, human papilloma virus with cervical cancer, and hepatitis B and C viruses with hepatocellular carcinoma (Karin et al., 2006). Nowadays, a role of inflammation in tumorigenesis is generally accepted, and it has becoming evident that an inflammatory microenvironment is an essential component of all tumors, including those in which a direct causal relationship with inflammation has not been proven (Mantovani, Allavena, Sica, & Balkwill, 2008).

Cytokines and chemokines, the inflammation mediators, are molecular messengers that enable cells of the immune system to coordinate a self-limiting immune response. In patients with normal immune system function, a complex cytokine network mediates an effective response. Cytokines and chemokines are released by both tumor and host cells (such as stromal, endothelial, and/or immune cells) within the TME.

Several interleukines (ILs) have been linked with inflammation and subsequent cancer development and progression, in particular IL-1, IL-6, IL-8, and IL-17. Given that these ILs, concomitantly with several cytokines, such as TNF, and transcription factors (AP-1, NF-κB, STAT3) are critical for both inflammation and tumor growth, they control hubs of protumorigenic signaling that may be targeted to counteract both tumor-associated inflammation and tumor growth. Indeed, pharmacological interference with cytokine signaling decreases tumorigenesis as well as cancer growth (Grivennikov et al., 2010) and may therefore serve as a starting point for preventive and therapeutic approaches.

Countless examples of causal relationship between ILs expression and tumor progression can be done. IL-1α, which is expressed in both normal tissue and several tumor cells, is a regulatory cytokine that can induce the activation of transcription factors, including NF-κB and AP-1, and promote the expression of various genes involved in cell survival, proliferation, and angiogenesis (Di Paolo & Shayakhmetov, 2016). In addition, direct evidence for the role of IL-1β in human cancer has been found in multiple myeloma: when released from myeloma cells, IL-1β is able to induce the production of IL-6 by bone marrow stromal cells and function as an autocrine growth factor for myeloma cells (Hope et al., 2014). IL-1β also induces proneoplastic transcription factor HIF-1α through a classical inflammatory signaling pathway involving NF-κB and COX-2 and culminating in upregulation of vascular endothelial growth factor (VEGF), a potent angiogenic factor required for tumor growth and metastasis (Lauta, 2003).

IL-6 is a major proinflammatory cytokine, one of the major players of inflammation-associated carcinogenesis. This cytokine can interact with membrane-bound (IL-6R) or soluble (sIL-6R) IL-6 receptor and through gp130 receptor subunit activates Janus kinases (JAK)/STAT, MAPK, and phosphatidylinositol 3-kinases/protein kinase B (PI3K/Akt) signaling pathways. The critical protumorigenic cascade leads to phosphorylation and activation of nuclear transcription factor STAT3 that exhibits nonredundant effects on tumor cells and TME, finally modulating the expression of genes involved in proliferation, survival, and angiogenesis (Bharti, Dey, & Mandal, 2016).

Analysis of biopsies from inflammation-associated gastric cancers showed that IL-1β and IL-6 levels are highly elevated in tumors compared with adjacent normal mucosa (Kai et al., 2005). An overproduction of IL-6 has also been found in 37% of multiple myeloma patients at diagnosis and is associated with disease aggressiveness, myeloma-cell proliferation, and poor prognosis (Lauta, 2003). Increased serum levels of IL-6 have been also found to be positively correlated with tumor burden in colorectal cancer patients, this indicating that inflammatory markers paralleled clinical outcome (Cammarota et al., 2010). Evidence obtained in breast cancer systems showed that IL-6 gene expression is upregulated in mammosphere models obtained from aggressive ductal breast carcinomas and that IL-6 regulates a Notch-3-dependent signaling pathway that promotes self-renewal and the invasive potential of normal and tumor mammosphere (Sansone et al., 2007). Moreover, in human epidermoid cell skin carcinoma, IL-6/STAT3 signaling has been correlated to resistance to anti-VEGF therapy, so that IL-6-blocking antibodies are able to revert this phenotype (Eichten et al., 2016).

Constitutive expression of IL-8 mRNA and secreted IL-8 protein has been observed in various tumor cell lines and animal models, thus suggesting that IL-8 could be a key factor involved in proliferation, angiogenesis, and metastasis of cancer cells (Grivennikov et al., 2010). Neutralizing antibodies to IL-8 can inhibit angiogenesis, tumor growth, and metastasis of human melanoma, suggesting the potential utility of anti-IL-8 to treat melanoma and other solid tumors either alone or in combination with conventional chemotherapy or other antitumor agents (Huang et al., 2016). Moreover, IL-8, CXCR1 (CXC chemokine receptor 1), and CXCR2 mRNA expression are found upregulated in response to hypoxia in a time-dependent manner in prostate cancer cells. Furthermore, inhibition of IL-8 signaling potentiates etoposide-induced cell death in hypoxic cells, thus suggesting that IL-8 signaling confers a survival advantage to hypoxic prostate cancer cells, and therefore strategies to inhibit IL-8 signaling may sensitize hypoxic tumor cells to conventional treatments (Maxwell et al., 2013).

In a recent study, tumor microvessel density was positively associated with IL-17, IL-6, IL-8, and VEGF expression in human lung adenocarcinoma tissues, and it was increased in tumor tissues of nude mice injected with lung adenocarcinoma cells-bearing IL-17. Importantly, positive correlations were also detected between IL-17 expression and IL-6, IL-8, and VEGF expression in the same tissues. This pathway is mediated by STAT1 activation, as demonstrated by the evidence that STAT1 knockdown attenuated IL-17-mediated increases in IL-6, IL-8, and VEGF expression in lung cancer cells (Huang et al., 2016). Similarly, in OC, IL-6 (Nilsson, Langley, & Fidler, 2005) and IL-8 (Wang et al., 2012) have been proven to be highly proangiogenic.

Altogether, cytokine production by immune and inflammatory cells is an important tumor-promoting mechanism that provides malignant cells with a continuous supply of growth and survival signals in an initially hostile microenvironment. In most cases, tumor-promoting cytokines act in a paracrine manner, yet several types of cancer cells produce their own cytokines to achieve the same effect (Bharti et al., 2016). This pathway is relevant in metastatic lung cancer, where a strong correlation between phospho-STAT3 and activating mutations in the tyrosine kinase domain of EGFR gene has been observed. Gao et al. (2007) demonstrated that STAT3 activation in mutant EGFR-expressing cell lines occurs, in part, through mutant EGFR-mediated transcriptional upregulation of the IL-6 gene, which, in turn, activates the gp130/JAK signaling pathway. Blockade of this pathway inhibits cell-cycle progression and cell growth and represses tumorigenesis. Several cell lines expressing EGFR are able to produce very high levels of IL-6, rapidly secreted into culture medium, that could stimulate STAT3 phosphorylation in vitro (Gao et al., 2007).

## 2.3 Inflammation-Induced EMT: An Alliance to Sustain Tumor Progression and Drug Resistance

A large body of literature demonstrated that inflammatory mediators have direct effects on tumor cell function and contribute to several hallmarks of cancer, e.g., stimulating the EMT and favoring the metastatic process (Hanahan & Weinberg, 2000). EMT is a key process in tumor metastatic cascade characterized by the loss of cell–cell junctions and cell polarity, traditionally considered important for the acquisition of migratory and invasive properties (Kalluri & Weinberg, 2009). However, in a novel and intriguing perspective, recent papers have highlighted a new role for EMT, whose function in cell spread and metastasis is becoming controversial: it has been shown that EMT is emerging as one of several mechanisms conferring resistance to cancer therapies and favoring the switch toward a nonproliferative, stem-like cell phenotype. Fischer et al. (2015) demonstrated a link between EMT features and chemoresistance rather than metastatic potential; in fact, EMT cells significantly contribute to recurrent disease after chemotherapy, thus suggesting a rationale for an EMT-targeting therapeutic strategy, coupled with conventional chemotherapies in breast cancer. In a parallel study, Zheng et al. (2015) suggested that EMT is not essential for metastatic spread in pancreatic cancer as well, since its suppression does not alter the emergence of invasive cells, systemic dissemination, or metastasis in primary tumors, but, conversely, suppression of EMT leads to increased cancer cell proliferation and enhanced sensitivity to chemotherapy.

Concomitantly, many studies have shown that the tumor surrounding milieu represents a source of molecular inputs not only for cell proliferation but, remarkably, also for drug resistance (Ricciardi et al., 2015). This mechanism is also achieved by paracrine secretion of cytokines, that in turn generates feed-forward signals to amplify survival networks, even upon drug treatment. In this context, inflammatory pathways and EMT collaborate in maintaining aggressive phenotypes in tumors, thus contributing to the onset of chemoresistance and tumor spreading (Ricciardi et al., 2015).

Recently, Bakker, Qattan, Mutti, Demonacos, and Krstic-Demonacos (2016) demonstrated that incubation with bone marrow-derived stromal cells alters the expression of some components of ATP-binding cassette (ABC) transporters, whose increased expression often causes chemoresistance in myeloid leukemia cell lines and that this process is dependent on insulin-like growth factor, thus supporting the microenvironmental effects on drug resistance. Notably, transfection of IL-6 into U-2OS osteosarcoma cells induces paclitaxel and doxorubicin resistance, in contrast to IL-8 transfection that has no effect (Duan, Lamendola, Penson, Kronish, & Seiden, 2002). High expression of IL-6 in oral squamous cell carcinoma cells predicts poor response to chemoradiotherapy and unfavorable prognosis in patients (Jinno et al., 2015). Accordingly, Zhang, Wu, et al. (2015) showed that stromal pancreatic stellate cells contribute to the chemoresistance of pancreatic cancer cells to gemtabicine through the activation of paracrine SDF-1α/CXCR4 signaling and, subsequently, the IL-6 autocrine loop. Dabkeviciene et al. (2015) showed that 5-fluorouracil treatment increases expression of IL-8 and its receptor CXCR2, thus resulting in upregulation of IL-8 in chemoresistant cell lines. Consequently, the blockage of CXCR2 resulted in a decreased expression of tumor growth-promoting cytokines, including IL-7 and EGF, as well as an increased expression of cancerogenesis-preventing cytokines IL-18 and CCL25, thus making IL-8 receptor an attractive target for eradicating chemoresistant cancers. IL-6 and IL-8 have been reported as overexpressed in paclitaxel and doxorubicin-resistant cell lines (Duan et al., 2002). In addition, several investigations have reported that elevated serum IL-6 or IL-8 levels in cancer patients are correlated with advanced disease and shortened survival time (Lane, Matte, Rancourt, & Piché, 2011).

At present, however, a substantial amount of evidence still links EMT and metastatic spread too. Interestingly, chemokine receptor expression was also found to be strongly correlated with organ-specific metastasis in different cancer types, thus providing to inflammatory pathways a significant role in tumor dissemination, besides drug resistance. Recently, extensive research suggested that IL-6 is involved in the metastatic process through induction of EMT in lung, head and neck, colon, ovarian, and cervical cancers, shedding light on how inflammation could trigger prometastatic signals (Korneev et al., 2017). For example, it has been shown that the induction of inflammation by lipopolysaccaride (LPS) is sufficient to induce an EMT program mediated by the transcription factor ZEB1, triggering metastatic dissemination in lung cancer (De Cock et al., in press). Accordingly, IL-17 treatment upregulates EMT phenotypic markers vimentin and E-cadherin in lung cancer cells and increased cell migration through ZEB1, whose expression and nuclear localization is induced by IL-17. Consistently, suppression of ZEB1 abrogates IL-17-stimulated vimentin expression and cell migration through the canonical NF-κB signaling (Gu et al., 2015). This reveals that TME can promote cancer metastasis through ZEB1 expression and indicates that targeting IL-17-induced ZEB1 expression may offer an effective therapeutic strategy for lung cancer treatment. Overall, ZEB1 can be considered a novel oncogenic factor required for tumor initiation, cancer cell plasticity, and drug resistance in carcinomas. Richard et al. (2016) demonstrated that high levels of ZEB1 expression are associated with inherent resistance to MAPK inhibitors in *BRAFV600E*-mutated cell lines and tumors. ZEB1 levels are indeed elevated in melanoma cells with acquired resistance to BRAF inhibition and in biopsies from relapsed patients, indicating that ZEB1 is a major driver of melanoma cell plasticity, yielding drug adaptation and phenotypic resistance to MAPK inhibitors.

The ZEB1-mediated EMT, however, does not represent an isolated case, as for inflammation-induced EMT. It was reported that THP-1 cells derived from macrophage promote EMT in lung adenocarcinoma cells (Che et al., 2014): coculture of THP-1-derived macrophage and lung adenoma cells triggered IL-6 secretion in cancer cells that promote EMT altering E-cadherin and vimentin expression. Moreover, IL-6 in combination with chemokine CCL2 induced EMT in small-cell lung carcinoma and, accordingly, expression level of IL-6 and E-cadherin/vimentin were correlated in lung cancer tissue samples (Zhao et al., 2014). This phenotype is mediated by STAT3 and can be reversed through inhibition of IL-6-induced STAT3 tyrosine phosphorylation by metformin, which also inhibited distant metastases formation in tumor-bearing nude mice (Zhao et al., 2014). In gastric cancer, the *H. pylori* infection leads to an active inflammatory microenvironment that confers to the cells the ability to metastasize to lymph nodes by undergoing EMT through alterations in signaling pathways such as activation of NF-κB or STAT3. For this reason, inflammation is accepted as a major driver of gastric carcinogenesis and metastatization (Ma, Liu, & Liang, 2016). In addition, in human bladder cancer cells, TNF-α stimulates the secretion of matrix metalloproteinase-9, which has been implicated in tumor invasion and metastasis (Lee et al., 2008). Similarly, Cohen et al. (2013) demonstrated that inflammatory breast cancer (IBC), the most insidious breast cancer subtype, is highly dependent on surrounding microenvironment for EMT activation. TNF-α production by peripheral blood T cells was indeed correlated with detection of circulating tumor cells expressing EMT markers. Accordingly, in IBC-model cell lines, soluble factors from activated T cells induce expression of EMT markers such as vimentin and ZEB1, thus increasing their invasive properties. Moreover, combination of TNF-α, IL-6, and TGF-β recapitulates EMT induction in IBC, and neutralizing antibodies against these factors reversed EMT activation, thus highlighting these molecules as potential mediators of immune-IBC interaction.

Accordingly, it has been shown that IL-6 levels in breast tumor increase with tumor grade, and elevated serum IL-6 correlates with poor breast cancer patient survival (Sullivan et al., 2009). In addition, there is evidence that tumorigenesis is initiated by IL-6 in prostate cancer via activation of insulin-like growth factor-1R and IL-6 induces EMT in benign prostate epithelial cells, as confirmed by downregulation of E-cadherin and upregulation of vimentin, N-cadherin, and Snail (Rojas et al., 2011). Remarkably, one the main EMT drivers, Snail, is stabilized by the inflammatory cytokine TNF-α through the activation of the NF-κB pathway, and its expression correlates with activation of NF-κB in cancer cell lines and metastatic tumor samples. Interestingly, Snail is responsible for the acquisition of a chemoresistant phenotype in colorectal cancer cells (Fan et al., 2012).

These recent observations unravel novel molecular pathways that drive chemoresistance and, intriguingly, may establish novel cross talks among apparently unrelated molecular pathways. Indeed, all these studies suggest the potential for an EMT-targeting strategy in conjunction with conventional chemotherapies for resistant cancer treatment but also suggest that EMT and inflammation run on parallel axes to induce drug resistance, tumor dissemination, and aggressiveness, thus opening new scenarios for combinatorial therapies in specific tumor settings.

## 2.4 Inflammation and Metabolism in Cancer: An Emerging Cross talk

Increasing evidence suggests that the sequential course of inflammation is linked with metabolism. This implies a radical change in biochemical machineries and signaling molecules of cross talks between tumor cells and the surrounding environment. Indeed, metabolism is a mediator of immune cell function in the supply of their nutritional requirements during inflammatory response. In this context, here a selection of data supporting the strict correlation between inflammatory environment and metabolic conditions is presented. Palmieri et al. (in press) described for the first time a metabolic mechanism in microglial response to a proinflammatory stimulus, based on glutamine synthase activity as a master modulator of immune cell function, thus indicating a potential therapeutic target. Further studies clarify glutamine role in metabolic-mediated inflammation, as chronic activation of macrophages with LPS causes an increase in intracellular succinate by glutamine-dependent anaplerosis. Intriguingly, succinate acts as an endogenous danger signal to stabilize the proneoplastic transcription factor HIF-1α, which in turn specifically regulates gene expression of IL-1b and other HIF-1α-dependent genes leading to protein succinylation. HIF-1α is normally stabilized under hypoxia; however, this signaling could occur under normoxia since elevated succinate has the same effect on HIF-1α stability as hypoxia; in addition, patients affected with cancers harboring mutations in succinate dehydrogenase (SDH) have increased HIF-1α activity and circulating succinate. The succinate-dependent activation of HIF-1α in tumors suggests a tumorigenic effect of the inflammatory process, and the identification of succinate as an inflammatory signal may therefore clarify the intricate cross talk between these cellular processes (Tannahill et al., 2013). In keeping with these data, it was demonstrated that: (i) the enhanced production of succinate is a critical mechanism responsible for the proinflammatory response to LPS, both through the generation of ROS by the mitochondrial ETC and HIF-1α stabilization; (ii) antiinflammatory gene expression is decreased upon succinate oxidation; (iii) germline mutations in SDH isoforms were observed in patients with hereditary paragangliomas and phaeochromocytomas (Mills & O'Neill, 2014). All these findings intimately connect mitochondrial metabolism alterations and tumorigenesis.

In recent years, several studies supported the role of intermediate metabolites as inflammation-promoting factors. For example, an elevation in citrate and a decrease in isocitrate were observed in LPS-activated macrophages, further supporting the role of this metabolite in LPS-induced responses (Iacobazzi & Infantino, 2014). Although a direct correlation between citrate and inflammatory response is lacking, citrate likely acts as a signal to produce key reactive oxygen intermediates for host defense and inflammation, as well as prostaglandins, which are also important mediators of inflammation (Iacobazzi & Infantino, 2014). Interestingly, nonimmunological disorders with a strong metabolic component, such as obesity and type 2 diabetes, have been linked to immune dysregulation, suggesting that metabolic alterations can be induced by, or be a consequence of, an altered state of immune tolerance (Johnson, Milner, & Makowski, 2012). In addition, immune-mediated disorders, such as multiple sclerosis, display severe metabolic alterations, and experimental evidence shows that metabolic imbalances (i.e., overweight and obesity) can increase the risk of developing such diseases (Johnson et al., 2012).

An important link between mitochondrial metabolism and inflammation in cancer has been demonstrated by Vats et al. (2006). These authors describe a feed-forward loop, in which PGC-1β, the master coordinator of gene expression for mitochondrial proteins, is induced by and coactivates STAT6, amplifying the metabolic, biogenic, and antiinflammatory programs of alternatively activated macrophages. Constitutive activation of PGC-1β in macrophages downregulates macrophage inflammatory responses induced by LPS, including IL-6 and IL-12 production, and increases IL-4-induced alternative macrophage activation markers. Consistently, PGC-1β knockdown reduced the effects of IL-4 on metabolic pathways and inflammatory cytokine production. This demonstrates an existing link between oxidative metabolism and antiinflammatory responses and highlights the important ability of transcriptional activators to facilitate cross talk between diverse cellular programs, such as metabolic and immune responses.

Moreover, recent evidence strongly supports the role of the microenvironment in metabolic shift of tumor cells. In particular, p62 has been recently identified as tumor suppressor in prostate cancer, since its deletion leads to downregulation of glucose and amino acid metabolism through repression on mTOR pathway, ROS production, increase in IL-6 synthesis, and consequent tumor growth (Valencia et al., 2014). In keeping with these findings, the antidiabetic drug metformin, widely used as anticancer agent and potent inhibitor of mitochondrial respiratory complex I (Owen, Doran, & Halestrap, 2000), counteracts several processes in the inflammatory pathway necessary for cell transformation and cancer stem cells formation, as nuclear translocation of NF-κB and phosphorylation of STAT3 (Hirsch, Iliopoulos, & Struhl, 2013). Furthermore, metformin reverts EMT and decreases IL-6 signaling activation in lung cancer cells resistant to anti-EGFR therapies, while IL-6 readdition bypasses metformin sensitizing effects (Li et al., 2014). This is further confirmed by the evidence that metformin-based combinatorial therapy efficiently blocked tumor growth in xenografts of TKI-resistant cancer cells, feature associated with decreased IL-6 secretion and expression, EMT reversal, and decreased IL-6 signaling activation in vivo (Li et al., 2014). It has been demonstrated that pretreatment with metformin of endothelial cells inhibited TNF-α-induced IL-6 production, phosphorylation of IκB kinase α/β, and IκB-α degradation (Huang et al., 2009), and metformin treatment inhibits pro-IL-1β induction in response to LPS in mouse models (Kelly, Tannahill, Murphy, & O'Neill, 2015). Interestingly, inhibition of complex I mimics the effects of metformin on LPS-induced cytokines, while AMPK inhibition does not (Kelly et al., 2015). These observations demonstrate that metformin antiinflammatory activity is due to its inhibitory role on OXPHOS-mediated ROS production. Moreover, metformin is able to block IL-8, IL-1β, ICAM, and IGFBP-1 expression in human endometrial stromal cells (Kelly et al., 2015).

Intriguingly, a novel link between OXPHOS and immune response through development of long-lived T-memory cells has been demonstrated, due to a mechanism of immune activation guided by the newly discovered protein LEM (Okoye et al., 2015). Through interaction with CR6 interacting factor, LEM controlled the levels of OXPHOS complexes and respiration, resulting in the production of proproliferative mitochondrial ROS. These findings provide a link between immune activation and expansion of protective CD8+ T cells driven by OXPHOS and represents a pathway for the restoration of long-term protective immunity based on metabolically modified cytotoxic CD8+ T cells.

# 3. TRAP1 AS A NOVEL PLAYER IN OVARIAN CANCER BIOLOGY

## 3.1 Ovarian Cancer: The Importance of Deciphering Metabolic and Inflammatory Features of a Lethal Disease

The puzzling scenario presented earlier could open unprecedented perspectives for tumors such as OC, which is characterized by a complex immune environment and almost unexplored metabolic features.

Epithelial OC is characterized by a very unfavorable outcome, mainly due to late diagnosis accompanied by a high tumor burden at first presentation. Local tumor spread and local tumor recurrence of finally chemotherapy-resistant tumor cells in the peritoneal cavity and associated clinical complications are the main causes of lethality. High Grade Serous OC (HGSOC) is the most malignant subtype of OC and accounts for up to 70% of all OC cases. Currently, survival rates of patients with OC remain low and OC continues to be a poorly understood disease (Siegel, Ma, Zou, & Jemal, 2014). However, recent advances are shedding light on the metabolic features of this disease and how these affect its outcome.

Yang et al. (2014) demonstrated that glutamine, rather than glucose, plays a critical role in high-invasive OC growth and in overall patient survival and that dependence on glutamine is strongly correlated with cancer invasiveness. A significant correlation exists between patient survival and gene expression in the glutaminolysis/glycolytic metabolic network: genes involved in glutaminolysis and TCA cycle are highly expressed in patients with poor survival, whereas glycolytic genes are associated with better patient survival (Yang et al., 2014). Considering that increased glutamine metabolism is an expected consequence of upregulation of oxidative metabolism (in order to replenish intermediates for the TCA cycle), it is consistent that OXPHOS is enhanced in highly invasive OC cells, and a concomitant pyruvate uptake is increased in highly invasive OC cells under anoikis conditions and required for anaplerosis, mitochondrial function, and migration (Yang et al., 2014).

In agreement with these data, the metabolic signature of cancer stem cells from patients affected by epithelial OC shows that OC-derived stem cells resist glucose deprivation and are strongly dependent on oxidative metabolism, with a biochemical pattern consistent with increase in mitochondrial metabolism (Pastò et al., 2014). This evidence is further supported by the current use of metformin as anticancer agent in OC to reduce tumor size in nude mice, diminish metastatic nodules in lungs compared with untreated mice, decrease cell proliferation and angiogenesis, and potentiate cisplatin-induced cytotoxicity (Rattan, Graham, Maguire, Giri, & Shridhar, 2011).

So far, OC has been nevertheless considered a genetic disease (Lech et al., 2013); however, epigenetic mechanisms affecting gene expression without genome modification are highly sensitive to external factors, such as infection, inflammation, oxidative stress, and diet. Inflammation has been suggested to act as an intermediary between extrinsic factors, such as infection, physical or chemical stimuli, and carcinogenesis. Sustained exposure to inflammatory mediators that aim to degrade aberrant stimuli, for instance cytokines and immune cells, could lead to disruption of homoeostasis, genomic instability, and, consequently, to abnormal proliferation. Indeed, an abundant network of chemokines is found in both the ovarian tumors and associated ascites (Kulbe et al., 2012). For this cancer, three key cytokine/chemokine mediators of cancer-related inflammation, TNF, CXCL12, and IL-6, are involved in an autocrine cytokine network, the "TNF network." This network has paracrine actions on angiogenesis, infiltration of myeloid cells, and NOTCH signaling in both murine xenografts and human OC biopsies. Neutralizing antibodies or interfering RNA to individual members of this TNF network reduced angiogenesis, myeloid cell infiltration, and experimental peritoneal OC growth (Kulbe et al., 2012). Gene expression profile analysis of ovarian transformed cells showed increased expression of several cytokines, including IL-6 and IL-8, which are upregulated by the NF-κB pathway. Furthermore, OC produce IL-6, and large amounts of IL-6 have been detected in sera and ascites from OC patients and their overexpression predicts poor clinical outcome (Reuter, Gupta, Chaturvedi, & Aggarwal, 2010). Accordingly, surgical removal of the tumor and resolution of ascites directly associated with decrease in serum levels of IL-6 (Wang et al., 2012). A highly significant increase in IL-6 and cIAP-2 mRNA and protein levels upon treatment with cisplatin has also been shown (Cohen et al., 2013). Furthermore, IL-6 by itself significantly increased cIAP-2 levels in OC cells, and cytotoxicity assays showed sensitization to cisplatin following the addition of IL-6 and cIAP-2 inhibitors. In conclusion, cisplatin treatment of OC cells upregulates IL-6 and cIAP-2 levels, while their inhibition significantly sensitizes them to cisplatin. Therefore, cIAP-2 could be considered as a novel inducer of platinum resistance in OC cells and suggest an axis induced by the exposure of tumor cells to cisplatin, mediated sequentially by IL-6 and cIAP-2, resulting in cisplatin resistance (Cohen et al., 2013).

Elevated IL-6 levels in serum and ascites of epithelial OC patients have been correlated with chemoresistance in several ways. Using a monoclonal antibody that specifically blocks IL-6 signaling (siltuximab), Guo et al. (2010) demonstrated that the combination of siltuximab with paclitaxel increased the sensitivity of OC cells to paclitaxel. In vitro studies by Wang et al. (2010) first demonstrated that autocrine production of IL-6 by OC cell lines is inversely associated with their response to cisplatin and paclitaxel. Moreover, the chemoresistance caused by IL-6 is associated with increased expression of both multidrug resistance-related genes (such as MDR1) and apoptosis inhibitory proteins (Bcl-2, Bcl-xL, and XIAP), as well as activation of Ras/MEK/ERK and PI3K/Akt signaling. Therefore, modulation of IL-6 expression or its related signaling pathway may be a promising strategy of treatment for drug-resistant OC.

IL-6 is not the only cytokine involved in OC development and progression. Stronach et al. (2015) showed that, in HSGC, IL-8 and IL-8 receptor expression by tumor cells increases upon cisplatin treatment, whereas IL-8 knockdown increases cells' sensitivity to cisplatin. The effect of IL-8 on platinum sensitivity appears to be mediated by an effect of IL-8 on antiapoptotic proteins Bcl-2 and Bad. This supports preclinical and clinical assessment of IL-8 and IL-8 receptors as therapeutic targets in conjunction with platinum-based therapy.

An important proof of the link between inflammation and platinum resistance was provided in a relevant paper by Wang et al. (2016). The authors described a role for cancer-associated fibroblasts in mediating platinum resistance of HSGOC through GSH increase and consequently depletion of platinum content from tumor cells. CD8+ T cells can abolish fibroblast-mediated platinum resistance via IFNγ, by promoting extracellular GSH degradation and decreasing fibroblast cysteine generation through downregulation of system cysteine–glutamate antiporter (xCT). Furthermore, IFNγ-activated STAT1 binds to specific promoter region of xCT and rapidly inhibits xCT gene transcription through JAK–STAT1 signaling in fibroblasts but not in tumor cells. Therefore, the metabolism of cancer-associated fibroblasts is significantly modulated by CD8+ T cells and IFNγ that selectively target stromal fibroblasts, shape their thiol metabolism, and affect tumor chemotherapeutic response. Effective immunotherapy may revert chemoresistance and revitalize tumor sensitivity to chemotherapy by exploiting the described cross talk between TME and cancer cells. Interestingly, Yang et al. (2016) described a novel mechanism whereby metabolism of reactive stromal cells is reprogrammed through an upregulated glutamine anabolic pathway of tumor core. This dysfunctional stromal metabolism confers atypical metabolic flexibility and adaptive mechanisms in stromal cells, allowing them to use noncanonical sources to synthesize glutamine in nutrient-deprived conditions existing in TME. In this perspective, TME-tumor cross talk could be a mechanism through which OC reinforces drug-resistant phenotypes.

Taken together, all these data demonstrate that both metabolism and inflammation are strongly involved in OC progression and chemoresistance; however, much effort is needed to understand how their connection is important in tumor biogenesis, to develop more effective drugs for OC treatment.

## 3.2 TRAP1 at the Crossroad of Stress-Adaptive Response in Cancer

The tumor environment produces multiple stress conditions that cancer cells must face by implementing integrated strategies of adaptation to sustain their growth and proliferation. Among the molecules involved in these survival processes, tumor necrosis factor receptor-associated protein 1 (TRAP1) was functionally identified in a differential display analysis among genes upregulated in Saos-2 osteosarcoma cells upon adaptation to mild oxidant conditions (Montesano Gesualdi et al., 2007). TRAP1 attracted increasing interest for its homology to HSP90—that has been long pursued as an attractive potential target for anticancer therapy—coupled with a distinct subcellular localization. Consequently, the majority of the studies focused on the expression of this protein in tumors, showing that TRAP1 is strongly expressed in tumor cells of adenocarcinomas of pancreas, breast, colon, and lung, whereas normal matched epithelia contain very low levels (Amoroso et al., 2014). Moreover, TRAP1 expression is upregulated in approximately 60% of human colorectal cancers and correlates with a multidrug-resistant phenotype in colon carcinoma cells (Costantino et al., 2009). In several neoplasms, such as prostate, colorectal, and breast cancers, elevation of TRAP1 levels correlates with malignant progression and metastasis (Agorreta et al., 2014; Leav et al., 2010; Zhang, Wang, et al., 2015).

In vitro studies demonstrate that TRAP1 levels are elevated in tumor cells resistant to 5-FU and to platin derivatives, thus suggesting this chaperone as a molecular link between resistance to antitumor agents and adaptation to oxidative stress. Accordingly, TRAP1 interference, as well as the use of dominant-negative mutants of TRAP1, sensitizes resistant cells to cell death inducers, thus supporting the hypothesis of common mechanisms shared by chemoresistance and adaptation to oxidative stress and providing evidence for the important role played by TRAP1 in the development and the maintenance of these phenotypes. In fact, TRAP1 hyperexpressing cells show both decreased cleavage of the apoptotic markers Caspase 3 and PARP, and increased levels of the scavenging tripeptide GSH (Costantino et al., 2009; Montesano Gesualdi et al., 2007).

The mechanistic role of TRAP1 in the maintenance of tumor phenotypes was strongly corroborated by the discovery of a novel mitochondrial network, only found in tumor cells, that regulates mitochondrial transition pore opening through the formation of a molecular complex containing TRAP1, its homologue HSP90 and the immunophilin cyclophilin D, that results stabilized, thus preventing apoptotic events (Kang et al., 2007). This network has been considered an attractive target for the design of anticancer drugs, finally leading to the development of a mitochondrial-targeted variant of 17-AAG (a potent and selective inhibitor of HSP90) carrying the Antennapedia peptide, called Shepherdin, that efficiently accumulates inside the mitochondria, binds mitochondrial TRAP1 and HSP90, and inhibits their chaperone activity via an ATP competition mechanism, thus resulting in cyclophilin D-mediated cell death (Kang et al., 2009). Following studies gave further relevance to this mitochondrial network not only in the immediate apoptotic response but also in the maintenance of mitochondrial homeostasis by buffering organelle-specific proteotoxic stress and mediating cytoprotective responses through integrating signals in different cellular compartments. These processes may further increase the apoptotic threshold of tumor cells undergoing mitochondriotoxic stress, resulting in better survival (Chae et al., 2012).

Different approaches contributed to expand the TRAP1 world: the identification and functional characterization of an antiapoptotic complex formed by TRAP1 with a novel mitochondrial isoform of the calcium-binding protein Sorcin (Landriscina et al., 2010) further reinforced the TRAP1 cytoprotective network inside tumor mitochondria. More importantly, a novel localization for TRAP1 has been demonstrated at the interface between mitochondria and endoplasmic reticulum, where it interacts with the regulatory particle of 19S proteasomal subunit, called TBP7 (Amoroso et al., 2012). This interaction is crucial in maintaining cellular proteostasis through the regulation of cotranslational ubiquitination of selective mitochondria-destined proteins (F1-ATPase, Sorcin) (Matassa et al., 2013), or oncogenes such as BRAF (Condelli et al., 2014, 2015), thus playing a quality control role by linking protein synthesis and degradation and shifting mRNA translation from the canonical cap-dependent mechanism to the IRES-mediated one, this favoring the synthesis of stress-responsive and oncogenic proteins (Matassa et al., 2014).

Recently, substantial amount of evidence have revealed the centrality of TRAP1 in metabolic regulation of cancer cells. Indeed, TRAP1 directly interacts with the mitochondrial subunit B of SDH (SDHB) and inhibits mitochondrial respiratory chain, thus providing survival to cancer cells by favoring the shift toward a Warburg phenotype. The respiratory downregulation elicited by TRAP1 interaction with SDH promotes tumorigenesis by priming the succinate-dependent stabilization of the proneoplastic transcription factor HIF1α, independently of hypoxic conditions (Sciacovelli et al., 2013). Consequently, TRAP1-deficient cells show high levels of mitochondrial ROS, elevated complex II and IV activity, and mitochondrial ATP production (Yoshida et al., 2013). Moreover, metabolomic analysis upon TRAP1 interference shows marked reduction in the steady-state levels of several glycolytic metabolites, but the levels of several TCA cycle metabolites (e.g., α-ketoglutarate and citrate) were increased, as well as fatty acid oxidation pathway and the NAD+/NADH ratio, consistent with increased oxygen consumption and suggesting increased metabolic flux through the TCA cycle of TRAP1 silenced cells (Yoshida et al., 2013). Interestingly, the same study reported that a specific set of cancers (renal, bladder, cervical) shows decreased TRAP1 expression compared to normal counterparts, as opposed to most previous literature. Of note, cervical carcinoma is among those cancers whose predominant energy metabolism is via OXPHOS, not glycolysis, thus promoting the reconsideration of TRAP1 role in carcinogenesis as uniformly prooncogenic (Yoshida et al., 2013) and suggesting that it could be dependent on the metabolic characteristic of the tumor tissue.

The fascinating hypothesis that, in some settings, and depending on the cell type context/metabolic features, TRAP1 may act as a tumor suppressor is partially supported by a study on 208 patients affected by OC. Immunohistochemical analyses showed that high TRAP1 expression correlated significantly with favorable chemotherapy response and showed a significantly positive impact on overall survival (Aust et al., 2012). In the same patients, TRAP1 expression was also found significantly correlated with estrogen receptor α levels. Consistently, the expression of TRAP1 was previously shown as significantly increased in cancers responsive to letrozole (an aromatase inhibitor used for the treatment of hormonally responsive cancers) compared to letrozole insensitive (O'Donnell, Macleod, Burns, Smyth, & Langdon, 2005). Considering that TRAP1 has been reported as an estrogen-regulated gene (Walker et al., 2007), this opens a new scenario on TRAP1 function in gynecological cancers, thus raising the hypothesis of a tumor-suppressing activity for specific cancers in a metabolic-dependent fashion.

## 3.3 TRAP1 in Ovarian Cancer: The Egg of Columbus in the Molecular Connection Between Oxidative Metabolism, Inflammation, EMT, and Platinum Resistance

Up to date, although OC is one of the most lethal cancers worldwide, its metabolic features are far to be elucidated. Surprisingly, and in agreement with the data published by Aust et al. (2012), recent studies show that TRAP1 inversely correlated to tumor stage, grade and directly associated to overall survival following analyses on large datasets of OC patients (Matassa et al., 2016). Accordingly, TRAP1 is lost by genetic deletion in late stages of the disease, suggesting that this could be one of the driving mechanisms of TRAP1 decrease in OC (Amoroso et al., 2016). In turn, such decrease could both be the cause or simply part of a selection process for the metabolic rewiring of the tumor.

In keeping with database analysis, in cellular models of HSGOC isogenic-matched paired cell lines derived from the same patient before and after platinum-based therapy, platinum-resistant cell lines show decreased TRAP1 levels compared to their sensitive counterparts. Moreover, public gene expression databases of cervical cancer display that TRAP1 expression also inversely correlates with tumor stage in this tumor type, that, interestingly, prevalently rely on OXPHOS for its energy metabolism (Yoshida et al., 2013). Therefore, TRAP1 could act as a tumor suppressor in a context in which OXPHOS is the preferential pathway for ATP source. This hypothesis is supported by metabolic analysis performed by measuring glycolytic activity and oxygen consumption rate in living cells, revealing that platinum-resistant OC cells rely on mitochondrial function and that TRAP1 knockdown shifted the metabolism toward OXPHOS. Accordingly, high grade tumors show a metabolic profile consistent with mitochondrial metabolism. As a result, TRAP1 silencing induces resistance to platinum treatment, and pharmacological inhibition of mitochondrial metabolism by metformin/oligomycin reverts platinum resistance, providing a causal nexus between TRAP1 expression, oxidative metabolism, and platinum resistance. Accordingly, a mitochondrial drug that uncouples respiration from ATP production reverts cisplatin apoptotic effect (Matassa et al., 2016). These pathways are most likely linked by mitochondrial-mediated ROS generation, due to elevated OXPHOS activity, since antioxidant treatment with *N*-acetyl cysteine sensitizes drug-resistant cells to cisplatin-induced cell death. Intriguingly, glucose deprivation leads to increase in platinum resistance, and this is probably due to metabolic shift toward OXPHOS by which cells escape nutrient deprivation (Matassa et al., 2016).

Gene expression analysis in OC cells upon glucose withdrawal revealed, among the five most relevant regulated biofunctions, the aryl hydrocarbon receptor and the glucocorticoid receptor signaling, two pathways involved in cellular immune response and inflammatory activation. These results recapitulate autocrine production of ILs and cytokines in OC and are consistent with recent studies associating metabolic reprogramming to IL signaling, which has been involved in cisplatin resistance (Cohen et al., 2013). In support of the hypothesis that metabolic shift could be responsible for inflammatory activation and that this phenotype could be reproduced by TRAP1 silencing, expression data obtained combining TRAP1 silencing, uncoupling mitochondrial respiration, and glucose starvation—three conditions pushing cells toward OXPHOS—have shown that IL-6 and CSF2, two important mediators of inflammatory response, were upregulated in all these experimental conditions (Matassa et al., 2016). A significant inverse correlation between IL-6 and TRAP1 expression in RNAseq data set (The Cancer Genome Atlas) has also been demonstrated. Notably, treatment with human recombinant IL-6 was sufficient to induce resistance to cisplatin in sensitive cells (Matassa et al., 2016). These data indicate that inhibition of glycolysis and increased oxidative metabolism in OC cells lead to the production of proinflammatory molecules, which may be responsible for cisplatin resistance in OC systems.

At a molecular level, such inflammatory activation induces increased expression of MDR family members. In particular, TAP1 was identified as a main mediator of platinum resistance in OC cells (Matassa et al., 2016). Incidentally, gene expression analysis in resistant cell lines vs sensitive controls revealed upregulation of TAP1 in platinum-resistant cells (Stronach et al., 2011). In agreement, TRAP1 silencing leads to increase in TAP1 expression, and TAP1 silencing in chemoresistant cells is sufficient to revert, at least in part, the platinum-resistant phenotype (Matassa et al., 2016).

Taken together, these results unravel a novel role for TRAP1 in OC, identifying its metabolic control as a key event in disease progression, platinum response, and inflammatory activation. This is of great interest because, as reported earlier, elevated levels of IL-6 in the sera and ascites predict poor clinical outcome and platinum resistance in HGSOC, and IL-8 and its receptor have been already proposed as potential prognostic biomarkers and therapeutic targets in OC (Stronach et al., 2015), but no evidence were given of a direct link between oxidative metabolism and inflammation in OC. Interestingly, metabolic analyses show that A2780 wild-type cells have increased oxidative metabolism than A2780 cisplatin-resistant cells, in which platinum resistance has been induced in vitro, and that, oppositely, have higher glycolytic activity than the respective counterpart (Fig. 1A and B). Accordingly, no difference in TRAP1 expression were detected in sensitive and resistant A2780 cells (Fig. 1C and D), as opposed to what observed in the matched sensitive/resistant cell lines with clinically acquired platinum resistance. It is worth noting that A2780 cells are unable to express a wide panel of chemokines when compared with other OC cells (Son, Kabir, Dong, Lee, & Adunyah, 2012), and a recent systematic genomic comparison of cell lines and tumors demonstrated that A2780 are among the OC cell lines that less resemble HGSOC tumors (Domcke, Sinha, Levine, Sander, & Schultz, 2013). Data presented in Fig. 1 further support these observations and point out the relevant difference between in vivo developed drug resistance mechanisms and those induced in cell cultures. These data highlight the link between inflammation and platinum resistance, as well as the role of TRAP1 in the metabolic choice, and confirm that the processes are strictly intertwined and interdependent.

It has been proposed that platinum-resistant HGSOC clones may arise from low-abundance subclones within the sensitive tumor through a selection process, rather than through adaptation (Cooke et al., 2010). Metabolic characterization of similar subpopulations of tumorigenic and multidrug-resistant cells in other cancer types revealed their dependency on mitochondrial respiration, which is consistent with the data presented earlier. In this regard, the antidiabetic metformin, recently reported as an effective antitumor agent in OC models, is gaining great importance in combined therapies for OXPHOS-addicted cancers (Li et al., 2015). These observations support the notion that metabolic features of tumor cells in the peculiar environment of OC can predict cisplatin sensitivity and that TRAP1, bioenergetic index and proinflammatory molecules can serve as biomarkers predictive of therapeutic outcome. In this context, oxidative

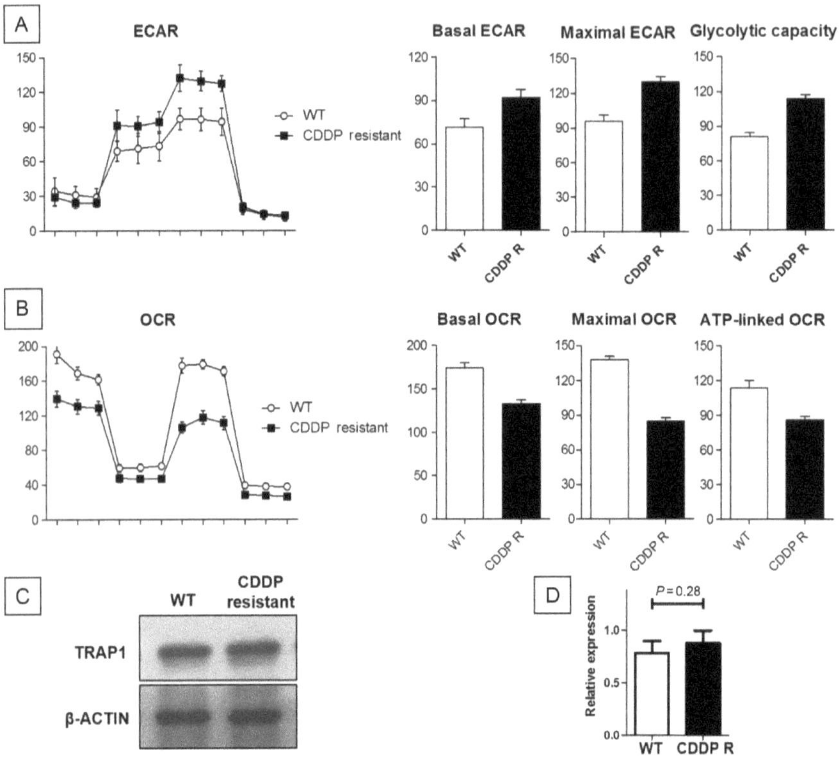


**Fig. 1** Metabolic analysis of sensitive (WT) and cisplatin-resistant (CDDP R) A2780 cells. Kinetic profile of (A) extracellular acidification rate (ECAR) and (B) oxygen consumption rate (OCR) was measured in real-time, under basal conditions and in response to glucose, oligomycin, and 2-DG for ECAR, or in response to mitochondrial drugs oligomycin, FCCP, Antimycin A, and Rotenone for OCR. Basal ECAR, maximal ECAR, and glycolytic capacity are shown as indices of glycolytic pathway activation; basal OCR, ATP-linked OCR, and maximal OCR are shown as indices of mitochondrial respiratory function. (C) Total lysates obtained from WT and CDDP-resistant A2780 cells were separated by SDS-PAGE and immunoblotted with the indicated antibodies. Images are representative of three independent experiments. (D) Real-time RT-PCR analysis of TRAP1 mRNAs expression in WT and CDDP-resistant A2780 cells. Data are expressed as mean ± SEM from three independent experiments with technical triplicates each. Numbers above bars indicate the statistical significance (*P*-value), based on the two-tailed Student's *t* test.

metabolism appears a suitable target to potentiate cisplatin activity in tumors highly dependent on mitochondrial function.

Consistently, the more aggressive phenotype of low-TRAP1-expressing tumors is associated to an increased ability to spread and disseminate in peritoneal sites distant from the tumor origin (Amoroso et al., 2016). This phenotype in OC has previously been associated to increased expression of the kinase p70S6K, a major regulator of EMT process and cell migration (Pon, Zhou, Cheung, Ngan, & Wong, 2008). Of note, p70S6K is also a major player in protein synthesis regulation and specifically regulates glutamine metabolism, being a key metabolic regulator downstream PI3K pathway (Csibi et al., 2013). Moreover, glutamine is an important biosynthetic amino acid source in cells relying on OXPHOS, and it has been shown that highly invasive OC cells remodel their bioenergetics to escape glycolysis dependence, thus shifting their metabolism toward glutaminolysis (Yang et al., 2014). As previously described in other cellular models (Agliarulo et al., 2015; Matassa et al., 2014), it has been found in OC that TRAP1 silencing leads to upregulation of p70S6K expression and activity at post-transcriptional level; moreover, an inverse correlation between TRAP1 and p70S6K phosphorylation is present in patients (Amoroso et al., 2016). TRAP1 downregulation in OC cells causes an increase in cell motility and activation of EMT program with increased expression of the main EMT-related transcription factors, Snail and Slug, and consequent decrease of E-cadherin levels. These regulations are conserved in a large cohort of OC patients, in which decrease in TRAP1 expression is associated to higher metastatic capacity. p70S6K pharmacological inhibition reverts the high motility phenotype of TRAP1 low expressing cells, but it is not able to revert EMT program induced by TRAP1 downmodulation (Amoroso et al., 2016). A model integrating connection between TRAP1 expression, p70S6K, metabolism, and EMT is represented in Fig. 2.

Notably, as discussed earlier, recent studies report that EMT is most likely responsible for resistance to chemotherapy rather than cell migration and metastatic spread (Fischer et al., 2015; Zheng et al., 2015). The evidence that downregulation of TRAP1 induces drug resistance in OC cells through metabolic remodeling are in keeping with these data, suggesting that TRAP1-mediated regulation of EMT could be linked to drug-resistant phenotypes, more than metastatic spread. Taken together, these observations support the hypothesis that the metabolic phenotype of the tumor and the nutrient availability in the surrounding microenvironment strongly influence migratory and invasive properties of OC cell, and responsiveness to therapy. On the basis of these observations, it remains unexplored if TRAP1 regulation of metabolism, that influences inflammatory response and chemoresistance, drives EMT through, for example, IL-6/8 induction. It would be fascinating the hypothesis that inflammation process is also responsible for EMT induction and that this could reinforce the resistant

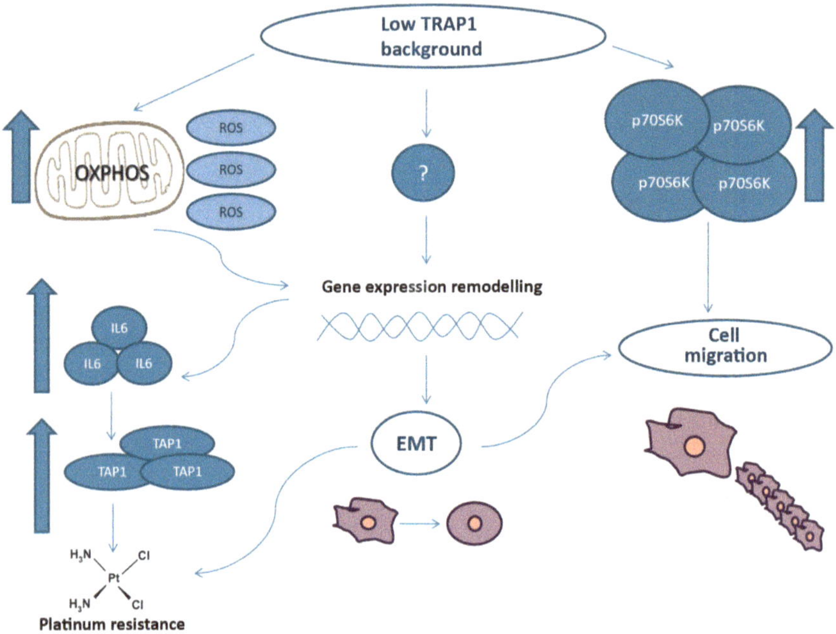


**Fig. 2** TRAP1 deletion causes metabolic remodeling, inflammation, drug resistance, and EMT in OC. TRAP1 downregulation in OC increases mitochondrial activity, with consequent ROS production. This elicits inflammatory response, with a consequent gene expression remodeling with increased IL production that, in turn, triggers expression of prosurvival genes such as the ABC transporter TAP1, involved in platinum resistance. On the other hand, low expression of TRAP1 leads to increased expression of p70S6K by translational regulation, thus favoring OC cell migration and spreading. Concomitantly, but in a p70S6K-independent and yet unknown manner, TRAP1 favors EMT activation by gene expression modifications, thus contributing to drug-resistant phenotypes.

phenotypes upon TRAP1 downregulation. Consequently, TRAP1-related signature could be a significant indicator for patients prognosis or, at least in part, may help in the selection of the best anticancer treatment fitted on tumor metabolic addiction.

## 4. CONCLUDING REMARKS

In the puzzling and heterogeneous context of human tumors, joining apparently separated molecular processes is challenging, and this is particularly true in the case of ovarian tumors, that are still poorly understood, with survival rates changed little over the past 30 years. However, the novelty represented by the introduction of TRAP1 in the molecular connection between oxidative metabolism, inflammation, EMT, and platinum resistance provides a tool to recapitulate different phenotypes of previously unrelated observations.

Tumors continuously face stress conditions during their establishment and development, due to increased energy demand in a nutrient-deprived environment. Their ability to survive to these adverse conditions determines their successful growth and spreading. During last years, metabolism and inflammation are gaining increasing relevance in tumor biology, thus becoming appealing targets for anticancer therapy. In this perspective, many efforts are focused on the discovery of molecular links between these different processes; indeed, reliance on OXPHOS may represent a previously unexpected "druggable" vulnerability of (multi)drug-resistant cancers, and inflammation has been recognized a key driver of tumorigenesis in different settings. Nowadays, it is becoming clear how these processes are strictly interconnected, thus rendering attractive the hypothesis that hub molecules could become the best targets to disrupt cancer homeostasis. In this context, there is an increasing need of tailored therapies, based on specific tumor types that correspond to peculiar metabolic settings and microenvironmental conditions.

The molecular chaperone TRAP1 has raised interest for anticancer drug development in many tumor types, but more recent advances have placed TRAP1 at the interconnection between different cellular functions. New anticancer drugs would desirably target not just a single molecule, but an entire network; in this perspective, an important aim will be the identification of crucial hubs to disrupt entire integrated circuitries or to identify specific disease settings in which an already known drug could be more effective. In this perspective, TRAP1 molecular and bioenergetic signature will provide novel biomarkers to identify the best therapy on the basis of metabolic background of each tumor type.

## REFERENCES

Agliarulo, I., Matassa, D. S., Amoroso, M. R., Maddalena, F., Sisinni, L., Sepe, L., et al. (2015). TRAP1 controls cell migration of cancer cells in metabolic stress conditions: Correlations with AKT/p70S6K pathways. *Biochimica et Biophysica Acta*, *1853*, 2570–2579.

Agorreta, J., Hu, J., Liu, D., Delia, D., Turley, H., Ferguson, D. J., et al. (2014). TRAP1 regulates proliferation, mitochondrial function, and has prognostic significance in NSCLC. *Molecular Cancer Research*, *12*, 660–669.

Alvarez-Calderon, F., Gregory, M. A., Pham-Danis, C., DeRyckere, D., Stevens, B. M., Zaberezhnyy, V., et al. (2015). Tyrosine kinase inhibition in leukemia induces an altered

metabolic state sensitive to mitochondrial perturbations. *Clinical Cancer Research*, *21*(6), 1360–1372. http://dx.doi.org/10.1158/1078-0432.CCR-14-2146.
Amoroso, M. R., Matassa, D. S., Agliarulo, I., Avolio, R., Lu, H., Sisinni, L., et al. (2016). TRAP1 downregulation in human ovarian cancer enhances invasion and epithelial–mesenchymal transition. *Cell Death and Disease*, 7(12), e2522.
Amoroso, M. R., Matassa, D. S., Laudiero, G., Egorova, A. V., Polishchuk, R. S., Maddalena, F., et al. (2012). TRAP1 and the proteasome regulatory particle TBP7/Rpt3 interact in the endoplasmic reticulum and control cellular ubiquitination of specific mitochondrial proteins. *Cell Death and Differentiation*, *19*, 592–604.
Amoroso, M. R., Matassa, D. S., Sisinni, L., Lettini, G., Landriscina, M., & Esposito, F. (2014). TRAP1 revisited: Novel localizations and functions of a 'next-generation' biomarker (review). *International Journal of Oncology*, *45*, 969–977.
Aust, S., Bachmayr-Heyda, A., Pateisky, P., Tong, D., Darb-Esfahani, S., Denkert, C., et al. (2012). Role of TRAP1 and estrogen receptor alpha in patients with ovarian cancer—A study of the OVCAD consortium. *Molecular Cancer*, *11*, 69.
Bakker, E., Qattan, M., Mutti, L., Demonacos, C., & Krstic-Demonacos, M. (2016). The role of microenvironment and immunity in drug response in leukemia. *Biochimica et Biophysica Acta*, *1863*(3), 414–426.
Balkwill, F. (2004). Cancer and the chemokine network. *Nature Reviews. Cancer*, *4*(7), 540–550.
Bharti, R., Dey, G., & Mandal, M. (2016). Cancer development, chemoresistance, epithelial to mesenchymal transition and stem cells: A snapshot of IL-6 mediated involvement. *Cancer Letters*, *375*(1), 51–61.
Cammarota, R., Bertolini, V., Pennesi, G., Bucci, E., Gottardi, O., Garlanda, C., et al. (2010). The tumor microenvironment of colorectal cancer: Stromal TLR-4 expression as a potential prognostic marker. *Journal of Translational Medicine*, *8*, 112.
Chae, Y. C., Caino, M. C., Lisanti, S., Ghosh, J. C., Dohi, T., Danial, N. N., et al. (2012). Control of tumor bioenergetics and survival stress signaling by mitochondrial HSP90s. *Cancer Cell*, *22*(3), 331–344.
Che, D. H., Pan, B., Tian, Q., Shang, L. H., Liu, F., Jin, S., et al. (2014). Enhanced invasion of lung adenocarcinoma cells after co-culture with THP-1-derived macrophages via the induction of EMT by IL-6. *Immunology Letters*, *160*, 1–10.
Cohen, S., Bruchim, I., Graiver, D., Evron, Z., Oron-Karni, V., Pasmanik-Chor, M., et al. (2013). Platinum-resistance in ovarian cancer cells is mediated by IL-6 secretion via the increased expression of its target cIAP-2. *Journal of Molecular Medicine (Berlin, Germany)*, *91*(3), 357–368.
Condelli, V., Maddalena, F., Sisinni, L., Lettini, G., Matassa, D. S., Piscazzi, A., et al. (2015). Targeting TRAP1 as a downstream effector of BRAF cytoprotective pathway: A novel strategy for human BRAF-driven colorectal carcinoma. *Oncotarget*, *6*, 22298–22309.
Condelli, V., Piscazzi, A., Sisinni, L., Matassa, D. S., Maddalena, F., Lettini, G., et al. (2014). TRAP1 is involved in BRAF regulation and downstream attenuation of ERK phosphorylation and cell-cycle progression: A novel target for BRAF-mutated colorectal tumors. *Cancer Research*, *74*, 6693–6704.
Cooke, S. L., Ng, C. K. Y., Melnyk, N., Garcia, M. J., Hardcastle, T., Temple, J., et al. (2010). Genomic analysis of genetic heterogeneity and evolution in high-grade serous ovarian carcinoma. *Oncogene*, *29*, 4905–4913.
Costantino, E., Maddalena, F., Calise, S., Piscazzi, A., Tirino, V., Fersini, A., et al. (2009). TRAP1, a novel mitochondrial chaperone responsible for multi-drug resistance and protection from apoptotis in human colorectal carcinoma cells. *Cancer Letters*, *279*, 39–46.
Csibi, A., Fendt, S. M., Li, C., Poulogiannis, G., Choo, A. Y., Chapski, D. J., et al. (2013). The mTORC1 pathway stimulates glutamine metabolism and cell proliferation by repressing SIRT4. *Cell*, *153*(4), 840–854.

Dabkeviciene, D., Jonusiene, V., Zitkute, V., Zalyte, E., Grigaitis, P., Kirveliene, V., et al. (2015). The role of interleukin-8 (CXCL8) and CXCR2 in acquired chemoresistance of human colorectal carcinoma cells HCT116. *Medical Oncology*, *32*(12), 258.

De Cock, J.M., Shibue, T., Dongre, A., Keckesova, Z., Reinhardt, F., Weinberg, R.A., Inflammation triggers Zeb1-dependent escape from tumor latency, Cancer Research, in press, pii: canres.0608.2016.

Di Paolo, N. C., & Shayakhmetov, D. M. (2016). Interleukin 1α and the inflammatory process. *Nature Immunology*, *17*(8), 906–913.

Domcke, S., Sinha, R., Levine, D. A., Sander, C., & Schultz, N. (2013). Evaluating cell lines as tumour models by comparison of genomic profiles. *Nature Communications*, *4*, 2126.

Duan, Z., Lamendola, D. E., Penson, R. T., Kronish, K. M., & Seiden, M. V. (2002). Overexpression of IL-6 but not IL-8 increases paclitaxel resistance of U-2OS human osteosarcoma cells. *Cytokine*, *17*(5), 234–242.

Eichten, A., Su, J., Adler, A. P., Zhang, L., Ioffe, E., Parveen, A. A., et al. (2016). Resistance to anti-VEGF therapy mediated by autocrine IL6/STAT3 signaling and overcome by IL6 blockade. *Cancer Research*, *76*(8), 2327–2339.

Fan, F., Samuel, S., Evans, K. W., Lu, J., Xia, L., Zhou, Y., et al. (2012). Overexpression of snail induces epithelial-mesenchymal transition and a cancer stem cell-like phenotype in human colorectal cancer cells. *Cancer Medicine*, *1*(1), 5–16.

Fischer, K. R., Durrans, A., Lee, S., Sheng, J., Li, F., Wong, S. T., et al. (2015). Epithelial-to-mesenchymal transition is not required for lung metastasis but contributes to chemoresistance. *Nature*, *527*(7579), 472–476.

Gao, S. P., Mark, K. G., Leslie, K., Pao, W., Motoi, N., Gerald, W. L., et al. (2007). Mutations in the EGFR kinase domain mediate STAT3 activation via IL-6 production in human lung adenocarcinomas. *The Journal of Clinical Investigation*, *117*(12), 3846–3856.

Gonda, T. A., Tu, S., & Wang, T. C. (2009). Chronic inflammation, the tumor microenvironment and carcinogenesis. *Cell Cycle*, *8*(13), 2005–2013.

Grivennikov, S. I., Greten, F. R., & Karin, M. (2010). Immunity, inflammation, and cancer. *Cell*, *140*(6), 883–899.

Gu, K., Li, M. M., Shen, J., Liu, F., Cao, J. Y., Jin, S., et al. (2015). Interleukin-17-induced EMT promotes lung cancer cell migration and invasion via NF-κB/ZEB1 signal pathway. *American Journal of Cancer Research*, *5*(3), 1169–1179.

Guo, Y., Nemeth, J., O'Brien, C., Susa, M., Liu, X., Zhang, Z., et al. (2010). Effects of siltuximab on the IL-6-induced signaling pathway in ovarian cancer. *Clinical Cancer Research*, *16*(23), 5759–5769.

Hanahan, D., & Weinberg, R. A. (2000). The hallmarks of cancer. *Cell*, *100*(1), 57–70.

Hanahan, D., & Weinberg, R. A. (2011). Hallmarks of cancer: The next generation. *Cell*, *144*(5), 646–674.

Haq, R., Shoag, J., Andreu-Perez, P., Yokoyama, S., Edelman, H., Rowe, G. C., et al. (2013). Oncogenic BRAF regulates oxidative metabolism via PGC1α and MITF. *Cancer Cell*, *23*(3), 302–315.

Hernandez-Davies, J. E., Tran, T. Q., Reid, M. A., Rosales, K. R., Lowman, X. H., Pan, M., et al. (2015). Vemurafenib resistance reprograms melanoma cells towards glutamine dependence. *Journal of Translational Medicine*, *13*, 210.

Hirsch, H. A., Iliopoulos, D., & Struhl, K. (2013). Metformin inhibits the inflammatory response associated with cellular transformation and cancer stem cell growth. *Proceedings of the National Academy of Sciences of the United States of America*, *110*(3), 972–977.

Hope, C., Ollar, S. J., Heninger, E., Hebron, E., Jensen, J. L., Kim, J., et al. (2014). TPL2 kinase regulates the inflammatory milieu of the myeloma niche. *Blood*, *123*(21), 3305–3315.

Huang, N. L., Chiang, S. H., Hsueh, C. H., Liang, Y. J., Chen, Y. J., & Lai, L. P. (2009). Metformin inhibits TNF-alpha-induced IkappaB kinase phosphorylation,

IkappaB-alpha degradation and IL-6 production in endothelial cells through PI3K-dependent AMPK phosphorylation. *International Journal of Cardiology*, *134*(2), 169–175.
Huang, Q., Duan, L., Qian, X., Fan, J., Lv, Z., Zhang, X., et al. (2016). IL-17 promotes angiogenic factors IL-6, IL-8, and vegf production via Stat1 in lung adenocarcinoma. *Scientific Reports*, *6*, 36551.
Iacobazzi, V., & Infantino, V. (2014). Citrate-new functions for an old metabolite. *Biological Chemistry*, *395*(4), 387–399.
Jacque, N., Ronchetti, A. M., Larrue, C., Meunier, G., Birsen, R., Willems, L., et al. (2015). Targeting glutaminolysis has antileukemic activity in acute myeloid leukemia and synergizes with BCL-2 inhibition. *Blood*, *126*(11), 1346–1356.
Jinno, T., Kawano, S., Maruse, Y., Matsubara, R., Goto, Y., Sakamoto, T., et al. (2015). Increased expression of interleukin-6 predicts poor response to chemoradiotherapy and unfavorable prognosis in oral squamous cell carcinoma. *Oncology Reports*, *33*(5), 2161–2168.
Johnson, A. R., Milner, J. J., & Makowski, L. (2012). The inflammation highway: Metabolism accelerates inflammatory traffic in obesity. *Immunological Reviews*, *249*(1), 218–238.
Kai, H., Kitadai, Y., Kodama, M., Cho, S., Kuroda, T., Ito, M., et al. (2005). Involvement of proinflammatory cytokines IL-1beta and IL-6 in progression of human gastric carcinoma. *Anticancer Research*, *25*(2A), 709–713.
Kalluri, R., & Weinberg, R. A. (2009). The basics of epithelial-mesenchymal transition. *The Journal of Clinical Investigation*, *119*(6), 1420–1428.
Kang, B. H., Plescia, J., Dohi, T., Rosa, J., Doxsey, S. J., & Altieri, D. C. (2007). Regulation of tumor cell mitochondrial homeostasis by an organelle-specific Hsp90 chaperone network. *Cell*, *131*(2), 257–270.
Kang, B. H., Plescia, J., Song, H. Y., Meli, M., Colombo, G., Beebe, K., et al. (2009). Combinatorial drug design targeting multiple cancer signaling networks controlled by mitochondrial Hsp90. *The Journal of Clinical Investigation*, *119*(3), 454–464.
Karin, M., Lawrence, T., & Nizet, V. (2006). Innate immunity gone awry: Linking microbial infections to chronic inflammation and cancer. *Cell*, *124*(4), 823–835.
Kelly, B., Tannahill, G. M., Murphy, M. P., & O'Neill, L. A. (2015). Metformin inhibits the production of reactive oxygen species from NADH: Ubiquinone oxidoreductase to limit induction of interleukin-1β (IL-1β) and boosts interleukin-10 (IL-10) in lipopolysaccharide (LPS)-activated macrophages. *The Journal of Biological Chemistry*, *290*(33), 20348–20359.
Korneev, K. V., Atretkhany, K. N., Drutskaya, M. S., Grivennikov, S. I., Kuprash, D. V., & Nedospasov, S. A. (2017). TLR-signaling and proinflammatory cytokines as drivers of tumorigenesis. *Cytokine*, *89*, 127–135. pii: S1043-4666(16)30022-9.
Kulbe, H., Chakravarty, P., Leinster, D. A., Charles, K. A., Kwong, J., Thompson, R. G., et al. (2012). A dynamic inflammatory cytokine network in the human ovarian cancer microenvironment. *Cancer Research*, *72*(1), 66–75.
Lagadinou, D., Sach, A., Callahan, K., Rossi, R., Neering, S., Minhajuddin, M., et al. (2013). BCL-2 inhibition targets oxidative phosphorylation and selectively eradicates quiescent human leukemia stem cells. *Cell Stem Cell*, *12*(3), 329–341.
Landriscina, M., Laudiero, G., Maddalena, F., Amoroso, M. R., Piscazzi, A., Cozzolino, F., et al. (2010). Mitochondrial chaperone Trap1 and the calcium binding protein Sorcin interact and protect cells against apoptosis induced by antiblasticagents. *Cancer Research*, *70*, 6577–6586.
Lane, D., Matte, I., Rancourt, C., & Piché, A. (2011). Prognostic significance of IL-6 and IL-8 ascites levels in ovarian cancer patients. *BMC Cancer*, *11*, 210.
Lauta, V. M. (2003). A review of the cytokine network in multiple myeloma: Diagnostic, prognostic, and therapeutic implications. *Cancer*, *97*(10), 2440–2452.

Leav, I., Plescia, J., Goel, H. L., Li, J., Jiang, Z., Cohen, R. J., et al. (2010). Cytoprotective mitochondrial chaperone TRAP-1 as a novel molecular target in localized and metastatic prostate cancer. *The American Journal of Pathology*, *176*, 393–401.

Lech, A., Daneva, T., Pashova, S., Gagov, H., Crayton, R., Kukwa, W., et al. (2013). Ovarian cancer as a genetic disease. *Frontiers in Bioscience*, *18*, 543–563.

Lee, S. J., Park, S. S., Cho, Y. H., Park, K., Kim, E. J., Jung, K. H., et al. (2008). Activation of matrix metalloproteinase-9 by TNF-alpha in human urinary bladder cancer HT1376 cells: The role of MAP kinase signaling pathways. *Oncology Reports*, *19*(4), 1007–1013.

Li, L., Han, R., Xiao, H., Lin, C., Wang, Y., Liu, H., et al. (2014). Metformin sensitizes EGFR-TKI-resistant human lung cancer cells in vitro and in vivo through inhibition of IL-6 signaling and EMT reversal. *Clinical Cancer Research*, *20*(10), 2714–2726.

Li, J., Hernanda, P. Y., Bramer, W. M., Peppelenbosch, M. P., van Luijk, J., & Pan, Q. (2015). Anti-tumor effects of metformin in animal models of hepatocellular carcinoma: A systematic review and meta-analysis. *PloS One*, *110*(6), e0127967.

Ma, H. Y., Liu, X. Z., & Liang, C. M. (2016). Inflammatory microenvironment contributes to epithelial-mesenchymal transition in gastric cancer. *World Journal of Gastroenterology*, *22*(29), 6619–6628.

Mantovani, A., Allavena, P., Sica, A., & Balkwill, F. (2008). Cancer-related inflammation. *Nature*, *454*(7203), 436–444.

Matassa, D. S., Agliarulo, I., Amoroso, M. R., Maddalena, F., Sepe, L., Ferrari, M. C., et al. (2014). TRAP1-dependent regulation of p70S6K is involved in the attenuation of protein synthesis and cell migration: Relevance in human colorectal tumors. *Molecular Oncology*, *8*, 1482–1494.

Matassa, D. S., Amoroso, M. R., Agliarulo, I., Maddalena, F., Sisinni, L., Paladino, S., et al. (2013). Translational control in the stress adaptive response of cancer cells: A novel role for the heat shock protein TRAP1. *Cell Death & Disease*, *4*, e851.

Matassa, D. S., Amoroso, M. R., Lu, H., Avolio, R., Arzeni, D., Procaccini, C., et al. (2016). Oxidative metabolism drives inflammation-induced platinum resistance in human ovarian cancer. *Cell Death and Differentiation*, *23*(9), 1542–1554.

Maxwell, P. J., Coulter, J., Walker, S. M., McKechnie, M., Neisen, J., McCabe, N., et al. (2013). Potentiation of inflammatory CXCL8 signalling sustains cell survival in PTEN-deficient prostate carcinoma. *European Urology*, *64*(2), 177–188.

Mills, E., & O'Neill, L. A. (2014). Succinate: A metabolic signal in inflammation. *Trends in Cell Biology*, *24*(5), 313–320.

Montesano Gesualdi, N., Chirico, G., Pirozzi, G., Costantino, E., Landriscina, M., & Esposito, F. (2007). Tumor necrosis factor-associated protein 1 (TRAP-1) protects cells from oxidative stress and apoptosis. *Stress*, *10*, 342–350.

Moreno-Sánchez, R., Rodríguez-Enríquez, S., Marín-Hernández, A., & Saavedra, E. (2007). Energy metabolism in tumor cells. *The FEBS Journal*, *274*, 1393–1418.

Nilsson, M. B., Langley, R. R., & Fidler, I. J. (2005). Interleukin-6, secreted by human ovarian carcinoma cells, is a potent proangiogenic cytokine. *Cancer Research*, *65*(23), 10794–10800.

Norberg E., Lako A., Chen P.H., Stanley I.A., Zhou F., Ficarro S.B., et al., Differential contribution of the mitochondrial translation pathway to the survival of diffuse large B-cell lymphoma subsets, *Cell Death and Differentiation*, in press, http://dx.doi.org/10.1038/cdd.2016.116, [Epub ahead of print].

O'Donnell, A. J., Macleod, K. G., Burns, D. J., Smyth, J. F., & Langdon, S. P. (2005). Estrogen receptor-alpha mediates gene expression changes and growth response in ovarian cancer cells exposed to estrogen. *Endocrine-Related Cancer*, *12*, 851–866.

Okoye, I., Wang, L., Pallmer, K., Richter, K., Ichimura, T., Haas, R., et al. (2015). T cell metabolism. The protein LEM promotes $CD8^+$ T cell immunity through effects on mitochondrial respiration. *Science*, *348*(6238), 995–1001.

Oliva, C. R., Nozell, S. E., Diers, A., McClugage, S. G., Sarkaria, J. N., Markert, J. M., et al. (2010). Acquisition of temozolomide chemoresistance in gliomas leads to remodeling of mitochondrial electron transport chain. *The Journal of Biological Chemistry*, *285*(51), 39759–39767.

Owen, M. R., Doran, E., & Halestrap, A. P. (2000). Evidence that metformin exerts its anti-diabetic effects through inhibition of complex 1 of the mitochondrial respiratory chain. *The Biochemical Journal*, *348*(Pt. 3), 607–614.

Palmieri E.M., Menga A., Lebrun A., Hooper D.C., Butterfield D.A., Mazzone M., et al., Blockade of glutamine synthetase enhances inflammatory response in microglial cells, *Antioxidants & Redox Signaling*, in press, http://dx.doi.org/10.1089/ars.2016.6715, [November, ahead of print].

Pastò, A., Bellio, C., Pilotto, G., Ciminale, V., Silic-Benussi, M., Guzzo, G., et al. (2014). Cancer stem cells from epithelial ovarian cancer patients privilege oxidative phosphorylation, and resist glucose deprivation. *Oncotarget*, *5*, 4305–4319.

Pavlova, N. N., & Thompson, C. B. (2016). The emerging hallmarks of cancer metabolism. *Cell Metabolism*, *23*(1), 27–47.

Pon, Y. L., Zhou, H. Y., Cheung, A. N., Ngan, H. Y., & Wong, A. S. (2008). p70 S6 kinase promotes epithelial to mesenchymal transition through snail induction in ovarian cancer cells. *Cancer Research*, *68*, 6524–6532.

Rattan, R., Graham, R. P., Maguire, J. L., Giri, S., & Shridhar, V. (2011). Metformin suppresses ovarian cancer growth and metastasis with enhancement of cisplatin cytotoxicity in vivo. *Neoplasia*, *13*(5), 483–491.

Reuter, S., Gupta, S. C., Chaturvedi, M. M., & Aggarwal, B. B. (2010). Oxidative stress, inflammation, and cancer: How are they linked? *Free Radical Biology & Medicine*, *49*(11), 1603–1616.

Ricciardi, M., Zanotto, M., Malpeli, G., Bassi, G., Perbellini, O., Chilosi, M., et al. (2015). Epithelial-to-mesenchymal transition (EMT) induced by inflammatory priming elicits mesenchymal stromal cell-like immune-modulatory properties in cancer cells. *British Journal of Cancer*, *112*(6), 1067–1075.

Richard, G., Dalle, S., Monet, M. A., Ligier, M., Boespflug, A., Pommier, R. M., et al. (2016). ZEB1-mediated melanoma cell plasticity enhances resistance to MAPK inhibitors. *EMBO Molecular Medicine*, *8*(10), 1143–1161.

Roesch, A. (2015). Tumor heterogeneity and plasticity as elusive drivers for resistance to MAPK pathway inhibition in melanoma. *Oncogene*, *34*(23), 2951–2957.

Rojas, A., Liu, G., Coleman, I., Nelson, P. S., Zhang, M., Dash, R., et al. (2011). IL-6 promotes prostate tumorigenesis and progression through autocrine cross-activation of IGF-IR. *Oncogene*, *30*, 2345–2355.

Sansone, P., Storci, G., Tavolari, S., Guarnieri, T., Giovannini, C., Taffurelli, M., et al. (2007). IL-6 triggers malignant features in mammospheres from human ductal breast carcinoma and normal mammary gland. *The Journal of Clinical Investigation*, *117*(12), 3988–4002.

Sciacovelli, M., Guzzo, G., Morello, V., Frezza, C., Zheng, L., Nannini, N., et al. (2013). The mitochondrial chaperone TRAP1 promotes neoplastic growth by inhibiting succinate dehydrogenase. *Cell Metabolism*, *17*, 988–999.

Siegel, R., Ma, J., Zou, Z., & Jemal, A. (2014). Cancer statistics, 2014. *CA: A Cancer Journal for Clinicians*, *64*, 364.

Skrtić, M., Sriskanthadevan, S., Jhas, B., Gebbia, M., Wang, X., Wang, Z., et al. (2011). Inhibition of mitochondrial translation as a therapeutic strategy for human acute myeloid leukemia. *Cancer Cell*, *20*(5), 674–688.

Son, D. S., Kabir, S. M., Dong, Y. L., Lee, E., & Adunyah, S. E. (2012). Inhibitory effect of tumor suppressor p53 on proinflammatory chemokine expression in ovarian cancer cells by reducing proteasomal degradation of IκB. *PloS One*, 7(12), e51116.

Stronach, E. A., Alfraidi, A., Rama, N., Datler, C., Studd, J. B., Agarwal, R., et al. (2011). HDAC4-regulated STAT1 activation mediates platinum resistance in ovarian cancer. *Cancer Research*, *71*(13), 4412–4422.

Stronach, E. A., Cunnea, P., Turner, C., Guney, T., Aiyappa, R., Jeyapalan, S., et al. (2015). The role of interleukin-8 (IL-8) and IL-8 receptors in platinum response in high grade serous ovarian carcinoma. *Oncotarget*, *6*(31), 31593–31603.

Sullivan, N. J., Sasser, A. K., Axel, A. E., Vesuna, F., Raman, V., Ramirez, N., et al. (2009). Interleukin-6 induces an epithelial-mesenchymal transition phenotype in human breast cancer cells. *Oncogene*, *28*(33), 2940–2947.

Tannahill, G. M., Curtis, A. M., Adamik, J., Palsson-McDermott, E. M., McGettrick, A. F., Goel, G., et al. (2013). Succinate is an inflammatory signal that induces IL-1β through HIF-1α. *Nature*, *496*(7444), 238–242.

Valencia, T., Kim, J. Y., Abu-Baker, S., Moscat-Pardos, J., Ahn, C. S., Reina-Campos, M., et al. (2014). Metabolic reprogramming of stromal fibroblasts through p62-mTORC1 signaling promotes inflammation and tumorigenesis. *Cancer Cell*, *26*, 121–135.

Vats, D., Mukundan, L., Odegaard, J. I., Zhang, L., Smith, K. L., Morel, C. R., et al. (2006). Oxidative metabolism and PGC-1β attenuate macrophage-mediated inflammation. *Cell Metabolism*, *4*(1), 13–24.

Vazquez, F., Lim, J. H., Chim, H., Bhalla, K., Girnun, G., Pierce, K., et al. (2013). PGC1α expression defines a subset of human melanoma tumors with increased mitochondrial capacity and resistance to oxidative stress. *Cancer Cell*, *23*(3), 287–301.

Vellinga, T. T., Borovski, T., de Boer, V. C., Fatrai, S., van Schelven, S., Trumpi, K., et al. (2015). SIRT1/PGC1α-dependent increase in oxidative phosphorylation supports chemotherapy resistance of colon cancer. *Clinical Cancer Research*, *21*(12), 2870–2879.

Viale, A., Corti, D., & Draetta, G. F. (2015). Tumors and mitochondrial respiration: A neglected connection. *Cancer Research*, *75*, 3685–3686.

Viale, A., Pettazzoni, P., Lyssiotis, C. A., Ying, H., Sánchez, N., Marchesini, M., et al. (2014). Oncogene ablation-resistant pancreatic cancer cells depend on mitochondrial function. *Nature*, *514*(7524), 628–632.

Walker, G., MacLeod, K., Williams, A. R., Cameron, D. A., Smyth, J. F., & Langdon, S. P. (2007). Estrogen-regulated gene expression predicts response to endocrine therapy in patients with ovarian cancer. *Gynecologic Oncology*, *106*, 461–468.

Wang, W., Kryczek, I., Dostál, L., Lin, H., Tan, L., Zhao, L., et al. (2016). Effector T cells abrogate stroma-mediated chemoresistance in ovarian cancer. *Cell*, *165*(5), 1092–1105.

Wang, Y., Li, L., Guo, X., Jin, X., Sun, W., Zhang, X., et al. (2012). Interleukin-6 signaling regulates anchorage-independent growth, proliferation, adhesion and invasion in human ovarian cancer cells. *Cytokine*, *59*(2), 228–236.

Wang, Y., Niu, X. L., Qu, Y., Wu, J., Zhu, Y. Q., Sun, W. J., et al. (2010). Autocrine production of interleukin-6 confers cisplatin and paclitaxel resistance in ovarian cancer cells. *Cancer Letters*, *295*(1), 110–123.

Warburg, O. (1956). On the origin of cancer cells. *Science*, *123*(3191), 309–314.

Weinberg, F., Hamanaka, R., Wheaton, W. W., Weinberg, S., Joseph, J., Lopez, M., et al. (2010). Mitochondrial metabolism and ROS generation are essential for Kras-mediated tumorigenicity. *Proceedings of the National Academy of Sciences of the United States of America*, *107*(19), 8788–8793.

Wheaton, W. W., Weinberg, S. E., Hamanaka, R. B., Soberanes, S., Sullivan, L. B., Anso, E., et al. (2014). Metformin inhibits mitochondrial complex I of cancer cells to reduce tumorigenesis. *eLife*, *3*, e02242.

Yang, L., Achreja, A., Yeung, T. L., Mangala, L. S., Jiang, D., Han, C., et al. (2016). Targeting stromal glutamine synthetase in tumors disrupts tumor microenvironment-regulated cancer cell growth. *Cell Metabolism*, *24*(5), 685–700.

Yang, L., Moss, T., Mangala, L. S., Marini, J., Zhao, H., Wahlig, S., et al. (2014). Metabolic shifts toward glutamine regulate tumor growth, invasion and bioenergetics in ovarian cancer. *Molecular Systems Biology*, *10*, 728.

Yoshida, S., Tsutsumi, S., Muhlebach, G., Sourbier, C., Lee, M. J., Lee, S., et al. (2013). Molecular chaperone TRAP1 regulates a metabolic switch between mitochondrial respiration and aerobic glycolysis. *Proceedings of the National Academy of Sciences of the United States of America*, *110*, 1604–1612.

Zhang, B., Wang, J., Huang, Z., Wei, P., Liu, Y., Hao, J., et al. (2015). Aberrantly upregulated TRAP1 is required for tumorigenesis of breast cancer. *Oncotarget*, *6*(42), 44495–44508.

Zhang, H., Wu, H., Guan, J., Wang, L., Ren, X., Shi, X., et al. (2015). Paracrine SDF-1α signaling mediates the effects of PSCs on GEM chemoresistance through an IL-6 autocrine loop in pancreatic cancer cells. *Oncotarget*, *6*(5), 3085–3097.

Zhao, Z. Q., Cheng, X. M., Wang, Y. B., Han, R., Li, L., & Xiang, T. (2014). Metformin inhibits the IL-6-induced epithelial-mesenchymal transition and lung adenocarcinoma growth and metastasis. *PloS One*, *9*(4), e95884.

Zheng, X., Carstens, J. L., Kim, J., Scheible, M., Kaye, J., Sugimoto, H., et al. (2015). Epithelial-to-mesenchymal transition is dispensable for metastasis but induces chemoresistance in pancreatic cancer. *Nature*, *527*(7579), 525–530.

CHAPTER EIGHT

# Molecular Targets of Ascochlorin and Its Derivatives for Cancer Therapy

**Jason Chua Min-Wen[*,†,2], Benjamin Chua Yan-Jiang[*,†,2], Srishti Mishra[‡,2], Xiaoyun Dai[‡], Junji Magae[§], Ng Shyh-Chang[†,1], Alan Prem Kumar[‡,¶,||,#,**,1], Gautam Sethi[§,††,1]**

[*]NUS Graduate School for Integrative Sciences and Engineering, National University of Singapore, Singapore, Singapore
[†]Stem Cell & Regenerative Biology, Genome Institute of Singapore, Agency for Science Technology and Research, Singapore, Singapore
[‡]Yong Loo Lin School of Medicine, National University of Singapore, Singapore, Singapore
[§]Magae Bioscience Institute, Tsukuba, Japan
[¶]Cancer Science Institute of Singapore, National University of Singapore, Singapore, Singapore
[||]Curtin Medical School, Faculty of Health Sciences, Curtin University, Perth, WA, Australia
[#]National University Cancer Institute, National University Health System, Singapore, Singapore
[**]University of North Texas, Denton, TX, United States
[††]School of Biomedical Sciences, Curtin Health Innovation Research Institute, Curtin University, Perth, WA, Australia
[1]Corresponding authors: e-mail address: ngsc1@gis.a-star.edu.sg; phcapk@nus.edu.sg; phcgs@nus.edu.sg

## Contents



## Abstract

Cancer is an extremely complex disease comprising of a multitude of characteristic hallmarks that continue to evolve with time. At the genomic level, random mutations leading to deregulation of diverse oncogenic signal transduction cascades and polymorphisms coupled with environmental as well as life style-related factors are major causative agent contributing to chemoresistance and the failure of conventional therapies as well as molecular targeted agents. Hence, there is an urgent need to identify novel alternative therapies based on alternative medicines to combat this dreaded disease. Ascochlorin (ASC), an isoprenoid antibiotic isolated initially from the fermented

[2] J.C.M.-W., B.C.Y.-J, and S.M. contributed equally to this chapter.

*Advances in Protein Chemistry and Structural Biology*, Volume 108
ISSN 1876-1623
http://dx.doi.org/10.1016/bs.apcsb.2017.01.001

broth of *Ascochyta viciae*, and its synthetic derivatives have recently demonstrated substantial antineoplastic effects in a variety of tumor cell lines and mouse models. The major focus of this review article is to briefly analyze the chemopreventive as well as therapeutic properties of ASC and its derivatives and to identify the multiple molecular targets modulated by this novel class of anticancer agent.

## ABBREVIATIONS

**ADP** adenosine diphosphate
**AF** ascofuranone
**AMP** Adenosine monophosphate
**AMPK** 5′ adenosine monophosphate-activated protein kinase
**AOX** alternative oxidase
**AP-1** activator protein 1
**AS-6** 4-*O*-carboxymethyl-ascochlorin
**ASC** ascochlorin
**ATF** activating transcription factor
**ATG5** autophagy protein 5
**ATM** ataxia telangiectasia mutated
**ATP** adenosine triphosphate
**ATR** ataxia telangiectasia and Rad3-related protein
**Bcl-2** B-cell lymphoma 2
**CHOP** CCAAT/enhancer binding protein homologous protein
**COX-2** cyclooxygenase-2
**CX3CL1** chemokine (C-X3-C motif) ligand 1
**DISC** death-inducing signaling complex
**DOX** doxorubicin
**ECM** extracellular matrix
**EGF** epidermal growth factor
**EGFR** epidermal growth factor receptor
**eIF4F** eukaryotic translation initiation factor
**EMT** epithelial-to-mesenchymal transition
**ER** endoplasmic reticulum
**ER** estrogen receptor
**ERK1/2** extracellular signal-regulated kinase 1/2
**ETC** electron transport chain
**FADD** Fas-associated protein with death domain
**FPTase** farnesyl-protein transferase
**GRP78** glucose-regulated protein 78
**HCC** hepatocellular carcinoma
**hDHODH** human dihydroorotate dehydrogenase
**HIF-1** hypoxia-inducible factor-1
**HIF-1α** hypoxia-inducible factor 1-alpha
**IFN-g** interferon gamma
**IL-2** interleukin-2
**ISP** iron–sulfur protein
**IκBα** nuclear factor of kappa light polypeptide gene enhancer in B-cell inhibitor, alpha

**JAK** Janus kinases
**JDP** Jun dimerization protein
**JNK1/2** c-Jun N-terminal kinases 1/2
**LC3-II** microtubule-associated protein 1A/1B-light chain 3
**MAC** 4-*O*-methyl-ascochlorin
**MAPK** mitogen-activated protein kinase
**MCP-1** monocyte chemotactic protein-1
**MDM2** E3 ubiquitin–protein ligase
**MMP** matrix metalloproteinase
**mTOR** mechanistic target of rapamycin
**mTORC1** mammalian target of rapamycin complex 1
**NF-κB** nuclear factor kappa-light-chain-enhancer of activated B cells
**oxLDL** oxidized low-density lipoprotein
**PARP** poly(ADP-ribose) polymerase
**PI3K** phosphatidylinositide 3-kinases
**PIAS3** protein inhibitor of activated STAT protein 3
**PKC** protein kinase C
**PMA** phorbol myristate acetate
**PPARγ** peroxisome proliferator-activated receptor-γ
**Q** ubiquinone
**ROS** reactive oxygen species
**SHAM** salicylhydroxamic
**STAT** signal transducer and activator of transcription
**TAO** Trypanosome alternative oxidase
**TGF-β** transforming growth factor beta
**UPR** unfolded protein response
**VCAM-1** vascular cell adhesion molecule-1
**VEGF** vascular endothelial growth factor

# 1. INTRODUCTION

Cancer is one of the primary causes of mortality worldwide. It is the second most common cause of death in developed nations and third most common cause of lethality in low-middle income nations. Worldwide, there will be 23.6 million new cases of cancer by 2030 (Society, 2015). The comprehensive research focused on cancer therapeutics till date has enhanced the survival rates of cancer patients through the conventional therapy, surgery, radiotherapy, and molecular targeted agents. However, the treatment options available so far have certain limitations such as high rates of relapse, huge toxicity levels of the chemotherapeutic agents, and development of resistance toward the anticancer agents. Consequently, there has been no significant reduction in the annual death rates due to cancer. Thus, novel strategies that can nullify the limitations of the current treatment options available for cancer are extremely vital (Chan, Supriady, Goh, & Kadir, 2015).

The remarkable progress in cancer research over the couple of decades has provided us important insights into the key hallmarks of cancer. Angiogenesis, metastasis, and evading apoptosis are some of the biological capabilities acquired during the multistep development of human tumors (Hanahan & Weinberg, 2011), thereby enabling identification of potential novel therapeutic agents to target these key events in cancer cells growth and survival. However, majority of pharmacological drugs employed for cancer therapy are generally associated with severe adverse effects, thereby making it imperative to explore the potential of alternative treatment strategies. Over the years, a multitude of natural compounds and its structures have been discovered, isolated, and characterized for their potential anticancer activities. Natural compounds as the name suggests are derived from naturally occurring living organisms such as fungi, molds, other botanical sources, microorganisms, marine organisms, and even terrestrial vertebrates and invertebrates (Newman & Cragg, 2016; Newman, Cragg, & Snader, 2000; Nobili et al., 2009; Shanmugam et al., 2016; Yang, Weng, Sethi, & Hu, 2013). Coupled with the abundant knowledge of biological and biochemical pathways and mechanisms of therapeutic actions, these have been employed as innovative therapeutic agents for treatment against various chronic ailments, including cancer (Bishayee & Sethi, 2016; Cragg, Grothaus, & Newman, 2014; Newman et al., 2000; Newman, Cragg, & Snader, 2003; Shanmugam, Kannaiyan, & Sethi, 2011; Tang, Sethi, & Kuo, 2014).

Alternatively, these compounds could also be used concurrently to increase the efficacy of existing chemotherapy and radiotherapy. Currently, there are multiple anticancer drugs that are made from natural compounds or which are synthetically similar to them (Hsieh, Yang, Sethi, & Hu, 2015; Newman & Cragg, 2007, 2016). Many plant-derived compounds are currently being used in cancer treatments. Some are even considered by the FDA as a "most required medicine in a basic health system." These include vincristine, irinotecan, etoposide, and paclitaxel, all of which exhibit diverse mechanism(s) of action in inducing apoptosis and preventing tumor growth (Jordan, 2002; Jordan & Wilson, 2004; Pommier, Leo, Zhang, & Marchand, 2010). In this review, we have briefly analyzed the anticancer effects of novel class of natural isoprenoid antibiotic, called ascochlorin (ASC) and highlighted its potential role in cancer prevention and therapy in different

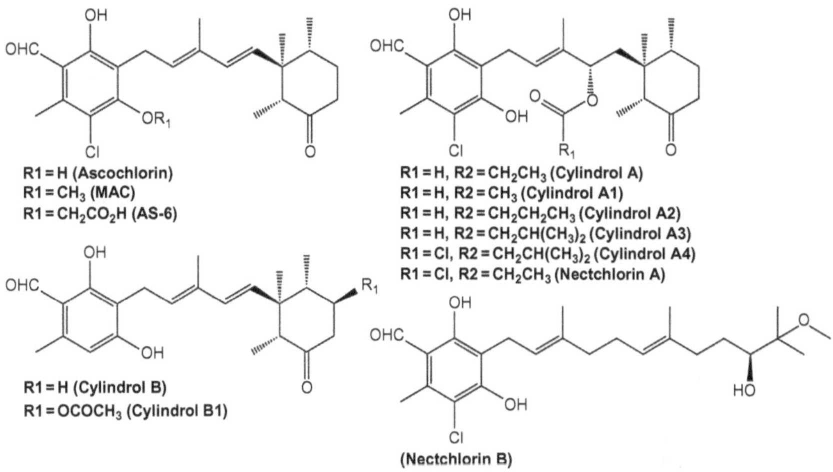


**Fig. 1** Chemical structure of ascochlorin and its synthetic derivatives.

tumor models. The molecular mechanisms underlying the pharmacological properties of ASC and its synthetic derivatives have also been discussed briefly (Fig. 1; Table 1).

## 2. PHARMACOLOGICAL PROPERTIES OF ASC AND ITS DERIVATIVES

The prenylphenol antibiotic, ASC was first isolated in the late 1960s by Tamara et al. from the filter cake of the fermented broth of *Ascochyta viciae* LIBERT (Tamura, Suzuki, Takatsuki, Ando, & Arima, 1968). Using the plaque inhibition method, it was observed that plaque formation by herpes simplex virus and Newcastle disease virus was inhibited, showing that ACS has antiviral activity against both DNA and RNA viruses (Tamura et al., 1968). ASC was found to inhibit the respiratory chain by targeting $Q_o$ and $Q_i$ sites of the mitochondrial cytochrome $bc_1$ complex, and cytochrome *bo* and trypanosome alternative oxidase in *Pichia anomala* and *Escherichia coli*, respectively (Berry et al., 2010; Mogi et al., 2009). It was suggested that the antiviral activity of ASC may be due to the inhibition of the mitochondrial respiration (Berry et al., 2010). ASC can selectively kill estrogen receptor (ER)-negative human breast cancer cell line, MX1, by suppressing the activator protein 1 (AP-1) activity of the cells (Sakaguchi et al., 2005). Treatment with ASC was found to elongate the lifespan of ER-negative mammary tumor-bearing mice (Nakajima et al., 2007). A proteomics analysis on ASC-treated human osteosarcoma

**Table 1** Summary of the Antineoplastic Properties of ASC and Its Derivatives

| Compound | Cancer Cell Type | Phenotypic Effects | Mechanisms | References |
|---|---|---|---|---|
| ASC | HepG2, Hep3B, Huh7 | Induces apoptosis<br>Suppresses migration | JAK/STAT3 | Dai et al. (2015) |
| ASC | HCCLM3, SNU38, SNU449, SK-HEP-1 | Induces apoptosis<br>Suppresses migration<br>Increases chemosensitization | JAK/STAT3 | Dai et al. (2016) |
| ASC | Cska-2, Chang, MDA-MB-231 | Suppresses migration | PKC/ERK | Hong et al. (2005) |
| ASC | MX-1, MDA231, ZR-75-1 | Induces apoptosis | AP-1 | Nakajima et al. (2007); Sakaguchi et al. (2005) |
| ASC | HS578T, BT-549, MDA-MB-435s, MDA-MB-231, MX-1 | Induces apoptosis | AP-1 | Nakajima et al. (2007); Sakaguchi et al. (2005) |
| AF | U2OS, Caki-1, MDA-MD-231 | Suppresses migration | ERK/AP-1 | Cho et al. (2007) |
| ASC/AF | U2OS, HCT116 | Activates p53 | p53 dependent | Jeong et al. (2009) |
| ASC | U2OS, HCT116 | Induces cell cycle arrest | c-Myc | Jeong and Chang (2010); Jeong et al. (2010) |
| MAC | K562 | Induces apoptosis | AMPK/mTOR | Shin et al. (2016) |
| AF | MCF-7, HepG2, U2OS, HCT116 | Induces cell cycle arrest | c-Myc | Jeong and Chang (2010); Jeong et al. (2010) |
| ASC | Caki-1 | Inhibits angiogenesis | EGFR/ERK/p70S6k | Jeong et al. (2012) |
| AF | MDA-MB-231, MCF-7 | Inhibits angiogenesis | Akt/mTOR/p70s6K | Jeong et al. (2013) |
| MAC | HB2–8, Jurkat cells | Induces apoptosis | Intrinsic/extrinsic | Tsuruga et al. (2004) |
| AS-6 | HepG2 | Induces apoptosis<br>Induces autophagy | PPARγ | Kang, Chang, and Maurizi (2012) |

cells revealed an increase in the tumor-suppressor protein, amphiphysin II, while a decrease in procancer proteins MMP-2 and prohibitin levels was observed (Kang et al., 2006).

A major problem associated with ASC in pharmacological studies was that it is poorly absorbed from the gastrointestinal tract due to its hydrophobicity. Hence, 4-O-carboxymethyl-ascochlorin (AS-6) was synthesized to enhance its physiochemical properties (Hosokawa, Ando, & Tamura, 1982). Oral treatment of AS-6 was found to reduce insulin resistance and ameliorates polydipsia, polyuria, and glycosuria in db/db genetically diabetic dyslipidemia mice (Hosokawa, Ando, & Tamura, 1985a, 1985b). Treatment of db/db mice with AS-6 was found not only to modulate the fatty acid composition (Hosokawa, Ando, & Tamura, 1985c) but also to enhance the binding of calcium ions on the plasma membrane of adipocytes (Hosokawa, Ando, & Tamura, 1985d).

AS-6 was found to be a potent agonist of nuclear hormone receptor, peroxisome proliferator-activated receptor-γ (PPARγ). AS-6 acts as a strong inducer of adipogenesis of mouse preadipocyte cell line 3T3-L1 by directly interacting with the PPARγ molecule (Togashi et al., 2002). Studies done on rat vascular smooth muscle cells showed that AS-6 inhibits NF-κB activity and several inflammatory molecule expressions, including vascular cell adhesion molecule-1 (VCAM-1), monocyte chemotactic protein-1 (MCP-1), and fractalkine (CX3CL1) (Park et al., 2006). Human hepatocellular cells (HepG2) treated with AS-6 undergo an autophagic response, resulting in an increased expression of beclin1, ATG5, and LC3-II. It also caused an increased expression of proteins involved in protein quality control, including glucose regulated protein 78 (GRP78/BiP) that can regulate endoplasmic reticulum (ER) stress responses, and transcription regulator CHOP, which can mediate ER stress-induced apoptosis (Kang et al., 2012). Human dihydroorotate dehydrogenase (hDHODH) is an inner mitochondrial membrane enzyme that can catalyze the fourth step of de novo biosynthesis of pyrimidine base, a crucial component of nucleotide and phospholipid (Munier-Lehmann, Vidalain, Tangy, & Janin, 2013). It catalyzes the oxidative conversion of dihydroorotate (DHO) to orotate with ubiquinone as the electron acceptor on the inner mitochondrial membrane. This pathway is crucial especially in proliferating cells and in modulating cell metabolism. Pharmacological inhibitors of hDHODH have been shown to be efficient in the treatment of inflammation, autoimmune disease, and cancer albeit with severe side effects (He et al., 2014; O'connor et al., 2006; Shen et al., 2016). Due to its similar structure with ubiquinone,

ASC and ascofuranone (AF) can competitively bind to hDHODH at the site of coenzyme Q substrate and inhibit it with $IC_{50}$ of 0.95 and 0.18 μM, respectively (Shen et al., 2016). ASC showed notable antiinflammatory and immunosuppressive effects in vitro and in vivo through reversible inhibition of hDHODH. ASC can also decrease concanavalin A-induced secretion of IL-2 and IFN-g by T-lymphocyte by suppressing its proliferation. In vivo, ASC was able to significantly suppress ear swelling, increase mean graft survival time from 12 to 16 days, and reduce circumference of ankle joint, indicating its beneficial antiarthritis effect (Shen et al., 2016). This immunosuppressive activity was demonstrated where 4-*O*-methyl-ascochlorin (MAC) increased the survival time of allogeneic skin implanted on flank mice (Tsuruga, Nakajima, & Magae, 2007). In addition, ASC also exerts its antiinflammatory effects via inhibition of the NF-κB pathway in LPS-stimulated RAW 264.7 macrophage cells without affecting cell viability. Mechanistically, ASC can suppress nuclear translocation and DNA binding affinity of NF-κB, concomitantly reducing the degradation of its negative regulator, IκBα. This can lead to the subsequent downregulation of its downstream factors such as iNOS, COX2, IL-1β, and IL6 that are actively involved in the inflammatory response (Lee et al., 2016).

MAC is a synthetic methylated derivative of ASC. Studies conducted in rat and mice to observe the hypocholesterolemic property of MAC demonstrated that it can inhibit intestinal absorption of cholesterol. However, the mechanism by which it occurred differed between the two species. In mice, increased fecal excretion of neutral sterols caused the inhibition (Hosokawa, Sawada, Ando, & Tamura, 1980), while in rats, enhanced synthesis of hepatic bile acids and increased biliary and fecal excretion of bile acids were the cause (Hosokawa, Sawada, Ando, & Tamura, 1981). MAC treatment notably retarded the hypertension, hyperlipidemia, and hypokalemia in deoxycorticosterone acetate-loaded uninephrectomized rats (Hosokawa, Okutomi, Sawada, Ando, & Tamura, 1981). While AS-6 can directly interact with PPARγ and enhance adipocyte differentiation, MAC suppressed adipocyte differentiation by inhibiting PPARγ expression through the modulation of AMP-activated protein kinase (AMPK) and PI3K/mTOR signaling pathways (Kim et al., 2015).

MAC has also been reported to act as an immunosuppressant and increased the survival time of the allogeneic skin grafted on mice by strongly suppressing the killer T-cell activity (Tsuruga et al., 2007). MAC was found to be capable of inducing apoptosis through diverse mechanism(s), one of which is via the activation of caspase/poly(ADP ribose) polymerase

(PARP) (Tsuruga et al., 2004), while another is through the downregulation of c-Myc expression leading to apoptosis via AMPK/mTOR-dependent mechanisms (Shin et al., 2016).

MAC can also stabilize hypoxia-inducible factor-1α (HIF-1α) via AMPK activation. Activation of AMPK can be triggered not by ROS generation but through the lowered intracellular ATP level. This activation can result in the increased glucose uptake and expression of VEGF which is a major target gene of HIF-1α (Jeong et al., 2011). As AMPK activation is directly associated with MAC and not ASC and AS-6, this may be indicative of the importance of the 4-*O*-methyl group (Jeong et al., 2011).

Cylindrols are other important members of the ASC family. Seven cylindrols, A, A1, A2, A3, A4, B, and B1, have been isolated from *Cylindrocarpon lucidum*. These ASC derivatives exhibited inhibitory activity against farnesyl-protein transferase (FPTase), an enzyme responsible for the farnesylation of Ras protein (Singh et al., 1996). As farnesylation is involved in cell-transforming activity of tumors, inhibitors of FPTase activity can be a potential chemopreventive drug (Vilella et al., 2000). The structure of cylindrol as an inhibitor has a profound influence on the FTPase activity. Cylindrol A, having a C12 propionate, was most active among the cylindrols to inhibit FTPase, while the acetate side chain on cylindrol A1 exhibited the minimal activity (Singh et al., 1996). An additional ASC derivative, cylindrol A5, has been isolated from *Cylindrocarpon* sp. FKI-4602. Cylindrol A5 showed evidence of antimicrobial activity against *Bacillus subtilis*, *Kocuria rhizophila*, *Mycobacterium smegmatis*, and *Acholeplasma laidlawii* (Kawaguchi et al., 2013).

Recently, two new ASC derivatives, nectchlorins A and B, have been isolated from leafhopper pathogen *Microcera* sp. BCC 17074. These ACS derivatives exhibited cytotoxicity to an oral cavity cancer cell line, KB. However, their cytotoxic activities against cancer cell lines NCI-H187 and MCF-7, small-cell lung cancer and breast cancer, respectively, were weaker as compared to ASC (Isaka et al., 2015).

## 3. ANTICANCER EFFECTS OF ASC AND ITS DERIVATIVES

A number of studies since the 1980s have investigated the antineoplastic properties of ASC and its derivatives and found these agents to exhibit profound antitumorigenic activity against numerous transplantable tumors, abolishing the metastasis capacity of melanomas and lung carcinoma in murine experimental models (Magae et al., 1988; Magae, Hosokawa,

Ando, Nagai, & Tamura, 1982). Over the past three decades, there has been a remarkable progress in understanding the mechanistic underpinnings, thereby providing strong evidence of ASC and its derivatives as a potential drug candidate.

The aberrant activation of the Janus kinases (JAK) and the activator of transcription proteins (STAT) signaling pathway have been implicated in various types of cancers including hepatocellular carcinoma (Siveen et al., 2014; Subramaniam et al., 2013; Thomas, Snowden, Zeidler, & Danson, 2015). In particular, STAT3 plays numerous tumorigenic roles involving sustaining cell proliferation, survival, epithelial-to-mesenchymal transition (EMT), angiogenesis, anoikis resistance, and metastasis (Chai et al., 2016; Yu, Lee, Herrmann, Buettner, & Jove, 2014). The phosphorylation and subsequent dimerization of STAT3 are crucial for its translocation into the nucleus to function as a transcription factor, regulating the expression of tumorigenic genes such as *cyclin D1*, *VEGF*, and *MMP-9* (Chai et al., 2016; Kowshik et al., 2014; Siveen et al., 2014).

One of the recent studies highlights the integral role ASC as a novel blocker of against JAK–STAT3 pathway. Not only can ASC induce apoptosis in HCC cells, but it can also circumvent doxorubicin (DOX)-induced resistance via modulating STAT3 expression and activity (Dai et al., 2015, 2016). A novel mechanism by which ASC can abrogate STAT3 activation is through the increased expression of PIAS3, a protein inhibitor of activated STAT that can negatively regulate its activation. By binding to STAT3 DNA binding domain, PIAS3 is able to downregulate both constitutive and inducible STAT3 activation in HCC cell lines. In addition, ASC was found to inhibit phosphorylation at Tyr705 residue, preventing STAT3 dimerization and nuclear translocation. This multitier inhibition of JAK–STAT3 signaling cascade leads to the reduction of cell viability, induction of apoptosis, and inhibition of invasive potential of HCC cells. In vivo experiments further validated the antitumor potential of ASC, as it was found to significantly suppress HCC tumor growth in an orthotopic mouse model (Dai et al., 2015; Fig. 2).

DOX, a conventional cytotoxic drug, is administered as a standard treatment in patients with various types of solid tumors and leukemia. However, tumor cells may develop DOX resistance due to various factors such as the gain of mesenchymal properties, polymorphisms, and gene mutations that may render the treatment ineffective (Aas et al., 1996; Abolhoda et al., 1999). In another follow-up study, ASC in combination with DOX was found to overcome the resistance in mesenchymal HCC cell lines, namely

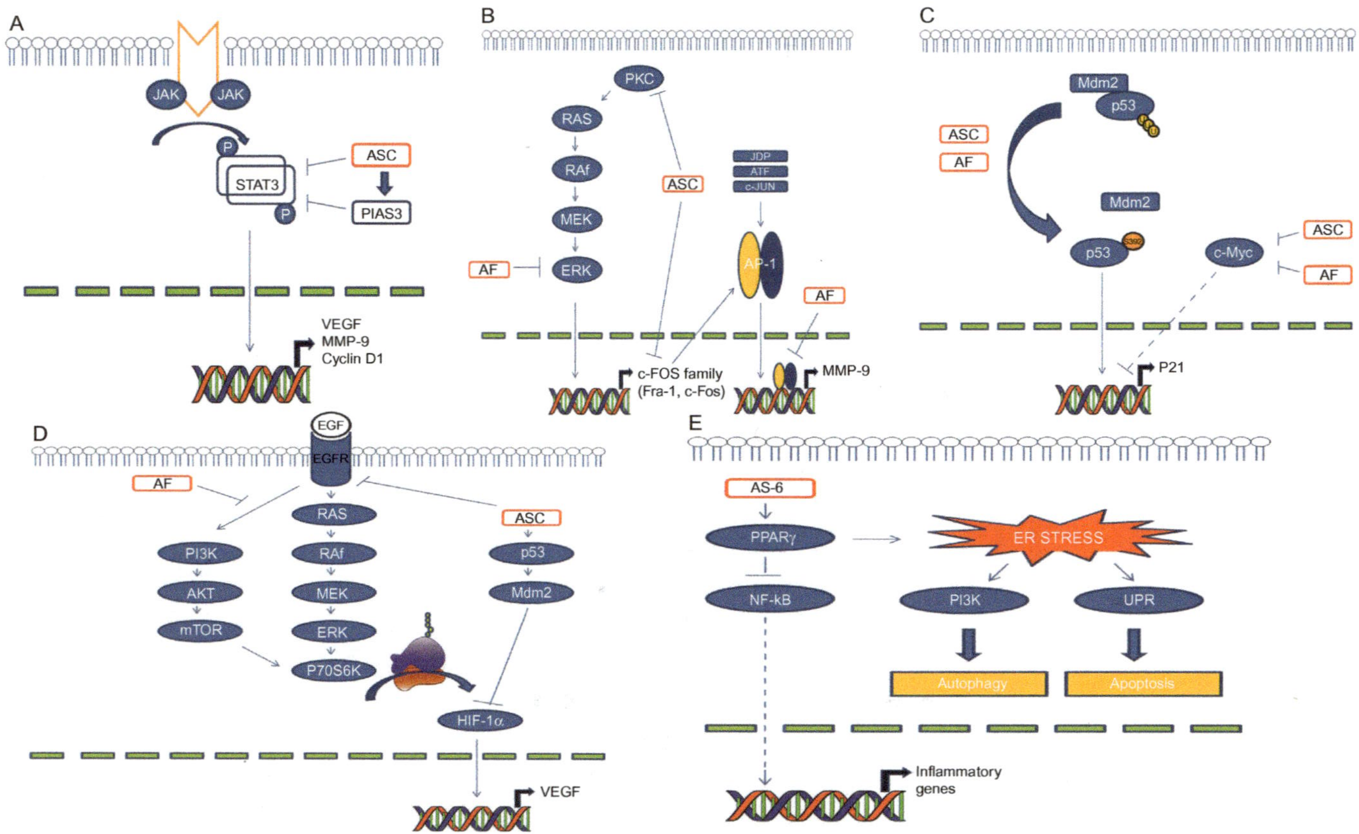


**Fig. 2** See legend on next page.

HCCLM3, SNU387, SNU449, and SK-HEP-1. In these cells, DOX alone was shown to activate the STAT3 pathway, leading to an elevated mRNA expression level of Snail and Slug (Bolós et al., 2003; Peinado, Ballestar, Esteller, & Cano, 2004). Both these proteins can repress and downregulate E-cadherin expression, which is an adherent protein responsible for the tight cell–cell interaction. Interestingly, in combination with DOX, ASC was able to reverse drug-induced resistance and cause chemosensitization. It was also observed that ASC in combination with DOX not only significantly suppressed tumor growth in HCCLM3 orthotopic mouse but also prevented detectable metastases to the lungs. Thus, targeting the STAT3 signaling pathway with ASC in combination with conventional therapy may be a unified strategy to circumvent chemoresistance in cancer cells (Dai et al., 2016).

Epithelial cells adhere tightly to each other most notably by E-cadherins and to the extracellular matrix (ECM) via integrins. This stable organization provides a barrier to the external environment, maintaining its morphology and structural conformation to carry out cellular functions effectively and for normal cell development (Engvall, 2003; Kalluri, 2003). Matrix metalloproteinase are a family of zinc-dependent endopeptidases and of preeminent importance is MMP-9. MMP-9 is secreted by tumor cells that degrade type IV collagen which is the main component of the basement membrane (Sand et al., 2013). Once the basement membrane is compromised, the tumor cells are able to dislodge and penetrate the ECM, secreting more MMPs to

**Fig. 2** Possible mechanism(s) by which ASC and its derivatives exert its antineoplastic properties. (A) ASC inhibits STAT3 phosphorylation at Tyr705, preventing dimerization and nuclear translocation. Additionally, ASC increases the expression of PIAS3 that negatively regulates STAT3. (B) ASC suppresses PKC activity, prevents AP-1 expression via the ERK pathway, and directly acts on AP-1 by decreasing the stability of c-FOS family proteins. AF inhibits the ERK pathway and blocked the binding of AP-1 protein to the binding site of the MMP-9 promoter. (C) Both ASC and AF can increase the transcription of p21 via a p53-dependent or a p53-independent pathway. In the P53-dependent pathway, p53 is activated via phosphorylation on serine 392, leading to increase transcription of p21. Alternatively, both compounds can release c-Myc inhibition on p21 most likely by target c-Myc for proteasome degradation via the p53-independent pathway. (D) ASC directly inhibits EGF-induced phosphorylation of EGFR at Tyr1068, suppressing synthesis of HIF-1$\alpha$ via the ERK pathway. Furthermore, ASC could target HIF-1$\alpha$ proteasome degradation via p53 activation. In contrast, AF can attenuate the synthesis of HIF-1$\alpha$ via the PI3K pathway. (E) AS-6 is a potent PPAR$\gamma$ agonist that negatively regulates NF-$\kappa$B expression. This impairment of NF-$\kappa$B response results in ER stress that forces the cells to undergo both autophagy and apoptosis.

further degrade the ECM, promoting tumor cell invasion and increased cell motility. Several studies have demonstrated a positive correlation between MMP-9 expression and tumor metastasis in many types of cancer (Wang et al., 2011; Zhang, Li, Lin, & Lin, 2003). A complex of signaling cascades regulate the expression of MMP-9, most notably AP-1, a transcription factor-regulated upstream by mitogen-activated protein kinase (MAPK), which includes p38, ERK1/2, and JNK1/2 (Cargnello & Roux, 2011; Peng et al., 2014). AP-1 is a heterodimer, which is formed by proteins from the c-FOS, c-JUN, ATF, and JDP families. AP-1 binds to the AP-1 site of the promoter region of MMP-9 and can positively regulate its activity (Ray, Bal, & Ray, 2005).

Numerous studies have demonstrated the ability of ASC and its derivative AF to target and inhibit AP-1 expression and activity. The subsequent downregulation of MMP-9 provides the basis by which ASC and AF could be effective to prevent MMP-9-induced invasion and metastasis. In human renal carcinoma (Caki-1) cells, ASC specifically attenuated ERK1/2 induction of AP-1 induced by phorbol myristate acetate (PMA) (Hong et al., 2005). Mechanistically, ASC suppression on protein kinase C (PKC) activity which is upstream of RAS/RAF/ERK ultimately can prevent the expression of c-FOS family proteins, most notably Fra-1 and c-fos, components of AP-1. As such, the invasive activity of Caki-1 induced by PMA was suppressed. ASC may also directly act on AP-1 by decreasing the stability of Fra-1 protein in particular. Similar results were obtained with AF whereby it can also suppress MMP-9 expression and activity simulated by PMA via inhibition of the ERK pathway in U2OS, Caki-1, and MDA-MD-231 cell lines. AF directly blocked the binding of AP-1 protein to the binding site of the consensus sequence of AP-1 of the MMP-9 promoter. Phosphorylation of upstream signal transducers was also suppressed by AF however not likely by inhibiting PKC as observed with ASC (Cho et al., 2007). Consistently, a similar mechanistic inhibition of MMP-9 was observed in human THP-1 macrophage whereby ASC can abrogate the activation of ERK pathway induced by oxidized low-density lipoprotein (oxLDL), thereby subsequently attenuating MMP-9 expression (Kang et al., 2007). Of note, in rat kidney fibroblast cells, ASC was shown to directly inhibit MMP-9 expression of heparin binding EGF-like growth factor, thereby suppressing the EGFR/MEK/ERK/TGF-B cascade, attenuating the fibrosis process (Cho et al., 2012). This indicates the possibility of ASC targeting MMP-9 directly in cancer cells, although the exact mechanisms remain to be completely understood.

For patients with ER-negative tumors, it is often associated with poor prognosis, highly malignant, resistant to conventional therapy and therefore no effective treatment regimen is readily available (Nakajima et al., 2007; Sakaguchi et al., 2005). Studies have shown that in ER-negative human breast cancer cells lines, targeting AP-1 by ASC could selectively kill these cells partly by induction of apoptosis (Nakajima et al., 2007; Sakaguchi et al., 2005). These cells express high levels of c-Jun, c-Fos, and Fra-1 compared to ER-positive cell lines; in particular, expression of Fra-1 is negatively correlated to ER expression (Nakajima et al., 2007). In vivo results demonstrated that ASC could suppress tumor growth and extend survival time of up to 60 days compared to 25 days of untreated mice implanted with ER-negative murine FM3A cells.

p53 tumor suppressor is one of the important protein orchestrating a series of processes to stabilize the genome and prevent neoplastic transformation (Greenblatt, Bennett, Hollstein, & Harris, 1994; Zilfou & Lowe, 2009). These include regulating DNA repair, halting cell cycle progression by p21, and inducing cell apoptosis if the degree of DNA damage is beyond repair by inducing proapoptotic proteins (Greenblatt et al., 1994; Zilfou & Lowe, 2009). Under normal physiological conditions, p53 is expressed at low level by the negative regulation of E3 ubiquitin–protein ligase MDM2. MDM2 not only binds to p53 and prevents its translocation into the nucleus; it can also cause its ubiquitination and tag it for degradation by proteasome (Brooks & Gu, 2006). Under stressful conditions such as oxidative stress and the damage of the DNA, p53 will be activated within minutes (Lakin & Jackson, 1999). Most notably, when there is a double-strand break, the ATM and ATR signaling cascades will be activated, phosphorylating serine 15 and 20 of p53, the former regarded as an initiating event in p53 activation (Lakin & Jackson, 1999). This will lead to the uncoupling of p53 from MDM2 and the accumulation of p53 in the nucleus. Conventional anticancer therapies target and damage the DNA (Helleday, Petermann, Lundin, Hodgson, & Sharma, 2008), activating the DNA damage response with p53 playing a pivotal role in regulating this process. The aberrant cell can undergo cell cycle arrest controlled by p21 to prevent defective replication or undergo apoptosis regulated by Bcl-2 family of proteins such as Bax and Puma (Childs, Baker, Kirkland, Campisi, & van Deursen, 2014).

ASC and AF can markedly increase the expression of p53 mainly by increasing protein stability against proteasome degradation and the subsequent transcription of its downstream targets such as p21 and hdm2 in human osteosarcoma cells and colon cancer cells (Jeong et al., 2009).

Interestingly, both compounds were found to selectively phosphorylate serine 392 with a negligible effect of serine 15 and 20 commonly induced by genotoxic drug. This serves to show that in contrast to the conventional cytotoxic drugs, both compounds do not induce DNA damage to activate the DNA damage response but most likely inhibit mitochondria respiration via oxidative phosphorylation to activate p53.

*c-Myc*, a proto-oncogene, is a key regulator of several vital cellular functions such as cell cycle, apoptosis, and metabolism (Hoffman & Liebermann, 2008). Interestingly, ASC and AF that are nongenotoxic compounds are able to induce G1 arrest via a p53-independent process in several human cancer cells such as U2OS and HCT116 (Jeong & Chang, 2010; Jeong et al., 2010). p21 can also prevent the hyperphosphorylation of Rb protein, thereby allowing it to bind to E2F. The role of E2F is to facilitate progression from G1 to S phase; hence, p21 inhibition of E2F may be responsible for the cell cycle arrest at the G1 phase, thereby preventing the progression to the S phase. The pronounced decline of c-Myc expression by MAC can efficiently convey signals to leukemic K652 cells to undergo sub-G1 arrest. MAC can also induce an increase in activity of caspase 3/7, leading to PARP-1 cleavage, which may eventually lead to apoptosis (Shin et al., 2016). Mechanistically, phosphorylation and activation of AMPK in concert with the suppression of mTORC1 by MAC can downregulate eukaryotic translation initiation factor (eIF4F) and p70S6K, both of which are crucial components in protein translation. Hence by suppressing c-Myc protein synthesis, cells cycle arrest is initiated and K562 leukemic cells ultimately undergo apoptosis. Interestingly, MAC-induced apoptosis was also specifically observed in the leukemic cells and not in normal peripheral blood mononuclear cell.

As the tumor grows in size, a more effective and efficient transport system is required for the cells to obtain sufficient nutrients and oxygen. The sheer size and volume of the tumor make normal diffusion inadequate as the limit between the cell and the blood vessel has to be less than 100 μm (Rieger & Welter, 2015). To circumvent this phenomenon for the lack of nutrients and a hypoxic microenvironment due to excessive cell growth, cancer cells express hypoxia-inducible factor-1 (HIF-1), a transcription factor that can regulate the process of angiogenesis, the formation of new blood vessel (Powis & Kirkpatrick, 2004). One protein that is positively regulated by HIF-1α is vascular endothelial growth factor (VEGF) that can stimulate angiogenesis in tumor cells (Büchler et al., 2003). EGF binding to its receptor EGFR, a hypoxia state, and AMPK and JAK–STAT signaling cascades are common pathways that can induce HIF-1α expression (Masoud & Li, 2015; Swinson & O'Byrne, 2006; Xu et al., 2005). In particular, EGFR, a receptor tyrosine kinase, is overexpressed in a wide range of cancers and unrestrained EGFR-mediated signaling is commonly observed in various cancers (Paul & Mukhopadhyay, 2004; Wieduwilt & Moasser, 2008). Aberrant expression of EGFR as well as HIF-1α has been found to correlate with poor prognosis and shorter survival, and demonstrate chemotherapy resistance in various cancers (Baba et al., 2010; Normanno, Maiello, & De Luca, 2003; Wieduwilt & Moasser, 2008). Hence, it would be efficacious to therapeutically target and attenuate both HIF-1α expression and EGFR signaling pathway to combat cancer progression.

Two studies have independently demonstrated the emerging potential of both ASC and AF in suppressing EGFR-induced HIF-1α expression in human cervical epidermoid carcinoma cells (CaSki) cells and MDA-MB-231 breast cancer cells, respectively, albeit by diverse mechanisms (Jeong et al., 2013, 2012). ASC was found to exert its inhibitory effects at multiple steps on this pathway. First, ASC can directly inhibit EGF-induced phosphorylation of EGFR at Tyr1068 and its subsequent downstream pathway via ERK/p70S6K and not the PI3K/AKT/mTORC1 pathway. This prevents the protein synthesis of HIF-1α downstream of ERK. Second, ASC was shown to degrade HIF-1α, most probably through the activation of p53 and its subsequent expression of mdm2 that marked HIF-1α for proteasomal degradation. In vivo experiment demonstrated that ASC could inhibit EGF-induced angiogenesis in CaSki-implanted matrigel plugs and a significant dose-dependent decrease in hemoglobin levels (Jeong et al., 2012). Conversely, AF was found to suppress EGF induction of HIF-1α by inhibition of the AKT/mTORC1/p70S6K pathway and not via the ERK pathway (Jeong et al., 2013). Additionally, the suppressive ability of ASC on EGFR was demonstrated where proteome analysis of human osteosarcoma cells suggested that ASC can cause a fourfold decrease in its activity (Kang et al., 2006).

Another upstream regulator of HIF-1α is AMPK, a serine/threonine protein kinase that is a vital modulator of the metabolism and autophagy process in a cell (Li, Saud, Young, Chen, & Hua, 2015; Wang, Song, & Zou, 2012). AMPK pathway is a survival mechanism activated under stressful conditions such as glucose starvation (low ATP/AMP levels) and hypoxia to counteract these conditions (Laderoute et al., 2006). Deregulation of the tightly controlled cellular energetics and imbalance in metabolic homeostasis is a characteristic hallmark of tumor cells; hence AMPK has been studied extensively in its role as a metabolic tumor suppressor (Li, Saud, et al., 2015). A number of distinct downstream signaling emerge from AMPK, such as the activation of p53, blocking of cell proliferation via mTORC1, and the negative regulation of COX-2, a proinflammatory enzyme associated with tumorigenesis (Li, Saud, et al., 2015). AMPK activation has been correlated with a better prognosis and a significant increase in overall survival in many types of cancer (William et al., 2012).

Paradoxically, in the early phase for MCF-7 cells to overcome the ATP-deprived condition induced by MAC, AMPK is activated, which in turn can increase the expression of HIF-1α as well as its target genes (Jeong et al., 2011), which might be counterintuitive for MAC, a tumorigenic drug to induce HIF-1α expression. It could be explained that for long-term exposure to MAC, cells will eventually undergo apoptosis due to the severe deprivation of ATP, as observed in human leukemic cell lines (Tsuruga et al., 2004). The mechanism through which MAC can induce substantial apoptosis is unique, in that it could possibly activate both the intrinsic and extrinsic cell death pathways. Interestingly, MAC could activate the death-inducing signaling complex (DISC) without the need for caspase-8 or FADD, components of the extrinsic pathway, possibly by the ER stressed-induced apoptosis.

Autophagy in the context of cancer can be a double-edged sword as it may enhance or prevent tumor progression depending on the signal it receives (Kenific, Thorburn, & Debnath, 2010). It is a cell survival mechanism, breaking down cellular components and organelles in the lysosome, and recycling and reallocating essential materials for vital processes under stressful conditions (Karam & Hsieh, 2009). Due to the inherent nature of the tumor microenvironment which is hypoxic and nutrient starved, and also the increased energetic demands of cancer cells, these cells may exploit this mechanism to proliferate rapidly and to invade neighboring cells (White, 2015). On the contrary, autophagy may release the inhibition of p53 protein, exerting its antitumorigenic capacity, and can serve as a barrier to tumorigenesis (White, 2016).

AS-6 was found to leverage on the autophagy mechanism to block sustained proliferation of human hepatocellular carcinoma cells (HepG2) (Kang et al., 2012). Mechanistically, AS-6 is a potent agonist of the PPARγ, which is a negative regulator of NF-κB. An NF-κB physiological role is to counteract and withstand stress via triggering the inflammatory response, promoting expression of antiapoptotic genes to ensure cell survival (Hoesel & Schmid, 2013; Li, Zhang, et al., 2015; Sethi & Tergaonkar,

2009). The impairment of NF-κB responsive pathways allows damage to accumulate inducing ER stress (Chai, Siveen, Shanmugam, Arfuso, & Sethi, 2015; Sethi, Shanmugam, Ramachandran, Kumar, & Tergaonkar, 2012; Sethi & Tergaonkar, 2009). Under ER stress, PI3K is activated, increasing the expression of beclin1, ATG5, and LC3-II, all crucial components that will lead to autophagosome formation. Second, ER stress was shown to activate the ER unfolded protein response (UPR), thereby increasing the expression of CHOP. CHOP can enhance the expression of proapoptotic protein BIM, thereby ultimately leading to apoptosis. Overall, the cytotoxic properties of AS-6 attributed to ER stress-induced autophagy and apoptosis via the PPARγ activation (Kang et al., 2012).

The antineoplastic properties of ASC and its derivatives could also be attributed to their role as potent, specific inhibitors of complex III of the electron transport chain (ETC). The ETC consists of five mitochondrial redox carriers responsible for generating a proton gradient which will be dissipated to drive the synthesis of ATP, requiring oxygen as the last electron acceptor. ASC and its derivative have been found to modulate cytochrome $bc_1$ complex, commonly known as complex III. A process known as the protonmotive Q cycle occurs at complex III. In summary, two fully reduced coenzyme Q ($QH_2$, ubiquinol) are needed to complete the reaction whereby one will be completed oxidized back to ubiquinone (Q) and the other regenerated back to $QH_2$, in the process pumping protons ($H^+$) from the matrix into the intermembrane space.

To basically understand that how ASC can function as a mitochondrial complex III inhibitor, the basic process of the Q cycle has to be understood (Bleier & Dröse, 2013; Cramer, Hasan, & Yamashita, 2011). First, complex III receives electrons from complexes I and II in the form of fully reduced ubiquinol, carrying two electrons. The first $QH_2$ binds to the $Q_o$ site of cytochrome *b* where bifurcated electron transfer reaction occurs. The first electron is passed on to the iron–sulfur protein (ISP), specifically to the Rieske center containing a 2Fe—2S cluster, which will be transported to the heme of cytochrome $c_1$, and finally to the heme of cytochrome *c*, which is bound to complex III at the surface of the intermembrane space. The reduced cytochrome *c* will then transport the electron to complex IV, allowing the ETC to progress. In this process, two protons diffuse into the intermembrane space. The second electron is transferred to heme $b_L$, followed by heme $b_H$, leaving ubiquinone (Q) fully oxidized which moves to the $Q_i$ site on cytochrome *b*, where it receives the second electron back from heme $b_H$.

This second electron does not bind to the Rieske center due to confirmation change in the ISP after the first electron binds to it. The next cycle begins where the second $QH_2$ similarly binds the $Q_o$ site, sending one electron to the Rieske center and ultimately to the cytochrome *c* via the same pathway, allowing two more $H^+$ diffused into the intermembrane space. The second electron will then be transferred to heme $b_L$, followed by $b_H$, and finally to the first partially reduced coenzyme Q bound at site $Q_i$. The fully oxidized ubiquinone will return to complex I or II for another pair of electrons, while the first coenzyme Q which is fully reduced by now will consume protons from the matrix and be released back. This Q cycle allows four protons to be transported from the matrix into the intermembrane space, creating a proton gradient that will be dissipated in complex V to drive the synthesis of ATP. Hence by inhibiting any of the pathways of the Q cycle, the formation of the proton gradient across the inner membrane will be disrupted, affecting the production of ATP.

ASC was shown to inhibit oxygen uptake activities with five different respiratory substrates much more strongly in *Hydrangea anomala* and rat liver mitochondria, even as potent with $IC_{50}$ comparable to antimycin A, a potent specific inhibitor at the $Q_i$ site (Berry et al., 2010). Interestingly, crystallographic studies demonstrated that ASC could be located at both the $Q_i$ and $Q_o$ sites; in particular the 4-OH is important for inhibition. It was elucidated that ASC binding at both $Q_i$ and $Q_o$ sites involves a halide bond with the Cl atom. ASC also immobilized the Rieske ISP in the cytochrome *b* position upon its binding to the $Q_o$, where the structural change of the ISP is essential for the electrons' bidirectional movement. In comparison to ASC and its derivative such as MAC, AF, and AS-6, ASC inhibits complex III much strongly with the lowest $IC_{50}$. Inhibitors that might inhibit both $Q_i$ and $Q_o$ active sites could be invaluable as the less likelihood of resistance to this dual action.

There exists another alternative route in which the electrons can bypass the conventional ETC for aerobic respiration known as the alternative oxidase (AOX) pathway. It is a nonprotonmotive ubiquinol oxidase that is found in mitochondria of all higher plants, algae, fungi, yeast, eubacteria, and protists (Elliott et al., 2014). The electrons from complexes I and II are able to bypass complexes III and IV, with AOX accepting the electron and catalyzing the four-electron reduction of dioxygen to water. AOX pathway protects against the accumulation of ROS when the ETC is blocked, for example, in the presence of cyanides which inhibit the activity of complex IV. Since complex I is the only remaining carrier pumping $H^+$, electron flow via AOX can still support a reduced ATP yield. As this unique metabolic difference is not seen in human, the AOX enzyme has been subjected for drug treatment in the disease human African trypanosomiasis, commonly known as sleeping sickness. The protozoan parasite *Trypanosoma brucei* can infect the human host and depends entirely on the TAO pathway for ATP production, hence an attractive target. Interestingly, both AF and ASC are also potent and specific inhibitors of ubiquinol oxidase activity of trypanosome mitochondrial AOX, with a much lower $IC_{50}$ than salicylhydroxamic acid (SHAM), the conventional inhibitor of AOX against native *Arum maculatum* and purified recombinant AOX (Saimoto, Kido, Haga, Sakamoto, & Kita, 2013).

## 4. CONCLUSION

ASC and its derivatives have been recently found to modulate diverse oncogenic cell signaling cascades in a wide variety of tumor models. This encompasses a remarkable diverse variety of anticancer effects including inhibition of aberrant proliferation, induction of selective apoptosis and autophagy, attenuation of metastasis, and promotion of chemosensitization. Hence, this preliminary knowledge provides a logical framework to complement conventional/molecular targeted therapies with ASC and its derivatives as a therapeutic strategy against diverse malignancies. Looking forward, an important agenda will be to verify their safety profile as well as pharmacokinetics, thereby ultimately refining and implementing these multitargeted natural agents for human clinical trials.

## ACKNOWLEDGMENTS

This work was supported by NUHS Basic seed grant (T1-BSRG 2015-02) and Ministry of Education Tier 1 grant to G.S. A.P.K. was supported by grants from National Medical Research Council of Singapore, NCIS Yong Siew Yoon Research Grant through donations from the Yong Loo Lin Trust, and the National Research Foundation Singapore and the Singapore Ministry of Education under its Research Centers of Excellence initiative to Cancer Science Institute of Singapore, National University of Singapore.

*Conflict of Interest*: The authors declare no conflict of interest.

## REFERENCES

Aas, T., Børresen, A.-L., Geisler, S., Smith-Sørensen, B., Johnsen, H., Varhaug, J. E., et al. (1996). Specific P53 mutations are associated with de novo resistance to doxorubicin in breast cancer patients. *Nature Medicine*, *2*(7), 811–814.

Abolhoda, A., Wilson, A. E., Ross, H., Danenberg, P. V., Burt, M., & Scotto, K. W. (1999). Rapid activation of MDR1 gene expression in human metastatic sarcoma after in vivo exposure to doxorubicin. *Clinical Cancer Research*, *5*(11), 3352–3356.

Baba, Y., Nosho, K., Shima, K., Irahara, N., Chan, A. T., Meyerhardt, J. A., et al. (2010). HIF1A overexpression is associated with poor prognosis in a cohort of 731 colorectal cancers. *The American Journal of Pathology*, *176*(5), 2292–2301.

Berry, E. A., Huang, L.-s., Lee, D.-W., Daldal, F., Nagai, K., & Minagawa, N. (2010). Ascochlorin is a novel, specific inhibitor of the mitochondrial cytochrome bc 1 complex. *Biochimica et Biophysica Acta (BBA)—Bioenergetics*, *1797*(3), 360–370.

Bishayee, A., & Sethi, G. (2016). Bioactive natural products in cancer prevention and therapy: Progress and promise. *Seminars in Cancer Biology*, *40–41*, 1–3. http://dx.doi.org/10.1016/j.semcancer.2016.08.006.

Bleier, L., & Dröse, S. (2013). Superoxide generation by complex III: From mechanistic rationales to functional consequences. *Biochimica et Biophysica Acta (BBA)—Bioenergetics*, *1827*(11), 1320–1331.

Bolós, V., Peinado, H., Pérez-Moreno, M. A., Fraga, M. F., Esteller, M., & Cano, A. (2003). The transcription factor slug represses E-cadherin expression and induces epithelial to mesenchymal transitions: A comparison with snail and E47 repressors. *Journal of Cell Science*, *116*(3), 499–511.

Brooks, C. L., & Gu, W. (2006). p53 ubiquitination: Mdm2 and beyond. *Molecular Cell*, *21*(3), 307–315.

Büchler, P., Reber, H. A., Büchler, M., Shrinkante, S., Büchler, M. W., Friess, H., et al. (2003). Hypoxia-inducible factor 1 regulates vascular endothelial growth factor expression in human pancreatic cancer. *Pancreas*, *26*(1), 56–64.

Cargnello, M., & Roux, P. P. (2011). Activation and function of the MAPKs and their substrates, the MAPK-activated protein kinases. *Microbiology and Molecular Biology Reviews*, *75*(1), 50–83.

Chai, E. Z., Shanmugam, M. K., Arfuso, F., Dharmarajan, A., Wang, C., Kumar, A. P., et al. (2016). Targeting transcription factor STAT3 for cancer prevention and therapy. *Pharmacology & Therapeutics*, *162*, 86–97. http://dx.doi.org/10.1016/j.pharmthera.2015.10.004.

Chai, E. Z., Siveen, K. S., Shanmugam, M. K., Arfuso, F., & Sethi, G. (2015). Analysis of the intricate relationship between chronic inflammation and cancer. *The Biochemical Journal*, *468*(1), 1–15. http://dx.doi.org/10.1042/BJ20141337.

Chan, C. K., Supriady, H., Goh, B. H., & Kadir, H. A. (2015). Elephantopus scaber induces apoptosis through ROS-dependent mitochondrial signaling pathway in HCT116 human colorectal carcinoma cells. *Journal of Ethnopharmacology*, *168*, 291–304. http://dx.doi.org/10.1016/j.jep.2015.03.072.

Childs, B. G., Baker, D. J., Kirkland, J. L., Campisi, J., & van Deursen, J. M. (2014). Senescence and apoptosis: Dueling or complementary cell fates? *EMBO Reports*, *15*(11), 1139–1153.

Cho, H.-J., Kang, J.-H., Jeong, J.-H., Jeong, Y.-J., Park, K.-K., Park, Y.-Y., et al. (2012). Ascochlorin suppresses TGF-β1-induced PAI-1 expression through the inhibition of phospho-EGFR in rat kidney fibroblast cells. *Molecular Biology Reports*, *39*(4), 4597–4603.

Cho, H.-J., Kang, J. H., Kwak, J.-Y., Lee, T.-S., Lee, I.-S., Park, N. G., et al. (2007). Ascofuranone suppresses PMA-mediated matrix metalloproteinase-9 gene activation through the Ras/Raf/MEK/ERK-and Ap1-dependent mechanisms. *Carcinogenesis*, *28*(5), 1104–1110.

Cragg, G. M., Grothaus, P. G., & Newman, D. J. (2014). New horizons for old drugs and drug leads. *Journal of Natural Products*, 77(3), 703–723. http://dx.doi.org/10.1021/np5000796.

Cramer, W. A., Hasan, S. S., & Yamashita, E. (2011). The Q cycle of cytochrome bc complexes: A structure perspective. *Biochimica et Biophysica Acta (BBA)—Bioenergetics, 1807*(7), 788–802.

Dai, X., Ahn, K. S., Kim, C., Siveen, K. S., Ong, T. H., Shanmugam, M. K., et al. (2015). Ascochlorin, an isoprenoid antibiotic inhibits growth and invasion of hepatocellular carcinoma by targeting STAT3 signaling cascade through the induction of PIAS3. *Molecular Oncology*, 9(4), 818–833. http://dx.doi.org/10.1016/j.molonc.2014.12.008.

Dai, X., Ahn, K. S., Wang, L. Z., Kim, C., Deivasigamni, A., Arfuso, F., et al. (2016). Ascochlorin enhances the sensitivity of doxorubicin leading to the reversal of epithelial-to-mesenchymal transition in hepatocellular carcinoma. *Molecular Cancer Therapeutics, 15*(12), 2966–2976.

Elliott, C., Young, L., May, B., Shearman, J., Albury, M. S., Kido, Y., et al. (2014). Purification and characterisation of recombinant DNA encoding the alternative oxidase from Sauromatum guttatum. *Mitochondrion, 19*, 261–268.

Engvall, E. (2003). Structure and function of basement membranes. *International Journal of Developmental Biology, 39*(5), 781–787.

Greenblatt, M., Bennett, W. P., Hollstein, M., & Harris, C. (1994). Mutations in the p53 tumor suppressor gene: Clues to cancer etiology and molecular pathogenesis. *Cancer Research, 54*(18), 4855–4878.

Hanahan, D., & Weinberg, R. A. (2011). Hallmarks of cancer: The next generation. *Cell, 144*(5), 646–674.

He, T., Haapa-Paananen, S., Kaminskyy, V., Kohonen, P., Fey, V., Zhivotovsky, B., et al. (2014). Inhibition of the mitochondrial pyrimidine biosynthesis enzyme dihydroorotate dehydrogenase by doxorubicin and brequinar sensitizes cancer cells to TRAIL-induced apoptosis. *Oncogene, 33*(27), 3538–3549.

Helleday, T., Petermann, E., Lundin, C., Hodgson, B., & Sharma, R. A. (2008). DNA repair pathways as targets for cancer therapy. *Nature Reviews Cancer, 8*(3), 193–204.

Hoesel, B., & Schmid, J. A. (2013). The complexity of NF-κB signaling in inflammation and cancer. *Molecular Cancer, 12*(1), 1.

Hoffman, B., & Liebermann, D. (2008). Apoptotic signaling by c-MYC. *Oncogene, 27*(50), 6462–6472.

Hong, S., Park, K.-K., Magae, J., Ando, K., Lee, T.-S., Kwon, T. K., et al. (2005). Ascochlorin inhibits matrix metalloproteinase-9 expression by suppressing activator protein-1-mediated Gene expression through the ERK1/2 signaling pathway: inhibitory effects of ascochlorin on the invasion of renal carcinoma cells. *Journal of Biological Chemistry, 280*(26), 25202–25209.

Hosokawa, T., Ando, K., & Tamura, G. (1982). An ascochlorin derivative, AS-6, potentiates insulin action in streptozotocin diabetic mice and rats. *Agricultural and Biological Chemistry, 46*(11), 2865–2869.

Hosokawa, T., Ando, K., & Tamura, G. (1985a). An ascochlorin derivative, AS-6, reduces insulin resistance in the genetically obese diabetic mouse, db/db. *Diabetes, 34*(3), 267–274.

Hosokawa, T., Ando, K., & Tamura, G. (1985b). Effect of oral treatment with a new hypoglycemic agent, AS-6, on the metabolic activities of adipocytes in dbdb mice: A comparative study. *Biochemical and Biophysical Research Communications, 126*(1), 471–476.

Hosokawa, T., Ando, K., & Tamura, G. (1985c). Modification of fatty acid composition in adipocyte plasma membranes by an oral treatment with a new antidiabetic agent, AS-6, in genetically obese diabetic mice, db/db. *Biochimica et Biophysica Acta, 834*(1), 130–133.

Hosokawa, T., Ando, K., & Tamura, G. (1985d). Treatment with an ascochlorin derivative, AS-6 increases $^{45}Ca^{2+}$ binding on the plasma membrane of adipocytes in db/db mice. *Biochemical and Biophysical Research Communications, 127*(1), 247–253.

Hosokawa, T., Okutomi, T., Sawada, M., Ando, K., & Tamura, G. (1981a). Unusual concentration of urine and prevention of polydipsia by fungal prenylphenols in DOCA hypertensive rats. *European Journal of Pharmacology, 69*(4), 429–438.

Hosokawa, T., Sawada, M., Ando, K., & Tamura, G. (1980). Enhanced excretion of fecal neutral sterols and the hypocholesterolemic property of 4-*O*-methylascochlorin in mice. *Agricultural and Biological Chemistry, 44*(10), 2461–2468.

Hosokawa, T., Sawada, M., Ando, K., & Tamura, G. (1981b). Alteration of cholesterol metabolism by 4-O-methylascochlorin in rats. *Lipids, 16*(6), 433–438.

Hsieh, Y. S., Yang, S. F., Sethi, G., & Hu, D. N. (2015). Natural bioactives in cancer treatment and prevention. *BioMed Research International, 2015*, 182835. http://dx.doi.org/10.1155/2015/182835.

Isaka, M., Yangchum, A., Supothina, S., Laksanacharoen, P., Luangsa-ard, J. J., & Hywel-Jones, N. L. (2015). Ascochlorin derivatives from the leafhopper pathogenic fungus Microcera sp. BCC 17074. *The Journal of Antibiotics, 68*(1), 47–51.

Jeong, J.-H., & Chang, Y.-C. (2010). Ascochlorin, an isoprenoid antibiotic, induces G1 arrest via downregulation of c-Myc in a p53-independent manner. *Biochemical and Biophysical Research Communications, 398*(1), 68–73.

Jeong, J.-H., Kang, J. H., Hwang, S.-L., Cho, H.-J., Park, K.-K., Park, Y.-Y., et al. (2011). 4-*O*-Methylascochlorin, methylated derivative of ascochlorin, stabilizes HIF-1α via AMPK activation. *Biochemical and Biophysical Research Communications, 406*(3), 353–358.

Jeong, J.-H., Kang, S.-S., Park, K.-K., Chang, H.-W., Magae, J., & Chang, Y.-C. (2010). p53-independent induction of G1 arrest and p21WAF1/CIP1 expression by ascofuranone, an isoprenoid antibiotic, through downregulation of c-Myc. *Molecular Cancer Therapeutics, 9*(7), 2102–2113.

Jeong, J. H., Jeong, Y. J., Cho, H. J., Shin, J. M., Kang, J. H., Park, K. K., et al. (2012). Ascochlorin inhibits growth factor-induced HIF-1α activation and tumor-angiogenesis through the suppression of EGFR/ERK/p70S6K signaling pathway in human cervical carcinoma cells. *Journal of Cellular Biochemistry, 113*(4), 1302–1313.

Jeong, J. H., Nakajima, H., Magae, J., Furukawa, C., Taki, K., Otsuka, K., et al. (2009). Ascochlorin activates p53 in a manner distinct from DNA damaging agents. *International Journal of Cancer, 124*(12), 2797–2803.

Jeong, Y.-J., Cho, H.-J., Magae, J., Lee, I.-K., Park, K.-G., & Chang, Y.-C. (2013). Ascofuranone suppresses EGF-induced HIF-1α protein synthesis by inhibition of the Akt/mTOR/p70S6K pathway in MDA-MB-231 breast cancer cells. *Toxicology and Applied Pharmacology, 273*(3), 542–550.

Jordan, M. (2002). Mechanism of action of antitumor drugs that interact with microtubules and tubulin. *Current Medicinal Chemistry. Anti-Cancer Agents, 2*(1), 1–17.

Jordan, M. A., & Wilson, L. (2004). Microtubules as a target for anticancer drugs. *Nature Reviews Cancer, 4*(4), 253–265.

Kalluri, R. (2003). Basement membranes: Structure, assembly and role in tumour angiogenesis. *Nature Reviews Cancer, 3*(6), 422–433.

Kang, J. H., Chang, Y.-C., & Maurizi, M. R. (2012). 4-*O*-Carboxymethyl ascochlorin causes ER stress and induced autophagy in human hepatocellular carcinoma cells. *Journal of Biological Chemistry, 287*(19), 15661–15671.

Kang, J. H., Kim, J. K., Park, W. H., Park, K. K., Lee, T. S., Magae, J., et al. (2007). Ascochlorin suppresses oxLDL-induced MMP-9 expression by inhibiting the MEK/ERK signaling pathway in human THP-1 macrophages. *Journal of Cellular Biochemistry, 102*(2), 506–514.

Kang, J. H., Park, K.-K., Lee, I.-S., Magae, J., Ando, K., Kim, C.-H., et al. (2006). Proteome analysis of responses to ascochlorin in a human osteosarcoma cell line by 2-D gel electrophoresis and MALDI-TOF MS. *Journal of Proteome Research, 5*(10), 2620–2631.

Karam, J. A., & Hsieh, J.-T. (2009). Anti-cancer strategy of transitional cell carcinoma of bladder based on induction of different types of programmed cell deaths. In G. G. Chen & P. B. S. Lai (Eds.), *Apoptosis in carcinogenesis and chemotherapy* (pp. 25–50). Springer.

Kawaguchi, M., Fukuda, T., Uchida, R., Nonaka, K., Masuma, R., & Tomoda, H. (2013). A new ascochlorin derivative from Cylindrocarpon sp. FKI-4602. *The Journal of Antibiotics, 66*(1), 23–29.

Kenific, C. M., Thorburn, A., & Debnath, J. (2010). Autophagy and metastasis: Another double-edged sword. *Current Opinion in Cell Biology, 22*(2), 241–245.

Kim, M., Cho, H.-J., Jeong, Y.-J., Chung, I.-K., Magae, J., & Chang, Y.-C. (2015). 4-*O*-Methylascochlorin suppresses differentiation of 3T3-L1 preadipocytes by inhibiting PPARγ expression through regulation of AMPK/mTOR signaling pathways. *Archives of Biochemistry and Biophysics, 583*, 79–86.

Kowshik, J., Baba, A. B., Giri, H., Deepak Reddy, G., Dixit, M., & Nagini, S. (2014). Astaxanthin inhibits JAK/STAT-3 signaling to abrogate cell proliferation, invasion and angiogenesis in a hamster model of oral cancer. *PLoS One, 9*(10). e109114. http://dx.doi.org/10.1371/journal.pone.0109114.

Laderoute, K. R., Amin, K., Calaoagan, J. M., Knapp, M., Le, T., Orduna, J., et al. (2006). 5′-AMP-activated protein kinase (AMPK) is induced by low-oxygen and glucose deprivation conditions found in solid-tumor microenvironments. *Molecular and Cellular Biology, 26*(14), 5336–5347.

Lakin, N. D., & Jackson, S. P. (1999). Regulation of p53 in response to DNA damage. *Oncogene, 18*(53), 7644–7655.

Lee, S. H., Kwak, C. H., Lee, S. K., Ha, S. H., Park, J., Chung, T. W., et al. (2016). Anti-inflammatory effect of ascochlorin in LPS-stimulated RAW 264.7 macrophage cells is accompanied with the down-regulation of iNOS, COX-2 and proinflammatory cytokines through NF-κB, ERK1/2, and p38 signaling pathway. *Journal of Cellular Biochemistry, 117*(4), 978–987.

Li, W., Saud, S. M., Young, M. R., Chen, G., & Hua, B. (2015b). Targeting AMPK for cancer prevention and treatment. *Oncotarget, 6*(10), 7365.

Li, F., Zhang, J., Arfuso, F., Chinnathambi, A., Zayed, M. E., Alharbi, S. A., et al. (2015a). NF-kappaB in cancer therapy. *Archives of Toxicology, 89*(5), 711–731. http://dx.doi.org/10.1007/s00204-015-1470-4.

Magae, J., Hayasaki, J., Matsuda, Y., Hotta, M., Hosokawa, T., Suzuki, S., et al. (1988). Antitumor and antimetastatic activity of an antibiotic, ascofuranone, and activation of phagocytes. *The Journal of Antibiotics, 41*(7), 959–965.

Magae, J., Hosokawa, T., Ando, K., Nagai, K., & Tamura, G. (1982). Antitumor protective property of an isoprenoid antibiotic, ascofuranone. *The Journal of Antibiotics, 35*(11), 1547–1552.

Masoud, G. N., & Li, W. (2015). HIF-1α pathway: Role, regulation and intervention for cancer therapy. *Acta Pharmaceutica Sinica B, 5*(5), 378–389.

Mogi, T., Ui, H., Shiomi, K., Ōmura, S., Miyoshi, H., & Kita, K. (2009). Antibiotics LL-Z1272 identified as novel inhibitors discriminating bacterial and mitochondrial quinol oxidases. *Biochimica et Biophysica Acta (BBA)—Bioenergetics, 1787*(2), 129–133.

Munier-Lehmann, H., Vidalain, P.-O., Tangy, F., & Janin, Y. L. (2013). On dihydroorotate dehydrogenases and their inhibitors and uses. *Journal of Medicinal Chemistry, 56*(8), 3148–3167.

Nakajima, H., Mizuta, N., Sakaguchi, K., Fujiwara, I., Mizuta, M., Furukawa, C., et al. (2007). Aberrant expression of Fra-1 in estrogen receptor-negative breast cancers and suppression of their propagation in vivo by ascochlorin, an antibiotic that inhibits cellular activator protein-1 activity. *Journal of Antibiotics, 60*(11), 682.

Newman, D. J., & Cragg, G. M. (2007). Natural products as sources of new drugs over the last 25 years⊥. *Journal of Natural Products, 70*(3), 461–477.

Newman, D. J., & Cragg, G. M. (2016). Natural products as sources of new drugs from 1981 to 2014. *Journal of Natural Products*, *79*(3), 629–661. http://dx.doi.org/10.1021/acs.jnatprod.5b01055.

Newman, D. J., Cragg, G. M., & Snader, K. M. (2000). The influence of natural products upon drug discovery. *Natural Product Reports*, *17*(3), 215–234.

Newman, D. J., Cragg, G. M., & Snader, K. M. (2003). Natural products as sources of new drugs over the period 1981-2002. *Journal of Natural Products*, *66*(7), 1022–1037.

Nobili, S., Lippi, D., Witort, E., Donnini, M., Bausi, L., Mini, E., et al. (2009). Natural compounds for cancer treatment and prevention. *Pharmacological Research*, *59*(6), 365–378.

Normanno, N., Maiello, M. R., & De Luca, A. (2003). Epidermal growth factor receptor tyrosine kinase inhibitors (EGFR-TKIs): Simple drugs with a complex mechanism of action? *Journal of Cellular Physiology*, *194*(1), 13–19.

O'connor, P., Li, D., Freedman, M., Bar-Or, A., Rice, G., Confavreux, C., et al. (2006). A phase II study of the safety and efficacy of teriflunomide in multiple sclerosis with relapses. *Neurology*, *66*(6), 894–900.

Park, K.-G., Lee, K.-M., Chang, Y.-C., Magae, J., Ando, K., Kim, K.-B., et al. (2006). The ascochlorin derivative, AS-6, inhibits TNF-α-induced adhesion molecule and chemokine expression in rat vascular smooth muscle cells. *Life Sciences*, *80*(2), 120–126.

Paul, M. K., & Mukhopadhyay, A. K. (2004). Tyrosine kinase-role and significance in cancer. *International Journal of Medical Sciences*, *1*(2), 101–115.

Peinado, H., Ballestar, E., Esteller, M., & Cano, A. (2004). Snail mediates E-cadherin repression by the recruitment of the Sin3A/histone deacetylase 1 (HDAC1)/HDAC2 complex. *Molecular and Cellular Biology*, *24*(1), 306–319.

Peng, H., Shi, M., Zhang, L., Li, Y., Sun, J., Zhang, L., et al. (2014). Activation of JNK1/2 and p38 MAPK signaling pathways promotes enterovirus 71 infection in immature dendritic cells. *BMC Microbiology*, *14*(1), 1.

Pommier, Y., Leo, E., Zhang, H., & Marchand, C. (2010). DNA topoisomerases and their poisoning by anticancer and antibacterial drugs. *Chemistry & Biology*, *17*(5), 421–433.

Powis, G., & Kirkpatrick, L. (2004). Hypoxia inducible factor-1α as a cancer drug target. *Molecular Cancer Therapeutics*, *3*(5), 647–654.

Ray, A., Bal, B. S., & Ray, B. K. (2005). Transcriptional induction of matrix metalloproteinase-9 in the chondrocyte and synoviocyte cells is regulated via a novel mechanism: Evidence for functional cooperation between serum amyloid A-activating factor-1 and AP-1. *The Journal of Immunology*, *175*(6), 4039–4048.

Rieger, H., & Welter, M. (2015). Integrative models of vascular remodeling during tumor growth. *Wiley Interdisciplinary Reviews: Systems Biology and Medicine*, 7(3), 113–129.

Saimoto, H., Kido, Y., Haga, Y., Sakamoto, K., & Kita, K. (2013). Pharmacophore identification of ascofuranone, potent inhibitor of cyanide-insensitive alternative oxidase of Trypanosoma brucei. *Journal of Biochemistry*, *153*(3), 267–273.

Sakaguchi, K., Nakajima, H., Mizuta, N., Furukawa, C., Ozawa, S., Ando, K., et al. (2005). Selective cytotoxicity of ascochlorin in ER-negative human breast cancer cell lines. *Biochemical and Biophysical Research Communications*, *329*(1), 46–50.

Sand, J. M., Larsen, L., Hogaboam, C., Martinez, F., Han, M., Larsen, M. R., et al. (2013). MMP mediated degradation of type IV collagen alpha 1 and alpha 3 chains reflects basement membrane remodeling in experimental and clinical fibrosis–validation of two novel biomarker assays. *PLoS One*, *8*(12). e84934.

Sethi, G., Shanmugam, M. K., Ramachandran, L., Kumar, A. P., & Tergaonkar, V. (2012). Multifaceted link between cancer and inflammation. *Bioscience Reports*, *32*(1), 1–15. http://dx.doi.org/10.1042/BSR20100136.

Sethi, G., & Tergaonkar, V. (2009). Potential pharmacological control of the NF-kappaB pathway. *Trends in Pharmacological Sciences*, *30*(6), 313–321. http://dx.doi.org/10.1016/j.tips.2009.03.004.

Shanmugam, M. K., Kannaiyan, R., & Sethi, G. (2011). Targeting cell signaling and apoptotic pathways by dietary agents: Role in the prevention and treatment of cancer. *Nutrition and Cancer, 63*(2), 161–173. http://dx.doi.org/10.1080/01635581.2011.523502.

Shanmugam, M. K., Lee, J. H., Chai, E. Z., Kanchi, M. M., Kar, S., Arfuso, F., et al. (2016). Cancer prevention and therapy through the modulation of transcription factors by bioactive natural compounds. *Seminars in Cancer Biology, 40-41*, 35–47. http://dx.doi.org/10.1016/j.semcancer.2016.03.005.

Shen, W., Ren, X., Zhu, J., Xu, Y., Lin, J., Li, Y., et al. (2016). Discovery of a new structural class of competitive hDHODH inhibitors with in vitro and in vivo anti-inflammatory, immunosuppressive effects. *European Journal of Pharmacology, 791*, 205–212.

Shin, J.-M., Jeong, Y.-J., Cho, H.-J., Magae, J., Bae, Y.-S., & Chang, Y.-C. (2016). Suppression of c-Myc induces apoptosis via an AMPK/mTOR-dependent pathway by 4-*O*-methyl-ascochlorin in leukemia cells. *Apoptosis, 21*(5), 657–668.

Singh, S. B., Ball, R. G., Bills, G. F., Cascales, C., Gibbs, J. B., Goetz, M. A., et al. (1996). Chemistry and biology of cylindrols: Novel inhibitors of Ras farnesyl-protein transferase from Cylindrocarpon lucidum. *The Journal of Organic Chemistry, 61*(22), 7727–7737.

Siveen, K. S., Sikka, S., Surana, R., Dai, X., Zhang, J., Kumar, A. P., et al. (2014). Targeting the STAT3 signaling pathway in cancer: Role of synthetic and natural inhibitors. *Biochimica et Biophysica Acta, 1845*(2), 136–154. http://dx.doi.org/10.1016/j.bbcan.2013.12.005.

Society, A. C. (2015). Global cancer facts & figures [Internet]. Atlanta: American Cancer Society.

Subramaniam, A., Shanmugam, M. K., Perumal, E., Li, F., Nachiyappan, A., Dai, X., et al. (2013). Potential role of signal transducer and activator of transcription (STAT)3 signaling pathway in inflammation, survival, proliferation and invasion of hepatocellular carcinoma. *Biochimica et Biophysica Acta, 1835*(1), 46–60. http://dx.doi.org/10.1016/j.bbcan.2012.10.002.

Swinson, D. E. B., & O'Byrne, K. J. (2006). Interactions between hypoxia and epidermal growth factor receptor in non-small-cell lung cancer. *Clinical Lung Cancer*, 7(4), 250–256.

Tamura, G., Suzuki, S., Takatsuki, A., Ando, K., & Arima, K. (1968). Ascochlorin, a new antibiotic, found by paper-disc agar-diffusion method. I. Isolation, biological and chemical properties of ascochlorin. (Studies on antiviral and antitumor antibiotics. I). *The Journal of Antibiotics, 21*(9), 539–544.

Tang, C. H., Sethi, G., & Kuo, P. L. (2014). Novel medicines and strategies in cancer treatment and prevention. *BioMed Research International, 2014*, 474078. http://dx.doi.org/10.1155/2014/474078.

Thomas, S. J., Snowden, J. A., Zeidler, M. P., & Danson, S. J. (2015). The role of JAK/STAT signalling in the pathogenesis, prognosis and treatment of solid tumours. *British Journal of Cancer, 113*(3), 365–371. http://dx.doi.org/10.1038/bjc.2015.233.

Togashi, M., Masuda, H., Kawada, T., Tanaka, M., Saida, K., Ando, K., et al. (2002). PPARgamma activation and adipocyte differentiation induced by AS-6, a prenyl-phenol antidiabetic antibiotic. *The Journal of Antibiotics, 55*(4), 417–422.

Tsuruga, M., Nakajima, H., & Magae, J. (2007). Immunosuppressive activity of 4-O-methylascochlorin. *Journal of Antibiotics, 60*(1), 20.

Tsuruga, M., Nakajima, H., Ozawa, S., Togashi, M., Chang, Y.-C., Ando, K., et al. (2004). Characterization of 4-*O*-methyl-ascochlorin-induced apoptosis in comparison with typical apoptotic inducers in human leukemia cell lines. *Apoptosis, 9*(4), 429–435.

Vilella, D., Sánchez, M., Platas, G., Salazar, O., Genilloud, O., Royo, I., et al. (2000). Inhibitors of farnesylation of Ras from a microbial natural products screening program. *Journal of Industrial Microbiology and Biotechnology, 25*(6), 315–327.

Wang, X., Lu, H., Urvalek, A. M., Li, T., Yu, L., Lamar, J., et al. (2011). KLF8 promotes human breast cancer cell invasion and metastasis by transcriptional activation of MMP9. *Oncogene*, *30*(16), 1901–1911.

Wang, S., Song, P., & Zou, M.-H. (2012). AMP-activated protein kinase, stress responses and cardiovascular diseases. *Clinical Science*, *122*(12), 555–573.

White, E. (2015). The role for autophagy in cancer. *The Journal of Clinical Investigation*, *125*(1), 42–46.

White, E. (2016). Autophagy and p53. *Cold Spring Harbor Perspectives in Medicine*, *6*(4), a026120.

Wieduwilt, M., & Moasser, M. (2008). The epidermal growth factor receptor family: Biology driving targeted therapeutics. *Cellular and Molecular Life Sciences*, *65*(10), 1566–1584.

William, W., Kim, J.-S., Liu, D., Solis, L., Behrens, C., Lee, J., et al. (2012). The impact of phosphorylated AMP-activated protein kinase expression on lung cancer survival. *Annals of Oncology*, *23*(1), 78–85.

Xu, Q., Briggs, J., Park, S., Niu, G., Kortylewski, M., Zhang, S., et al. (2005). Targeting Stat3 blocks both HIF-1 and VEGF expression induced by multiple oncogenic growth signaling pathways. *Oncogene*, *24*(36), 5552–5560.

Yang, S. F., Weng, C. J., Sethi, G., & Hu, D. N. (2013). Natural bioactives and phytochemicals serve in cancer treatment and prevention. *Evidence-based Complementary and Alternative Medicine*, *2013*, 698190. http://dx.doi.org/10.1155/2013/698190.

Yu, H., Lee, H., Herrmann, A., Buettner, R., & Jove, R. (2014). Revisiting STAT3 signalling in cancer: New and unexpected biological functions. *Nature Reviews. Cancer*, *14*(11), 736–746. http://dx.doi.org/10.1038/nrc3818.

Zhang, S., Li, L., Lin, J.-Y., & Lin, H. (2003). Imbalance between expression of matrix metalloproteinase-9 and tissue inhibitor of metalloproteinase-1 in invasiveness and metastasis of human gastric carcinoma. *World Journal of Gastroenterology*, *9*(5), 899–904.

Zilfou, J. T., & Lowe, S. W. (2009). Tumor suppressive functions of p53. *Cold Spring Harbor Perspectives in Biology*, *1*(5), a001883.

CHAPTER NINE

# Cardiokines as Modulators of Stress-Induced Cardiac Disorders

**Anna Planavila*[,†,1], Joaquim Fernández-Solà[‡], Francesc Villarroya*[,†]**

*Institut de Biomedicina de la Universitat de Barcelona (IBUB), Universitat de Barcelona, Barcelona, Spain
[†]CIBER Fisiopatología de la Obesidad y Nutrición (CIBEROBN), Barcelona, Spain
[‡]Hospital Clínic, Institut de Recerca Biomèdica August Pi i Sunyer (IDIBAPS), Universitat de Barcelona, Barcelona, Spain
[1]Corresponding author: e-mail address: aplanavila@ub.edu

## Contents



## Abstract

Almost 30 years ago, the protein, atrial natriuretic peptide, was identified as a heart-secreted hormone that provides a peripheral signal from the myocardium that communicates to the rest of the organism to modify blood pressure and volume under conditions of heart failure. Since then, additional peripheral factors secreted by the heart, termed cardiokines, have been identified and shown to coordinate this interorgan cross talk. In addition to this interorgan communication, cardiokines also act in an autocrine/paracrine manner to play a role in intercellular communication within the myocardium. This review focuses on the roles of newly emerging cardiokines that are mainly increased in stress-induced cardiac diseases. The potential of these cardiokines as clinical biomarkers for diagnosis and prognosis of cardiac disorders is also discussed.

*Advances in Protein Chemistry and Structural Biology*, Volume 108
ISSN 1876-1623
http://dx.doi.org/10.1016/bs.apcsb.2017.01.002

## 1. INTRODUCTION

Heart failure represents a major cause of morbidity and mortality, constituting a major disease burden worldwide (Nicholson, Gandra, Halbert, Richhariya, & Nordyke, 2016). It is a complex disorder in which the heart is unable to meet the metabolic demands of peripheral tissues. Current pharmacological approaches for treating heart failure seek to decrease ventricle pre- and afterload using diuretics, β-blockers, and angiotensin-converting enzyme inhibitors and improve cardiac contractile function by modifying neurohormonal signaling (Holland et al., 2005). Despite the health benefits of these drugs, up to 50% of patients still die from this disorder within 5 years of diagnosis (Linseman & Bristow, 2003). Therefore, there remains a need to develop additional therapeutic strategies and new prognostic biomarkers to improve the survival of heart failure patients.

Several studies have established the crucial importance of the inflammatory process in the development and progression of heart failure (Jiang & Liao, 2010; Kleinbongard, Schulz, & Heusch, 2011). Proinflammatory factors induce myocardial remodeling by promoting recruitment of inflammatory cells or by producing maladaptive effects in the heart, such as endothelial dysfunction and ventricular remodeling, thereby facilitating the development of cardiac hypertrophy and fibrosis (Ambardekar & Buttrick, 2011; Hill & Olson, 2008; Quttainah et al., 2015). However, antiinflammatory therapeutic strategies tested to date have been largely unsuccessful, causing no detectable effect or even worsening the outcome of heart failure treatment. These findings highlight the importance of identifying novel targets as part of an effort to more effectively treat cardiac disorders (Heymans et al., 2009).

A number of recent basic research studies have identified proteins produced by the heart, referred to as *cardiokines,* which may function analogously to adipokines in signaling either locally or peripherally and are deregulated during cardiac stress and inflammation. The scientific quest for new biomarkers for heart failure diagnosis and therapeutics has yielded a steadily increasing number of identified cardiokines, with conservative estimates yielding a total of 30–60 cardiokines, based on the presence of an N-terminal sequence predicted to target the protein for the classic secretory pathway (Frost & Engelhardt, 2007; Stastna, Chimenti, Marban, & Van Eyk, 2010). The identification of key cardiokines induced during cardiac disorders is important, as it may shed light on the development of novel or better diagnostic tools and treatment modalities for cardiac diseases.

## 2. THE HEART AS A SECRETORY ORGAN

Traditionally, the heart has been viewed simply as a mechanically beating organ unrelated to the endocrine system. Braunwald, Harrison, and Chidsey (1964) published an editorial providing the first description of the heart as being capable of releasing proteins into the bloodstream (Shimano, Ouchi, & Walsh, 2012), specifically citing the ability of the myocardium to supply and produce catecholamines for subsequent release. This concept lay dormant until 1985, when studies from a number of laboratories identified atrial natriuretic peptide (ANP), the prototypical cardiokine, prompting reconsideration of the concept of an endocrine function of the heart (Frohlich, 1985). Since then, a number of cardiokines have been identified, including the well-known examples of brain natriuretic peptide (BNP), tumor necrosis factor-α, and angiotensin II (AngII). Recently, a number of studies have focused on using gene expression array screening, cloning, and other techniques to identify new cardiokines that are regulated during cardiac stress and inflammation.

Cardiokines are secreted from all types of cells found in cardiac tissue, including myocytes, fibroblasts, and endothelial and vascular cells. These secreted proteins are required for the maintenance of normal cardiac function and also control pathological remodeling of the myocardium in response to injury through their ability to act in a paracrine/autocrine manner to modulate cardiomyocyte hypertrophy, survival, and apoptosis; fibroblast activation; and inflammation (Kakkar & Lee, 2010; Tian & Morrisey, 2012; Tirziu, Giordano, & Simons, 2010). Some of these factors also function throughout the body to exert systemic endocrine actions, influencing kidney function or cachectic processes (Nakamura & Sadoshima, 2014). Similarly, peripheral tissue-derived mediators can influence heart function, for instance, affecting the development and progression of heart failure (Jahng, Song, & Sweeney, 2016). Examples of well-known cardiokines include the two circulating cardiokines, ANP and BNP, which are synthesized mainly in cardiomyocytes in the myocardium and are upregulated in response to heart failure. The endocrine effects of ANP and BNP are thought to counteract heart failure by promoting vasodilation and inhibiting AngII and aldosterone. Their autocrine and paracrine effects include inhibition of cardiomyocyte hypertrophy and activation of fibroblasts

(Nishikimi, Maeda, & Matsuoka, 2006). Plasma BNP levels have also been used as the "gold standard" biomarker for detecting left ventricular hypertrophy (Maisel et al., 2008; McDonagh et al., 1998; Yamamoto et al., 1996). TGF-β1, a cardiokine elevated in pathological states, serves to induce inflammatory cell recruitment, cardiomyocyte hypertrophy, and/or fibroblast activation (Teekakirikul et al., 2010). Cardiac secretion of AngII, which is also induced in heart failure, promotes cardiomyocyte hypertrophy and fibrosis, as discussed earlier. Recent studies have shown that AngII also has endocrine catabolic effects in skeletal muscle, suggesting that it may contribute to muscle wasting in heart failure (Delafontaine & Akao, 2006; Sukhanov et al., 2011). New cardiokines with autocrine, paracrine, and/or endocrine functions that are regulated during cardiac stress are continuously being discovered (Doroudgar & Glembotski, 2011; Shimano et al., 2012). In this review, we will focus on the regulation and function of newly emerging cardiokines with promising near-term applications.

## 3. CARDIOKINES AND CARDIAC STRESS

Many cellular "conversations" occur in the heart and are coordinated to ensure that cellular cross talk directs integrated biological responses (Tirziu et al., 2010). At the physiological level, cardiokines are important in maintaining this aspect of the normal functioning of cardiac tissue. Paracrine/autocrine signaling within the heart is one of the most important communication channels for directing this process. However, in situations of cardiac stress, such as myocardial infarction or heart failure, the heart responds coordinately through cardiokine secretion in an attempt to compensate for the cardiac disturbance. Some of these molecules have pro- or antihypertrophic, fibrotic, and/or angiogenic properties that act coordinately to resolve the insult. Thus, a change in the repertoire of cardiokines plays a critical role in determining how the heart responds to a damaging stress (Doroudgar & Glembotski, 2011).

It has been shown that some transcription factors, such as activating transcription factor (ATF)-4 and -6 and X-box-binding protein 1 (XBP1), are activated in response to cardiac stress produced by, for example, ischemia or cardiac hypertrophy (Doroudgar & Glembotski, 2011). These factors increase the transcription of genes involved in protein folding and the secretory machinery (Ron & Walter, 2007), thus promoting protein folding and secretion. In addition, these factors control the expression of some cardiokines themselves (Belmont et al., 2008; Thuerauf et al., 2006). For instance, FSTL1 (follistatin-like 1) and FGF21 (fibroblast growth factor 21), cardiokines that will be described further in this review (see below), are transcriptionally regulated by ATF6 and ATF4, respectively, and are upregulated upon cardiac stress (De Sousa-Coelho, Marrero, & Haro, 2012; Doroudgar, Thuerauf, Marcinko, Belmont, & Glembotski, 2009). Thus, cardiokines released by the stressed heart are likely to play central roles in establishing whether the injured myocardium is amenable to repair.

## 4. CARDIOKINES: AUTOCRINE/PARACRINE SIGNALING

The main cell type in the human heart is the cardiac myocyte, accounting for an estimated 2–3 billion total cells; but this amount represents less than a third of the total number of cells present in the myocardium. Among the remaining additional cell types are fibroblasts, smooth muscle cells, endothelial cells, and immune system-related cells. These cell populations are not isolated from each other, but instead interact physically as well as via cardiokines (Fig. 1). The cross talk between myocytes and other cell types within the heart plays crucial and interdependent roles (Jahng et al., 2016; Tian & Morrisey, 2012; Tirziu et al., 2010). To fully understand the biology of the heart, we might consider this cellular cross talk as a window into the molecular mechanisms that govern the heart failure process. In this section, we will discuss the autocrine and paracrine roles of new emerging cardiokines.

### 4.1 The TGF-β Family

The TGF-β family of secreted cytokines is critical for a diverse array of cellular functions, including inflammation, extracellular matrix deposition, cell proliferation, differentiation, and growth (Aihara, Mogi, Shibata, Bishop-Bailey, & Ma, 2011; Dobaczewski, Chen, & Frangogiannis, 2011). In mammals, three different genes encode the three structurally similar TGF-β isoforms: TGF-β1, -β2, and -β3. TGF-β glycoproteins are synthesized and secreted into the extracellular matrix as immature precursor proteins that are maintained in binding-incompetent conformation through noncovalent binding of their latency-associated peptide (LAP) domain. The LAP domain prevents TGF-β from binding to its receptors until it is removed through the proteolytic activity of thrombospondin-1 (Murphy-Ullrich & Poczatek, 2000). Once released from the LAP, the active TGF-β molecule is able to bind with high affinity to its constitutively active serine/threonine type II TGF-β receptors, evoking a conformational

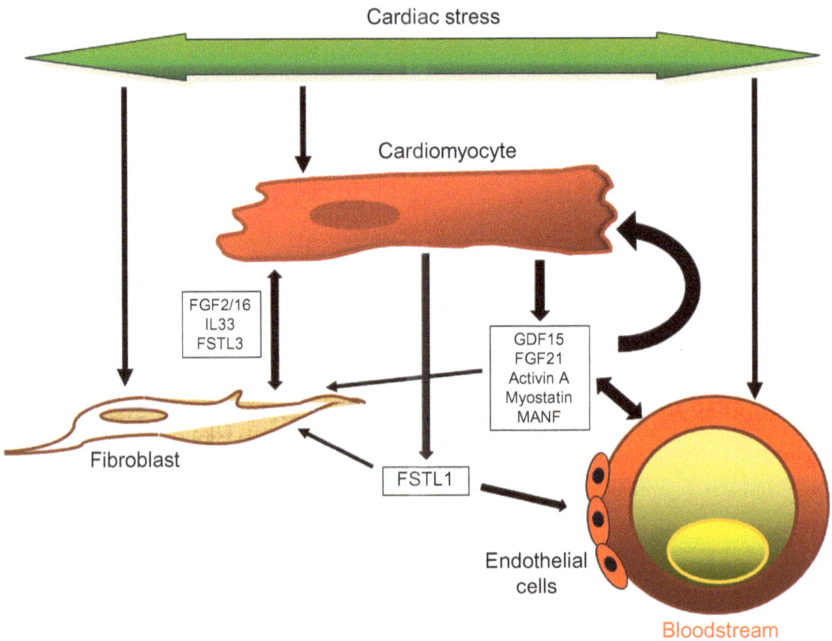


**Fig. 1** Autocrine/paracrine functions of cardiokines in the myocardium. Cardiokines are produced by different cell types in the myocardium, mainly cardiomyocytes and fibroblasts, and mediate cellular cross talk in the heart through autocrine/paracrine signaling. Perturbations in the balance of this cellular network during sustained hemodynamic stress lead to secretion of cardiokines with multiple autocrine/paracrine effects on cardiac hypertrophy, fibrosis, and/or vascular remodeling.

change that activates recruitment of the appropriate type I TGF-β receptor. Type II receptors then phosphorylate the glycine-/serine-rich domain of type I receptors to activate intracellular signaling (Dobaczewski et al., 2011). The signaling pathways downstream of TGF-β receptor type I activation include SMAD (Sma and Mad)-dependent and -independent pathways. The classic SMAD-dependent pathway begins with the activation of receptor-activated SMAD (R-SMAD)-2 and -3, which then recruit SMAD4 (co-SMAD) for translocation into the nucleus. Once inside the nucleus, SMAD2/3/4 complexes, together with their binding partners, activate or repress the transcription of target genes. Several studies have reported SMAD-independent pathways that function through the activation of extracellular signal-regulated kinase (ERK), c-Jun N-terminal kinase, p38 mitogen-activated kinase (MAPK), Abelson nonreceptor tyrosine kinase, and TGF-β-activated kinase 1 (Derynck & Zhang, 2003).

Although the three TGF-β isoforms have similar signaling activities and structures, recent evidence suggests that they have different expression patterns and nonoverlapping functions. TGF-β1 is the most ubiquitously expressed isoform in many organs and cell types. In the heart, TGF-β1 is markedly elevated in animal models of heart failure and is associated with cardiac hypertrophy and fibrosis. However, specific roles of the other TGF-β isoforms in modulating myocardial remodeling remain unknown. Only a few reports have shown elevated TGF-β2 or TGF-β3 in the stressed heart. For example, early upregulation of TGF-β2 and TGF-β1, together with delayed elevation of TGF-β3, has been reported in the infarct area in a rodent model of myocardial infarction (Deten, Holzl, Leicht, Barth, & Zimmer, 2001). Additionally, it has been shown that TGF-β2 and TGF-β1 expression are upregulated in myocardial biopsies from patients with dilated cardiomyopathy (Pauschinger et al., 1999). Studies employing knockout mice have also proved informative in assessing the importance of each individual TGF-β isoform. These studies have shown that approximately 50% of TGF-β1-null mice die in utero owing to defective yolk sac vasculogenesis and hematopoiesis, whereas animals that do survive to birth subsequently die due to multiorgan inflammation, primarily in the heart and lungs (Dickson et al., 1995; Kulkarni et al., 1993). TGF-β3-knockout mice exhibit defective epithelial-to-mesenchymal transition, leading to cleft palate and abnormal lung development (Proetzel et al., 1995). TGF-β2-null mice exhibit perinatal mortality and severe developmental organ defects, including congenital heart defects (Sanford et al., 1997). TGF-β2 is indispensable for the embryonic development of cardiac tissue, as evidenced by the cardiac valvular defects in TGF-β2-null mice (Azhar et al., 2011). Cardiac expression of TGF-β2 has been observed in myocardial, endocardial, neural crest, and vascular smooth muscle cell lineages (Azhar et al., 2011; Dickson, Slager, Duffie, Mummery, & Akhurst, 1993; Molin et al., 2002), suggesting the importance of the autocrine/paracrine activity of TGF-β2. However, because TGF-β2 knockout causes perinatal mortality, it has been difficult to study the role of TGF-β2 in adult cardiac diseases.

Under the larger umbrella of the TGF-β superfamily are the newly identified cardiokines, GDF15 (growth differentiation factor 15), myostatin, and activin A. These cardiokines all act through the SMAD2/3 receptor, to activate downstream signaling (Bartholin et al., 2002; Rebbapragada, Benchabane, Wrana, Celeste, & Attisano, 2003; Xu et al., 2006), and have both similar and nonoverlapping functions.

#### 4.1.1 GDF15

The gene encoding GDF15 was first identified through microarray analyses as a nitric oxide-inducible gene in cardiac myocytes (Bootcov et al., 1997). Subsequent studies have shown that GDF15 is also upregulated in response to ischemia and in the infarcted human heart, where it serves to protect cardiac cells from apoptosis (Kempf et al., 2006). GDF15 is also induced during cardiac hypertrophy in a pressure-overload murine model, and its cardiac-specific overexpression protects the heart from hypertrophic responses (Xu et al., 2006). More recently, it was demonstrated that ablation of GDF15 increases infiltration of polymorphonuclear leukocytes into the infarct area and thus enhances inflammation in the heart. By contrast, GDF15 reduces leukocyte recruitment by modulating chemokine signaling, thereby promoting resolution of the inflammatory process (Kempf et al., 2011). Collectively, the available evidence indicates that GDF15 functions as a stress-inducible cardiokine with antiinflammatory effects that protects the myocardium against cardiac remodeling.

#### 4.1.2 Myostatin

Myostatin, also known as GDF8 (growth differentiation factor 8), is known classically to act as an inhibitor of skeletal muscle growth (McPherron, Lawler, & Lee, 1997). The main site of myostatin expression is skeletal muscle, where its inactivation produces a dramatic increase in muscle mass in humans as well as in other animal species (Salzler et al., 2016). However, myostatin is also significantly expressed in the heart and adipose tissue (Dschietzig, 2014; Karlowatz et al., 2011). Evidence of a role for myostatin in heart pathophysiology has recently emerged. For example, myostatin is expressed in failing human hearts (George et al., 2010; Lenk et al., 2009) and is upregulated in the context of different heart pathologies, including ventricular hypertrophy (Rodgers et al., 2009), volume overload (Shyu, Lu, Wang, Sun, & Chang, 2006), chronic heart failure (Baan et al., 2015; George et al., 2010), and acute myocardial infarction (Castillero et al., 2015). We have found that myostatin upregulation is a prominent feature in chronic alcoholic consumption (Fernandez-Sola et al., 2011) as well as long-standing hypertension (Fernandez-Sola, Borrisser-Pairo, Antunez, & Tobias, 2015). The role of myostatin in regulating heart homeostasis is also relevant in the context of preventing cardiac hypertrophy and averting heart failure (Baan et al., 2015; George et al., 2010). Myostatin signaling represses adenosine monophosphate-activated protein kinase (AMPK) and modifies its interaction with key metabolic enzymes and proteins involved in cardiac hypertrophy (Kemaladewi et al., 2012). At present, myostatin may be considered a cardioprotective factor with antihypertrophic properties that is necessary for maintaining normal heart homeostasis (Biesemann et al., 2014).

#### 4.1.3 Activin A

In the heart, activin A expression is upregulated following ischemic injury and pressure overload (Oshima et al., 2009; Shimano, Ouchi, Nakamura, Oshima, et al., 2011). Activin A has been demonstrated to exert a protective effect in vitro by rescuing myocytes from apoptosis following hypoxia/reoxygenation stress (Oshima et al., 2009) and shown to prevent cardiac injury in vivo following systemic delivery after ischemia/reperfusion (Zimmers et al., 2002).

### 4.2 Follistatin Family

The follistatin family is a group of extracellular regulators that selectively bind to members of the TGF-β superfamily, including activin A and myostatin. The follistatin family consists of follistatin (FST) and follistatin-like 3 (FSTL3), which are structurally and functionally related glycoproteins, and the distantly related follistatin-like 1 (FSTL1), which has little or no functional overlap with the rest of the family. FST and FSTL3 have been identified on the basis of their antagonistic activities against activin A and myostatin (Schneyer et al., 2004; Sidis et al., 2006), whereas FSTL1 has been shown to antagonize the activity of BMP1 (bone morphogenetic protein 1) (Sylva et al., 2011). Although FST and FSTL3 are similar in many respects, their differential expression suggests nonoverlapping actions. FST has its highest expression in ovaries, testes, and kidneys, whereas FSTL3 is most highly expressed in the placenta, followed by the testis, pancreas, and heart (Schneyer et al., 2004).

#### 4.2.1 FSTL1

Although FSTL1 is also a secreted glycoprotein, it has low sequence homology to other members of the family, suggesting that it may have a limited ability to modulate TGF-β family proteins. FSTL1 was first identified as a secreted cardiac protein that is upregulated during physiological heart growth (Oshima et al., 2008; Schiekofer et al., 2006). However, FSTL1 expression also increases in response to cardiac ischemia or myocardial hypertrophy, leading to elevated plasma levels of this cardiokine (Oshima et al., 2008; Shimano, Ouchi, Nakamura, van Wijk, et al., 2011). FSTL1 upregulation also promotes heart repair mechanisms (Sanchis-Gomar, Perez-Quilis, & Lucia, 2016) and plays a protective role in the myocardium, where it has been shown to attenuate infarct size and apoptosis in a mouse model of ischemia–reperfusion through the activation of the phosphoinositide 3-kinase/AKT signaling pathway (Oshima et al., 2008). Moreover, FSTL1 exerts antihypertrophic effects in the heart (Shimano, Ouchi, Nakamura, van Wijk, et al., 2011) that are mediated by its promotion of AMPK signaling in cardiac myocytes. Overexpression or systemic delivery of FSTL1 results in a diminished cardiac hypertrophy response in a mouse model of pressure overload. In addition to its beneficial role in cardiomyocytes, FSTL1 exerts direct effects on endothelial cells, promoting cell growth in response to ischemia (Ouchi et al., 2010, 2008; Tanaka et al., 2010). The ability of FSTL1 to act on the vasculature may contribute to the beneficial effects on cardiac function observed after FSTL1 treatment. Therefore, FSTL1 appears to function as an injury-induced cardiokine that protects the heart against injury via autocrine/paracrine signaling. Very recent studies have demonstrated that FSTL1 acts through the suppression of inflammatory pathways to exert its cardioprotective effects. Specifically, FSTL1 inhibits myocardial inflammation by blocking the BMP4 proinflammatory signaling pathway (Ogura et al., 2012). Collectively, these findings indicate that FSTL1 is a cardiokine with antiinflammatory, antihypertrophic, and proangiogenic actions in cardiac tissue.

### 4.2.2 FSTL3

FSTL3 binds to activin A specifically in the heart and antagonizes its cardioprotective function. Increased FSTL3 has been reported in the myocardium of end-stage heart failure patients, where its expression is positively correlated with disease severity (Lara-Pezzi et al., 2008). In the same study, cardiomyocytes and endothelial cells were identified as the sources of elevated FSTL3 expression. Myocardial infarction (Oshima et al., 2009) and pressure overload (Shimano, Ouchi, Nakamura, Oshima, et al., 2011) in mice result in increased FSTL3 and activin A expression. The decrease in infarct size in cardiomyocyte-specific *Fstl3*-knockout mice subjected to ischemic injury highlights the cardioprotective role of activin A and the opposing force of FSTL3 (Oshima et al., 2009).

FSTL3 also plays a role in myocardial hypertrophy, as evidenced by the fact that cardiac-specific ablation of *Fstl3* attenuates cardiac hypertrophy in response to pressure overload (Shimano, Ouchi, Nakamura, Oshima, et al., 2011). In cultured neonatal myocytes, phenylephrine-stimulated hypertrophic responses were significantly attenuated by pretreatment with small interfering RNA (siRNA) targeting FSTL3 (Shimano, Ouchi, Nakamura, Oshima, et al., 2011). The contribution of myocyte-derived FSTL3 to cardiac fibroblast activation has been demonstrated in cultured neonatal fibroblasts treated with recombinant FSTL3, where full activation required the purported FSTL3-binding partner, connective tissue growth factor (Panse et al., 2012). Taken together, these observations clearly indicate that FSTL3 plays an important role in mediating both hypertrophy and fibrosis in the context of ischemic and pressure-overload stress. Moreover, activin A and FSTL3 are cardiokines with opposing actions that function to regulate the response to myocardial injury.

## 4.3 Fibroblast Growth Factor Family

Fibroblast growth factors (FGFs) are ~150–300-amino-acid proteins with a conserved core of ~120 amino acids that share 30%–60% identity among family members. They have diverse functions, mainly during development and in systemic metabolism. The family comprises 22 members, which can be classified phylogenetically into 7 subfamilies (Itoh & Ornitz, 2011), but are also categorized as intracellular, paracrine or endocrine based on their action mechanisms (Itoh & Ohta, 2013). Paracrine and endocrine FGFs are secreted proteins that act in an autocrine/paracrine or endocrine manner, and can be considered to be cardiokines when they are expressed and play a role in the myocardium. Recently, three members of the FGF family—FGF2, FGF16, and FGF21—have been shown to play pathophysiological roles in the heart. FGF2 and FGF16 belong to the paracrine FGF subfamily, whose members mediate their biological responses by binding FGF receptors (FGFRs) using heparin/heparan sulfate as a cofactor. FGF21 is an endocrine FGF that exerts its biological responses in an FGFR-dependent manner. However, it binds FGFR in a heparin sulfate-independent manner, instead requiring β-Klotho as a cofactor. This enables FGF21 to function over long distances in an endocrine manner; the main influence of this hormone-like signaling molecule is on metabolism (Itoh, 2010; Itoh & Ornitz, 2011; Long & Kharitonenkov, 2011). Among FGFRs, FGFR1c is the predominant form expressed in the heart (Fon et al., 2010). Paracrine and endocrine FGFs induce intracellular tyrosine phosphorylation of FGFRs, triggering activation of cytoplasmic signal transduction pathways that elicit different intracellular responses (Goetz & Mohammadi, 2013). These new findings of pathophysiological roles of FGFs in the heart may provide potential therapeutic strategies for heart failure patients.

### 4.3.1 FGF2

FGF2, a paracrine FGF, is released by damaged cells through exocytotic mechanisms independent of the endoplasmic reticulum–Golgi pathway (Nickel, 2010). FGF2 is broadly expressed in mice, and its deletion in *Fgf2*-null mice protects against cardiac hypertrophy and fibrosis after either myocardial infarction or isoproterenol-induced cardiac hypertrophy (House et al., 2010; Virag et al., 2007). FGF2 is mainly expressed in non-cardiomyocytes and is stored in these cells and released in response to hemodynamic stress (Kaye et al., 1996). Studies in vitro with cultured cardiomyocytes have shown that FGF2 acts through FGFR1c and causes cardiac hypertrophy by inducing MAPK signaling (Bogoyevitch et al., 1994; Kaye et al., 1996). Moreover, cardiac-specific overexpression of FGF2 exacerbates cardiac hypertrophy via a mechanism that involves the MAPK cascade (House et al., 2010). Collectively, these observations indicate that FGF2 acts on cardiac cells in a paracrine manner and promotes cardiac remodeling by activating MAPK signaling.

### 4.3.2 FGF16

Although FGF16 is also a paracrine FGF, it exerts effects opposite those of FGF2 in the myocardium. FGF16 is predominantly expressed in the heart (Fon et al., 2010) and is mainly present in cardiomyocytes (Hotta et al., 2008). FGF16 is efficiently secreted by cardiomyocytes and acts as an autocrine signaling molecule to induce proliferation of embryonic mouse hearts (Hotta et al., 2008). In adulthood, FGF16 antagonizes FGF2 signaling by competing for a binding site in FGFR1c, which is predominantly expressed in the heart (Lu, Sontag, Detillieux, & Cattini, 2008). This FGF16-mediated suppression of FGF2 signaling prevents cardiac hypertrophy and fibrosis, as evidenced by the fact that mice lacking FGF16 exhibit enhanced AngII-induced cardiac hypertrophy and fibrosis (Matsumoto et al., 2013). Consistent with this, AngII induces cardiac expression of FGF2 in vivo, followed by increased expression of FGF16, and induces FGF2-dependent expression of FGF16 in cultured cardiomyocytes. These findings indicate a possible mechanism by which FGF16 prevents cardiac hypertrophy, suggesting that it competes with FGF2 for binding to FGFR1c (Matsumoto et al., 2013).

### 4.3.3 FGF21

FGF21 is a secreted protein that has been shown to act as a metabolic regulator that plays a role in controlling glucose homeostasis, insulin sensitivity, and ketogenesis (Badman et al., 2007; Galman et al., 2008). The liver is considered the main site of FGF21 production and is released into the blood (Badman et al., 2007; Inagaki et al., 2007). Extrahepatic tissues, such as white and brown adipose tissues and skeletal muscle, also express FGF21 (Hondares et al., 2011; Izumiya et al., 2008; Muise et al., 2008). As noted earlier, the action of FGF21 on target cells requires FGFRs and β-Klotho, a single-pass transmembrane protein that functions as an obligate cofactor for FGF21 signaling (Kharitonenkov et al., 2008; Kurosu et al., 2007). The heart was originally not considered an FGF21 target or source, primarily because it only modestly expresses mRNA for β-Klotho, the obligate co-receptor for cellular responsiveness to FGF21 (Fon et al., 2010). However, a recent study from our laboratory provided the first report of the cardioprotective effects of FGF21 (Planavila et al., 2013). In this report, we showed that significant amounts of both the FGF21 receptor, FGFR1, and the cofactor, β-Klotho, are present at the protein level in cardiac cells. Moreover, treatment of cardiomyocytes in culture with FGF21 was found to activate the ERK signaling pathway, which is considered the main intracellular pathway responsible for FGF21 intracellular actions. In vivo, hearts of *Fgf21*-knockout mice exhibit an increase in relative weight and develop enhanced enlargement potential. In response to isoproterenol infusion, a standard model used to induce cardiac hypertrophy, heart size, and cardiomyocyte volume is increased to a larger extent in *Fgf21*-knockout mice than in control mice. Furthermore, FGF21 treatment reverses cardiac hypertrophy development, enhances fatty acid oxidation, and prevents the induction of proinflammatory pathways in the heart, confirming the antihypertrophic properties of FGF21 (Planavila et al., 2013). Studies have also reported that FGF21 exerts protection after myocardial infarction by inhibiting cardiomyocyte apoptosis (Liu et al., 2013, 2011), attenuating pathological myocardial remodeling, and reducing infarct size (Joki et al., 2015; Patel et al., 2014). In addition, FGF21 modulates oxidative stress, which plays a role in the pathogenesis of heart failure (Planavila, Redondo-Angulo, et al., 2015). Cardiac FGF21 regulates genes involved in antioxidant pathways, thus preventing the production of reactive oxygen species by cardiac cells. Furthermore, FGF21 is expressed in and secreted by cells of the heart in response to different cardiac stress stimuli, such as cardiac hypertrophy and myocardial infarction. An analysis of different cell populations isolated from the heart showed that FGF21 is mainly produced by cardiomyocytes (Planavila et al., 2013). Importantly, this cardiomyocyte-generated FGF21 acts locally in an autocrine manner to prevent cardiac hypertrophy, metabolic dysregulation, and activation of proinflammatory pathways. Thus, in addition to the protective role exerted by systemic, circulating FGF21 on the heart, local secretion of FGF21 in the context of cardiac damage may serve as an endogenous autocrine cardioprotective signaling pathway (Fernandez-Sola & Planavila, 2016).

## 4.4 Other Cardiokines

### *4.4.1 IL33*

IL33 is a member of the IL1 family of cytokines and is a ligand for the membrane receptor, ST2 (Miller & Liew, 2011). IL33 is widely expressed in many tissues, including the heart (Schmitz et al., 2005), and is mainly produced and secreted by cardiac fibroblasts after mechanical stretch, although other cell types in the heart, such as myocytes, can produce it (Kakkar, Hei, Dobner, & Lee, 2012; Sanada et al., 2007). IL33 signaling through the ST2 receptor (also known as IL1RL1-β) activates the MAPK signaling cascade (Schmitz et al., 2005). After binding to the ST2 receptor, IL33 exerts antihypertrophic and antifibrotic effects in the heart. IL33 suppresses hypertrophic responses in vitro in cardiac myocytes and, when administered in vivo, attenuates cardiac hypertrophy and fibrosis after pressure overload in wild-type mice, but not in mice lacking ST2 (Sanada et al., 2007). Moreover, IL33 treatment attenuates cardiomyocyte apoptosis and protects against myocardial infarction through ST2-dependent mechanisms in mouse models (Seki et al., 2009). Collectively, these findings indicate that the IL33–ST2 axis represents a paracrine cardioprotective system in which cardiac fibroblasts produce IL33, which acts on cardiomyocytes through ST2 to prevent cardiac hypertrophy.

### *4.4.2 Mesencephalic Astrocyte-Derived Neurotrophic Factor*

Mesencephalic astrocyte-derived neurotrophic factor (MANF) is an endoplasmic reticulum stress-inducible cardiokine originally identified as a secreted factor that protects rat dopaminergic neurons in vitro and prevents neuron degeneration in Parkinson's disease (Lindholm et al., 2007; Petrova et al., 2003). Specifically in the heart, MANF protects against cardiac hypertrophy and myocardial infarction. In cardiac myocytes, MANF is induced in response to endoplasmic reticulum stress in a manner that depends on ATF6, a regulator of the unfolded protein response (Tadimalla et al., 2008). Myocardial infarction also leads to induction of MANF in all cell types in the heart. Moreover, treatment of cardiomyocytes with MANF reduces cell death in response to ischemia/reperfusion, whereas the lack of MANF promotes cell death in the same model (Tadimalla et al., 2008). Systemic administration of recombinant MANF protein is cardioprotective in mice after ischemia/reperfusion (Glembotski et al., 2012). Finally, MANF also inhibits hypertrophy in cultured cardiac myocytes (Glembotski, 2011), suggesting an autocrine effect of MANF in this cell type.

## 5. ENDOCRINE FUNCTIONS OF CARDIOKINES

As mentioned earlier, the concept of the heart as an endocrine organ was first proposed in 1981 based on reports of the diuretic and natriuretic properties of ANP (de Bold, Borenstein, Veress, & Sonnenberg, 1981). Extensive subsequent research has led to the characterization of the other members of the natriuretic peptide family, BNP and C-type natriuretic peptide (Kangawa, Fukuda, Minamino, & Matsuo, 1984; Sudoh, Kangawa, Minamino, & Matsuo, 1988; Sudoh, Minamino, Kangawa, & Matsuo, 1990). These cardiokines are mainly synthesized in the myocardium and are upregulated in response to myocardial stretching. ANP and BNP act via endocrine mechanisms to regulate electrolyte and water excretion in the kidney by functionally antagonizing the "salt-sparing" role of the renin–angiotensin–aldosterone system; they also regulate vascular cell growth and tone to promote vasodilation (Nishikimi et al., 2006). The natriuretic peptides activate their intracellular responses by binding to natriuretic peptide receptor A (NPRA) in the plasma membrane. Moreover, natriuretic peptides are removed from the circulation through internalization via NP receptor C (NPRC)—the clearance receptor. The relative level of NPRA and NPRC is the major determinant of the activity of natriuretic peptides.

Despite the well-recognized endocrine functions of natriuretic peptides on renal and vascular tissues, the recent (re)discovery of potential endocrine functions of natriuretic peptides on adipose tissue was reinforced by publication of new data linking natriuretic peptides with obesity (Collins, 2014; Thoonen, Hindle, & Scherrer-Crosbie, 2016; Fig. 2). Almost 20 years ago, natriuretic peptides receptors were found to be expressed in adipose tissue in humans and rats, and studies in vitro showed a strong lipolytic effect of ANP and BNP on human adipocytes (Sarzani, Dessi-Fulgheri, Paci, Espinosa, &

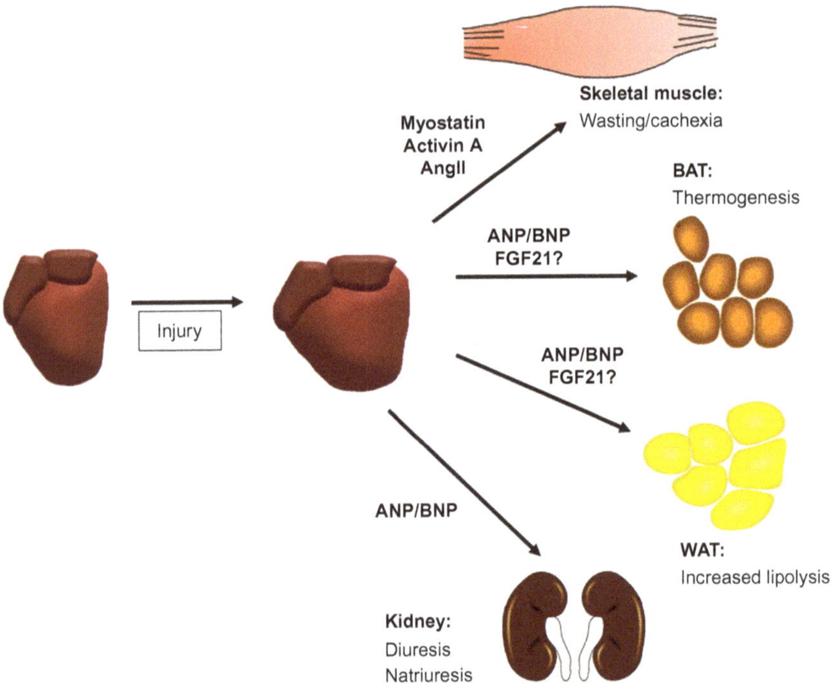


**Fig. 2** Cardiokines in interorgan communication. Cardiac injury leads to the systemic production of cardiokines that act in an endocrine manner in several tissues. Natriuretic peptides act on the kidney to regulate diuresis and natriuresis and on the vascular system to regulate blood pressure. Together with FGF21, they can also act on adipose tissues to regulate energy metabolism. Finally, cardiokines such as myostatin, activin A, and AngII influence muscle wasting and cachexia.

Rappelli, 1996; Sengenes, Berlan, De Glisezinski, Lafontan, & Galitzky, 2000). Moreover, human obesity appears to be associated with decreased circulating levels of natriuretic peptides and altered expression of their receptors (Dessi-Fulgheri et al., 1997; Wang et al., 2004). Unfortunately, this interorgan communication between the heart and adipose tissue was not further explored at the time. However, in 2012, it was shown that natriuretic peptides are capable of activating brown fat to increase energy expenditure, also referred to as the thermogenic program, an action that suggests a therapeutic approach for combating obesity (Bordicchia et al., 2012). Moreover, it has been recently reported that the expression of adipose tissue natriuretic peptide receptors in humans is related to insulin sensitivity in obesity and diabetes (Kovacova et al., 2016). These studies support the hypothesis that the cardiac natriuretic peptide system is dysregulated in obesity, insulin resistance, and type 2 diabetes, and suggest a tissue connection between the cardiac natriuretic peptide pathway and the control of whole-body energy metabolism.

Another cardiokine with potential effects on energy metabolism is FGF21, reflecting its well-known effects as a metabolic regulator when produced by other tissues in the organism. Transgenic mice with cardiac-specific FGF21 overexpression show increased circulating FGF21 levels and altered body mass composition (Brahma et al., 2014), suggesting that FGF21 released by the heart has the potential to act at a distance in an endocrine manner (Planavila, Redondo-Angulo, & Villarroya, 2015). However, additional studies are needed to elucidate possible endocrine actions of cardiac FGF21 and assess its actual contribution to whole-body homeostasis.

The most recently described endocrine function of cardiokines is their influence on cachexic processes (Fig. 2). Cachexia, the development of muscle wasting disease, is associated with an increased mortality risk in heart failure patients (Anker et al., 1997). As mentioned earlier, cardiac AngII has catabolic effects on skeletal muscle, contributing to muscle wasting in congestive heart failure (Delafontaine & Akao, 2006; Sukhanov et al., 2011). Myostatin has also been shown to exert an inhibitory action on skeletal muscle growth by inhibiting phosphorylation of AKT (protein kinase B) (Trendelenburg et al., 2009). Because of this inhibitory effect on skeletal muscle growth, myostatin plays a critical role in muscle wasting during cachexia. An elegant recent report identified myostatin as a cardiokine that mediates skeletal muscle wasting during heart failure (Heineke et al., 2010). Genetic deletion of myostatin specifically in the heart has been shown to prevent muscle atrophy during heart failure. Similarly, injection of a myostatin-blocking antibody in a model of preexisting heart failure maintains skeletal muscle mass. By contrast, cardiac-specific overexpression of myostatin promotes muscle wasting. Finally, activin A, through the activation of the activin IIB receptor, has been suggested to be an even more potent inducer of cachexia (Chen et al., 2014). Activin A-haploinsufficient mice exhibit increased skeletal muscle mass, indicating that heart-derived activin A influences the cardiac cachexic state by acting on skeletal muscle in an endocrine manner (Lee et al., 2010). Collectively, these data indicate that myostatin and activin A are potential targets for the prevention and/or treatment of cachexia during heart failure.

## 6. CARDIOKINES AS BIOMARKERS FOR CARDIAC DISORDERS

The fact that cardiokines are proteins released by the heart into the circulatory system makes them promising biomarker candidates with prognostic clinical value (Doroudgar & Glembotski, 2011). Circulating plasma levels of ANP and BNP are massively increased in many cardiac disorders and indicate the development of a deranged cardiorenal axis (D'Souza, Davis, & Baxter, 2004). Therefore, ANP and BNP are commonly measured in a clinical setting, especially for heart failure diagnosis and prognosis (Rademaker & Richards, 2005). However, there remains a need for more accurate biomarkers in cardiac disorders. The study of these new emerging cardiokines as useful prognostic biomarkers could provide better tools for monitoring disease severity and predicting cardiovascular risk more efficiently. In fact, recent studies have identified GDF15, FSTL1, myostatin, and FGF21 as potential biomarkers in heart disease. Circulating GDF15 and FSTL1 levels are elevated in patients with acute coronary syndromes (ACSs), are predictive of cardiovascular death in ACSs, and are independently and closely associated with one another in the same patient population (Eggers, Kempf, et al., 2010; Widera et al., 2012; Wollert, Kempf, Lagerqvist, et al., 2007; Wollert, Kempf, Peter, et al., 2007). Increased levels of GDF15 and FSTL1 are also associated with higher risk of death in patients with heart failure, and circulating levels of FSTL1 are positively correlated with left ventricular hypertrophy (Anand et al., 2010; El-Armouche et al., 2011). The identification of FSTl1 as a positive regulator of GDF15 (Widera et al., 2012) suggests that these two cardiokines function as a biomarker network and provide similar information about patient outcome at the clinical level. Myostatin has also recently emerged as an important biomarker in heart failure states. For example, cardiac biopsies of patients with severe congestive heart failure were shown to exhibit increased levels of myostatin (George et al., 2010). Moreover, plasma concentrations of myostatin were found to be significantly increased in heart failure patients and were demonstrated to correlate with biomarkers related to heart failure severity (Gruson, Ahn, Ketelslegers, & Rousseau, 2011). In addition, we have reported myostatin upregulation during the remodeling process in alcoholic and hypertensive human dilated cardiomyopathies (Fernandez-Sola et al., 2015, 2011). These studies point to myostatin as a promising biomarker in cardiac disorders (Fernandez-Sola & Planavila, 2016).

Several recent clinical studies in humans have explored the role of FGF21 in cardiovascular diseases. Elevated serum FGF21 levels have been recently reported in subjects with coronary heart disease or carotid artery plaques independently of established cardiovascular risk factors, suggesting the potential of serum FGF21 as a biomarker for atherosclerotic diseases (An et al., 2012; Chow et al., 2013; Lin et al., 2010). Serum FGF21 levels are also elevated in atrial fibrillation patients in association with atrial remodeling (Han et al., 2015). Interestingly, a recent study reported that serum FGF21 levels are independently associated with acute myocardial infarction (Zhang, Chu, Ding, & Wang, 2015), showing that serum FGF21 levels are markedly increased on the first day after the onset of myocardial infarction and remain high on days 3 and 7. Moreover, FGF21 levels were found to be closely related to those of BNP, a common marker of cardiac diseases. The authors of this study concluded that high levels of FGF21 might be related to the incidence of reinfarction within 30 days after onset. Collectively, these studies suggest that FGF21 is a potential new biomarker for cardiac diseases.

Finally, the ST2 receptor has recently emerged as a potential biomarker for cardiac disease (Shah & Januzzi, 2010). Alternative splicing of the ST2 gene results in formation of a truncated soluble receptor (sST2) that can be measured in the peripheral circulation (Kakkar & Lee, 2008), and several clinical studies have shown that blood concentrations of sST2 are enhanced in heart disease. In fact, high levels of sST2 are increased in acute myocardial infarction and ACS (Dhillon et al., 2011; Eggers, Armstrong, et al., 2010; Sabatine et al., 2008; Weir et al., 2010), as well as in heart failure (Bayes-Genis et al., 2012; Broch et al., 2012; Ky et al., 2012). Moreover, serial determination of sST2 in heart failure has shown value for both prognosis and biomarker-directed therapy (Boisot et al., 2008; Weinberg et al., 2003).

## 7. CONCLUSIONS

Several lines of evidence indicate that communication among cardiac cells via secreted factors may contribute to myocardial remodeling. The term *cardiokine* has emerged to describe proteins secreted by the heart that exert autocrine, paracrine, and/or endocrine functions crucial for the maintenance of cardiac function. Cardiokines can have either positive or negative effects in the heart, and an imbalance in their production may contribute

**Table 1** Summary of the Roles of Cardiokines in Cardiac Remodeling

| Cardiokine | Remodeling | Cell Type | Action Manner | Biomarker |
|---|---|---|---|---|
| GDF15 | Prevention | Cardiomyocytes | Autocrine/paracrine | Yes |
| Myostatin | Prevention | Cardiomyocytes | Autocrine/paracrine<br>Endocrine | Yes |
| Activin A | Prevention | Cardiomyocytes | Autocrine/paracrine<br>Endocrine | No |
| FSTL1 | Prevention | Cardiomyocytes | Autocrine/paracrine | Yes |
| FSTL3 | Promotion | Cardiomyocytes<br>Endothelial cells | Autocrine/paracrine | No |
| FGF2 | Promotion | Noncardiomyocytes | Paracrine | No |
| FGF16 | Prevention | Cardiomyocytes | Autocrine/paracrine | No |
| FGF21 | Prevention | Cardiomyocytes | Autocrine/paracrine<br>Endocrine? | Yes |
| IL33/ST2 | Prevention | Cardiomyocytes<br>Fibroblasts | Autocrine/paracrine | Yes (sST2) |
| MANF | Prevention | Cardiac cells | Autocrine/paracrine | No |

to cardiac disease outcome. Moreover, through their endocrine functions, cardiokines may act on distant tissues to affect whole-body homeostasis and skeletal muscle cachexia. Table 1 summarizes the roles of the new cardiokines that had been reviewed in this chapter during cardiac remodeling. Collectively, these findings reinforce the concept that the heart functions as a secretory organ. From a biomedical point of view, the possibility of preventing or even reversing pathological cardiac states, and thereby slowing the development of heart disease, is of utmost importance. In this sense, the identification and characterization of new cardiokines are of interest from pathogenic, diagnostic, and therapeutic perspectives.

## ACKNOWLEDGMENTS

This work has been partially supported by Grants from CIBEROBN (Fisiopatologia de la Obesitat i la Nutrició), Instituto de Salud Carlos III, Sociedad Española de Cardiología (SEC), Fundación BBVA and Fundació La Marató de TV3: 201533-30/31. A.P. is supported by a RyC-2014 fellowship and SAF2014-55702-JIN project from the Ministerio de Ciencia e Innovación, Spain and cofinanced by the European Regional Development Fund (ERDF).

## REFERENCES


Aihara, K., Mogi, M., Shibata, R., Bishop-Bailey, D., & Ma, X. L. (2011). Interactions between CKD and MetS and the development of CVD. *Cardiology Research and Practice, 2011*, 878065.

Ambardekar, A. V., & Buttrick, P. M. (2011). Reverse remodeling with left ventricular assist devices: A review of clinical, cellular, and molecular effects. *Circulation: Heart Failure, 4*, 224–233.

An, S. Y., Lee, M. S., Yi, S. A., Ha, E. S., Han, S. J., Kim, H. J., et al. (2012). Serum fibroblast growth factor 21 was elevated in subjects with type 2 diabetes mellitus and was associated with the presence of carotid artery plaques. *Diabetes Research and Clinical Practice, 96*, 196–203.

Anand, I. S., Kempf, T., Rector, T. S., Tapken, H., Allhoff, T., Jantzen, F., et al. (2010). Serial measurement of growth-differentiation factor-15 in heart failure: Relation to disease severity and prognosis in the valsartan heart failure trial. *Circulation, 122*, 1387–1395.

Anker, S. D., Ponikowski, P., Varney, S., Chua, T. P., Clark, A. L., Webb-Peploe, K. M., et al. (1997). Wasting as independent risk factor for mortality in chronic heart failure. *Lancet, 349*, 1050–1053.

Azhar, M., Brown, K., Gard, C., Chen, H., Rajan, S., Elliott, D. A., et al. (2011). Transforming growth factor Beta2 is required for valve remodeling during heart development. *Developmental Dynamics, 240*, 2127–2141.

Baan, J. A., Varga, Z. V., Leszek, P., Kusmierczyk, M., Baranyai, T., Dux, L., et al. (2015). Myostatin and IGF-I signaling in end-stage human heart failure: A qRT-PCR study. *Journal of Translational Medicine, 13*, 1.

Badman, M. K., Pissios, P., Kennedy, A. R., Koukos, G., Flier, J. S., & Maratos-Flier, E. (2007). Hepatic fibroblast growth factor 21 is regulated by PPARalpha and is a key mediator of hepatic lipid metabolism in ketotic states. *Cell Metabolism, 5*, 426–437.

Bartholin, L., Maguer-Satta, V., Hayette, S., Martel, S., Gadoux, M., Corbo, L., et al. (2002). Transcription activation of FLRG and follistatin by activin A, through Smad proteins, participates in a negative feedback loop to modulate activin A function. *Oncogene, 21*, 2227–2235.

Bayes-Genis, A., de Antoniao, M., Galán, A., Sanz, H., Urrutia, A., Cabanes, R., et al. (2012). Combined use of high-sensitivity ST2 and NTproBNP to improve the prediction of death in heart failure. *European Journal of Heart Failure, 14*, 32–38.

Belmont, P. J., Tadimalla, A., Chen, W. J., Martindale, J. J., Thuerauf, D. J., Marcinko, M., et al. (2008). Coordination of growth and endoplasmic reticulum stress signaling by regulator of calcineurin 1 (RCAN1), a novel ATF6-inducible gene. *The Journal of Biological Chemistry, 283*, 14012–14021.

Biesemann, N., Mendler, L., Wietelmann, A., Hermann, S., Schafers, M., Kruger, M., et al. (2014). Myostatin regulates energy homeostasis in the heart and prevents heart failure. *Circulation Research, 115*, 296–310.

Bogoyevitch, M. A., Glennon, P. E., Andersson, M. B., Clerk, A., Lazou, A., Marshall, C. J., et al. (1994). Endothelin-1 and fibroblast growth factors stimulate the mitogen-activated protein kinase signaling cascade in cardiac myocytes. The potential role of the cascade in the integration of two signaling pathways leading to myocyte hypertrophy. *The Journal of Biological Chemistry, 269*, 1110–1119.

Boisot, S., Beede, J., Isakson, S., Chiu, A., Clopton, P., Januzzi, J., et al. (2008). Serial sampling of ST2 predicts 90-day mortality following destabilized heart failure. *Journal of Cardiac Failure, 14*, 732–738.

Bootcov, M. R., Bauskin, A. R., Valenzuela, S. M., Moore, A. G., Bansal, M., He, X. Y., et al. (1997). MIC-1, a novel macrophage inhibitory cytokine, is a divergent member of the TGF-beta superfamily. *Proceedings of the National Academy of Sciences of the United States of America, 94*, 11514–11519.

Bordicchia, M., Liu, D., Amri, E. Z., Ailhaud, G., Dessi-Fulgheri, P., Zhang, C., et al. (2012). Cardiac natriuretic peptides act via p38 MAPK to induce the brown fat thermogenic program in mouse and human adipocytes. *The Journal of Clinical Investigation*, *122*, 1022–1036.

Brahma, M. K., Adam, R. C., Pollak, N. M., Jaeger, D., Zierler, K. A., Poecher, N., et al. (2014). FGF21 is induced upon cardiac stress and alters cardiac lipid homeostasis. *Journal of Lipid Research*, *55*, 2229–2241.

Braunwald, E., Harrison, D. C., & Chidsey, C. A. (1964). The heart as an endocrine organ. *The American Journal of Medicine*, *36*, 1–4.

Broch, K., Ueland, T., Nymo, S. H., Kjekshus, J., Hulthe, J., Muntendam, P., et al. (2012). Soluble ST2 is associated with adverse outcome in patients with heart failure of ischaemic aetiology. *European Journal of Heart Failure*, *14*, 268–277.

Castillero, E., Akashi, H., Wang, C., Najjar, M., Ji, R., Kennel, P. J., et al. (2015). Cardiac myostatin upregulation occurs immediately after myocardial ischemia and is involved in skeletal muscle activation of atrophy. *Biochemical and Biophysical Research Communications*, *457*, 106–111.

Chen, J. L., Walton, K. L., Winbanks, C. E., Murphy, K. T., Thomson, R. E., Makanji, Y., et al. (2014). Elevated expression of activins promotes muscle wasting and cachexia. *The FASEB Journal*, *28*, 1711–1723.

Chow, W. S., Xu, A., Woo, Y. C., Tso, A. W., Cheung, S. C., Fong, C. H., et al. (2013). Serum fibroblast growth factor-21 levels are associated with carotid atherosclerosis independent of established cardiovascular risk factors. *Arteriosclerosis, Thrombosis, and Vascular Biology*, *33*, 2454–2459.

Collins, S. (2014). A heart-adipose tissue connection in the regulation of energy metabolism. *Nature Reviews. Endocrinology*, *10*, 157–163.

de Bold, A. J., Borenstein, H. B., Veress, A. T., & Sonnenberg, H. (1981). A rapid and potent natriuretic response to intravenous injection of atrial myocardial extract in rats. *Life Sciences*, *28*, 89–94.

Delafontaine, P., & Akao, M. (2006). Angiotensin II as candidate of cardiac cachexia. *Current Opinion in Clinical Nutrition and Metabolic Care*, *9*, 220–224.

Derynck, R., & Zhang, Y. E. (2003). Smad-dependent and Smad-independent pathways in TGF-beta family signalling. *Nature*, *425*, 577–584.

De Sousa-Coelho, A. L., Marrero, P. F., & Haro, D. (2012). Activating transcription factor 4-dependent induction of FGF21 during amino acid deprivation. *The Biochemical Journal*, *443*, 165–171.

Dessi-Fulgheri, P., Sarzani, R., Tamburrini, P., Moraca, A., Espinosa, E., Cola, G., et al. (1997). Plasma atrial natriuretic peptide and natriuretic peptide receptor gene expression in adipose tissue of normotensive and hypertensive obese patients. *Journal of Hypertension*, *15*, 1695–1699.

Deten, A., Holzl, A., Leicht, M., Barth, W., & Zimmer, H. G. (2001). Changes in extracellular matrix and in transforming growth factor beta isoforms after coronary artery ligation in rats. *Journal of Molecular and Cellular Cardiology*, *33*, 1191–1207.

Dhillon, O. S., Narayan, H. K., Quinn, P. A., Squire, I. B., Davies, J. E., & Ng, L. L. (2011). Interleukin 33 and ST2 in non-ST-elevation myocardial infarction: Comparison with global registry of acute coronary events risk scoring and NT-proBNP. *American Heart Journal*, *161*, 1163–1170.

Dickson, M. C., Martin, J. S., Cousins, F. M., Kulkarni, A. B., Karlsson, S., & Akhurst, R. J. (1995). Defective haematopoiesis and vasculogenesis in transforming growth factor-beta 1 knock out mice. *Development (Cambridge, England)*, *121*, 1845–1854.

Dickson, M. C., Slager, H. G., Duffie, E., Mummery, C. L., & Akhurst, R. J. (1993). RNA and protein localisations of TGF beta 2 in the early mouse embryo suggest an involvement in cardiac development. *Development (Cambridge, England)*, *117*, 625–639.

Dobaczewski, M., Chen, W., & Frangogiannis, N. G. (2011). Transforming growth factor (TGF)-beta signaling in cardiac remodeling. *Journal of Molecular and Cellular Cardiology, 51*, 600–606.

Doroudgar, S., & Glembotski, C. C. (2011). The cardiokine story unfolds: Ischemic stress-induced protein secretion in the heart. *Trends in Molecular Medicine, 17*, 207–214.

Doroudgar, S., Thuerauf, D. J., Marcinko, M. C., Belmont, P. J., & Glembotski, C. C. (2009). Ischemia activates the ATF6 branch of the endoplasmic reticulum stress response. *The Journal of Biological Chemistry, 284*, 29735–29745.

Dschietzig, T. B. (2014). Myostatin-from the mighty mouse to cardiovascular disease and cachexia. *Clinica Chimica Acta, 433*, 216–224.

D'Souza, S. P., Davis, M., & Baxter, G. F. (2004). Autocrine and paracrine actions of natriuretic peptides in the heart. *Pharmacology & Therapeutics, 101*, 113–129.

Eggers, K. M., Armstrong, P. W., Califf, R. M., Simoons, M. L., Venge, P., Wallentin, L., et al. (2010). ST2 and mortality in non-ST-segment elevation acute coronary syndrome. *American Heart Journal, 159*, 788–794.

Eggers, K. M., Kempf, T., Lagerqvist, B., Lindahl, B., Olofsson, S., Jantzen, F., et al. (2010). Growth-differentiation factor-15 for long-term risk prediction in patients stabilized after an episode of non-ST-segment-elevation acute coronary syndrome. *Circulation. Cardiovascular Genetics, 3*, 88–96.

El-Armouche, A., Ouchi, N., Tanaka, K., Doros, G., Wittkopper, K., Schulze, T., et al. (2011). Follistatin-like 1 in chronic systolic heart failure: A marker of left ventricular remodeling. *Circulation: Heart Failure, 4*, 621–627.

Fernandez-Sola, J., Borrisser-Pairo, F., Antunez, E., & Tobias, E. (2015). Myostatin and insulin-like growth factor-1 in hypertensive heart disease: A prospective study in human heart donors. *Journal of Hypertension, 33*, 851–858.

Fernandez-Sola, J., Lluis, M., Sacanella, E., Estruch, R., Antunez, E., & Urbano-Marquez, A. (2011). Increased myostatin activity and decreased myocyte proliferation in chronic alcoholic cardiomyopathy. *Alcoholism, Clinical and Experimental Research, 35*, 1220–1229.

Fernandez-Sola, J., & Planavila, P. A. (2016). New treatment strategies for alcohol-induced heart damage. *International Journal of Molecular Sciences, 17*.

Fon, T. K., Bookout, A. L., Ding, X., Kurosu, H., John, G. B., Wang, L., et al. (2010). Research resource: Comprehensive expression atlas of the fibroblast growth factor system in adult mouse. *Molecular Endocrinology (Baltimore, Md.), 24*, 2050–2064.

Frohlich, E. D. (1985). The heart. An endocrine organ (revisited). *Archives of Internal Medicine, 145*, 809–811.

Frost, R. J., & Engelhardt, S. (2007). A secretion trap screen in yeast identifies protease inhibitor 16 as a novel antihypertrophic protein secreted from the heart. *Circulation, 116*, 1768–1775.

Galman, C., Lundasen, T., Kharitonenkov, A., Bina, H. A., Eriksson, M., Hafstrom, I., et al. (2008). The circulating metabolic regulator FGF21 is induced by prolonged fasting and PPARalpha activation in man. *Cell Metabolism, 8*, 169–174.

George, I., Bish, L. T., Kamalakkannan, G., Petrilli, C. M., Oz, M. C., Naka, Y., et al. (2010). Myostatin activation in patients with advanced heart failure and after mechanical unloading. *European Journal of Heart Failure, 12*, 444–453.

Glembotski, C. C. (2011). Functions for the cardiomyokine, MANF, in cardioprotection, hypertrophy and heart failure. *Journal of Molecular and Cellular Cardiology, 51*, 512–517.

Glembotski, C. C., Thuerauf, D. J., Huang, C., Vekich, J. A., Gottlieb, R. A., & Doroudgar, S. (2012). Mesencephalic astrocyte-derived neurotrophic factor protects the heart from ischemic damage and is selectively secreted upon sarco/endoplasmic reticulum calcium depletion. *The Journal of Biological Chemistry, 287*, 25893–25904.

Goetz, R., & Mohammadi, M. (2013). Exploring mechanisms of FGF signalling through the lens of structural biology. *Nature Reviews Molecular Cell Biology*, *14*, 166–180.
Gruson, D., Ahn, S. A., Ketelslegers, J. M., & Rousseau, M. F. (2011). Increased plasma myostatin in heart failure. *European Journal of Heart Failure*, *13*, 734–736.
Han, X., Chen, C., Cheng, G., Xie, C., Yang, M., Shou, X., et al. (2015). Serum fibroblast growth factor 21 levels are increased in atrial fibrillation patients. *Cytokine*, *73*, 176–180.
Heineke, J., Auger-Messier, M., Xu, J., Sargent, M., York, A., Welle, S., et al. (2010). Genetic deletion of myostatin from the heart prevents skeletal muscle atrophy in heart failure. *Circulation*, *121*, 419–425.
Heymans, S., Hirsch, E., Anker, S. D., Aukrust, P., Balligand, J. L., Cohen-Tervaert, J. W., et al. (2009). Inflammation as a therapeutic target in heart failure? A scientific statement from the Translational Research Committee of the Heart Failure Association of the European Society of Cardiology. *European Journal of Heart Failure*, *11*, 119–129.
Hill, J. A., & Olson, E. N. (2008). Cardiac plasticity. *The New England Journal of Medicine*, *358*, 1370–1380.
Holland, R., Battersby, J., Harvey, I., Lenaghan, E., Smith, J., & Hay, L. (2005). Systematic review of multidisciplinary interventions in heart failure. *Heart*, *91*, 899–906.
Hondares, E., Iglesias, R., Giralt, A., Gonzalez, F. J., Giralt, M., Mampel, T., et al. (2011). Thermogenic activation induces FGF21 expression and release in Brown adipose tissue. *The Journal of Biological Chemistry*, *286*, 12983–12990.
Hotta, Y., Sasaki, S., Konishi, M., Kinoshita, H., Kuwahara, K., Nakao, K., et al. (2008). Fgf16 is required for cardiomyocyte proliferation in the mouse embryonic heart. *Developmental Dynamics*, *237*, 2947–2954.
House, S. L., House, B. E., Glascock, B., Kimball, T., Nusayr, E., Schultz, J. E., et al. (2010). Fibroblast growth factor 2 mediates isoproterenol-induced cardiac hypertrophy through activation of the extracellular regulated kinase. *Molecular and Cellular Pharmacology*, *2*, 143–154.
Inagaki, T., Dutchak, P., Zhao, G., Ding, X., Gautron, L., Parameswara, V., et al. (2007). Endocrine regulation of the fasting response by PPARalpha-mediated induction of fibroblast growth factor 21. *Cell Metabolism*, *5*, 415–425.
Itoh, N. (2010). Hormone-like (endocrine) Fgfs: Their evolutionary history and roles in development, metabolism, and disease. *Cell and Tissue Research*, *342*, 1–11.
Itoh, N., & Ohta, H. (2013). Pathophysiological roles of FGF signaling in the heart. *Frontiers in Physiology*, *4*, 247.
Itoh, N., & Ornitz, D. M. (2011). Fibroblast growth factors: From molecular evolution to roles in development, metabolism and disease. *Journal of Biochemistry*, *149*, 121–130.
Izumiya, Y., Bina, H. A., Ouchi, N., Akasaki, Y., Kharitonenkov, A., & Walsh, K. (2008). FGF21 is an Akt-regulated myokine. *FEBS Letters*, *582*, 3805–3810.
Jahng, J. W., Song, E., & Sweeney, G. (2016). Crosstalk between the heart and peripheral organs in heart failure. *Experimental & Molecular Medicine*, *48*. e217
Jiang, B., & Liao, R. (2010). The paradoxical role of inflammation in cardiac repair and regeneration. *Journal of Cardiovascular Translational Research*, *3*, 410–416.
Joki, Y., Ohashi, K., Yuasa, D., Shibata, R., Ito, M., Matsuo, K., et al. (2015). FGF21 attenuates pathological myocardial remodeling following myocardial infarction through the adiponectin-dependent mechanism. *Biochemical and Biophysical Research Communications*, *459*, 124–130.
Kakkar, R., Hei, H., Dobner, S., & Lee, R. T. (2012). Interleukin 33 as a mechanically responsive cytokine secreted by living cells. *The Journal of Biological Chemistry*, *287*, 6941–6948.
Kakkar, R., & Lee, R. T. (2008). The IL-33/ST2 pathway: Therapeutic target and novel biomarker. *Nature Reviews. Drug Discovery*, *7*, 827–840.
Kakkar, R., & Lee, R. T. (2010). Intramyocardial fibroblast myocyte communication. *Circulation Research*, *106*, 47–57.

Sudoh, T., Minamino, N., Kangawa, K., & Matsuo, H. (1990). C-type natriuretic peptide (CNP): A new member of natriuretic peptide family identified in porcine brain. *Biochemical and Biophysical Research Communications*, *168*, 863–870.

Sukhanov, S., Semprun-Prieto, L., Yoshida, T., Michael, T. A., Higashi, Y., Galvez, S., et al. (2011). Angiotensin II, oxidative stress and skeletal muscle wasting. *The American Journal of the Medical Sciences*, *342*, 143–147.

Sylva, M., Li, V. S., Buffing, A. A., van Es, J. H., van den Born, M., van der Velden, S., et al. (2011). The BMP antagonist follistatin-like 1 is required for skeletal and lung organogenesis. *PLoS·One*, *6*. e22616

Tadimalla, A., Belmont, P. J., Thuerauf, D. J., Glassy, M. S., Martindale, J. J., Gude, N., et al. (2008). Mesencephalic astrocyte-derived neurotrophic factor is an ischemia-inducible secreted endoplasmic reticulum stress response protein in the heart. *Circulation Research*, *103*, 1249–1258.

Tanaka, M., Murakami, K., Ozaki, S., Imura, Y., Tong, X. P., Watanabe, T., et al. (2010). DIP2 disco-interacting protein 2 homolog a (drosophila) is a candidate receptor for follistatin-related protein/follistatin-like 1--analysis of their binding with TGF-beta superfamily proteins. *The FEBS Journal*, *277*, 4278–4289.

Teekakirikul, P., Eminaga, S., Toka, O., Alcalai, R., Wang, L., Wakimoto, H., et al. (2010). Cardiac fibrosis in mice with hypertrophic cardiomyopathy is mediated by non-myocyte proliferation and requires Tgf-beta. *The Journal of Clinical Investigation*, *120*, 3520–3529.

Thoonen, R., Hindle, A. G., & Scherrer-Crosbie, M. (2016). Brown adipose tissue: The heat is on the heart. *American Journal of Physiology. Heart and Circulatory Physiology*, *310*, H1592–H1605.

Thuerauf, D. J., Marcinko, M., Gude, N., Rubio, M., Sussman, M. A., & Glembotski, C. C. (2006). Activation of the unfolded protein response in infarcted mouse heart and hypoxic cultured cardiac myocytes. *Circulation Research*, *99*, 275–282.

Tian, Y., & Morrisey, E. E. (2012). Importance of myocyte-nonmyocyte interactions in cardiac development and disease. *Circulation Research*, *110*, 1023–1034.

Tirziu, D., Giordano, F. J., & Simons, M. (2010). Cell communications in the heart. *Circulation*, *122*, 928–937.

Trendelenburg, A. U., Meyer, A., Rohner, D., Boyle, J., Hatakeyama, S., & Glass, D. J. (2009). Myostatin reduces Akt/TORC1/p70S6K signaling, inhibiting myoblast differentiation and myotube size. *American Journal of Physiology—Cell Physiology*, *296*, C1258–C1270.

Virag, J. A., Rolle, M. L., Reece, J., Hardouin, S., Feigl, E. O., & Murry, C. E. (2007). Fibroblast growth factor-2 regulates myocardial infarct repair: Effects on cell proliferation, scar contraction, and ventricular function. *The American Journal of Pathology*, *171*, 1431–1440.

Wang, T. J., Larson, M. G., Levy, D., Benjamin, E. J., Leip, E. P., Wilson, P. W., et al. (2004). Impact of obesity on plasma natriuretic peptide levels. *Circulation*, *109*, 594–600.

Weinberg, E. O., Shimpo, M., Hurwitz, S., Tominaga, S., Rouleau, J. L., & Lee, R. T. (2003). Identification of serum soluble ST2 receptor as a novel heart failure biomarker. *Circulation*, *107*, 721–726.

Weir, R. A., Miller, A. M., Murphy, G. E., Clements, S., Steedman, T., Connell, J. M., et al. (2010). Serum soluble ST2: A potential novel mediator in left ventricular and infarct remodeling after acute myocardial infarction. *Journal of the American College of Cardiology*, *55*, 243–250.

Widera, C., Giannitsis, E., Kempf, T., Korf-Klingebiel, M., Fiedler, B., Sharma, S., et al. (2012). Identification of follistatin-like 1 by expression cloning as an activator of the growth differentiation factor 15 gene and a prognostic biomarker in acute coronary syndrome. *Clinical Chemistry*, *58*, 1233–1241.

Wollert, K. C., Kempf, T., Lagerqvist, B., Lindahl, B., Olofsson, S., Allhoff, T., et al. (2007). Growth differentiation factor 15 for risk stratification and selection of an invasive treatment strategy in non ST-elevation acute coronary syndrome. *Circulation*, *116*, 1540–1548.

Wollert, K. C., Kempf, T., Peter, T., Olofsson, S., James, S., Johnston, N., et al. (2007). Prognostic value of growth-differentiation factor-15 in patients with non-ST-elevation acute coronary syndrome. *Circulation*, *115*, 962–971.

Xu, J., Kimball, T. R., Lorenz, J. N., Brown, D. A., Bauskin, A. R., Klevitsky, R., et al. (2006). GDF15/MIC-1 functions as a protective and antihypertrophic factor released from the myocardium in association with SMAD protein activation. *Circulation Research*, *98*, 342–350.

Yamamoto, K., Burnett, J. C., Jr., Jougasaki, M., Nishimura, R. A., Bailey, K. R., Saito, Y., et al. (1996). Superiority of brain natriuretic peptide as a hormonal marker of ventricular systolic and diastolic dysfunction and ventricular hypertrophy. *Hypertension*, *28*, 988–994.

Zhang, W., Chu, S., Ding, W., & Wang, F. (2015). Serum level of fibroblast growth factor 21 is independently associated with acute myocardial infarction. *PLoS·One*, *10*. e0129791

Zimmers, T. A., Davies, M. V., Koniaris, L. G., Haynes, P., Esquela, A. F., Tomkinson, K. N., et al. (2002). Induction of cachexia in mice by systemically administered myostatin. *Science (New York, NY)*, *296*, 1486–1488.